Learning and Analytics in Intelligent Systems

Volume 18

Series Editors

George A. Tsihrintzis, University of Piraeus, Piraeus, Greece

Maria Virvou, University of Piraeus, Piraeus, Greece

Lakhmi C. Jain, Faculty of Engineering and Information Technology,
Centre for Artificial Intelligence, University of Technology Sydney, NSW,
Australia;
KES International, Shoreham-by-Sea, UK;
Liverpool Hope University, Liverpool, UK

The main aim of the series is to make available a publication of books in hard copy form and soft copy form on all aspects of learning, analytics and advanced intelligent systems and related technologies. The mentioned disciplines are strongly related and complement one another significantly. Thus, the series encourages cross-fertilization highlighting research and knowledge of common interest. The series allows a unified/integrated approach to themes and topics in these scientific disciplines which will result in significant cross-fertilization and research dissemination. To maximize dissemination of research results and knowledge in these disciplines, the series publishes edited books, monographs, handbooks, textbooks and conference proceedings.

More information about this series at http://www.springer.com/series/16172

George A. Tsihrintzis · Lakhmi C. Jain
Editors

Machine Learning Paradigms

Advances in Deep Learning-based
Technological Applications

 Springer

Editors
George A. Tsihrintzis
Department of Informatics
University of Piraeus
Piraeus, Greece

Lakhmi C. Jain
University of Technology
Sydney, NSW, Australia

Liverpool Hope University
Liverpool, UK

ISSN 2662-3447   ISSN 2662-3455   (electronic)
Learning and Analytics in Intelligent Systems
ISBN 978-3-030-49726-2   ISBN 978-3-030-49724-8   (eBook)
https://doi.org/10.1007/978-3-030-49724-8

This Springer imprint is published by the registered company Springer Nature Switzerland AG
The registered company address is: Gewerbestrasse 11, 6330 Cham, Switzerland

Dedicated to our loving and supporting families

George A. Tsihrintzis and Lakhmi C. Jain

Foreword

Indisputably, the paradigms of Machine Learning and deep learning with its plethora of applications and a diversity of technological pursuits have positioned themselves at the forefront of studies on intelligent systems. It is very likely that this interest will grow over time. Even for an experienced researcher, the broad spectrum of directions and a rapidly changing landscape of the discipline create a quest to navigate efficiently across current developments and gain a thorough insight into their accomplishments and limitations. This volume addresses the evident challenges.

Professors George A. Tsihrintzis and Lakhmi C. Jain did a superb job by bringing a collection of carefully selected and coherently structured contributions to studies on recent developments. The chapters have been written by active researchers who reported on their research agendas and timely outcomes of their investigations being positioned at the forefront of Machine Learning methodologies and its existing practices.

A set of 16 chapters delivers an impressive spectrum of authoritatively covered timely topics by being focused on deep learning to sensing (3 chapters), Social Media and Internet of Things (IoT) (2 chapters), health care (2 chapters) focused on medical imaging and electroencephalography, systems control (2 chapters), and forecasting and prediction (3 chapters). The last part of the book focuses on performance analysis of the schemes of deep learning (3 chapters).

In virtue of the organization of the contributions and the way the key ideas have been exposed, cohesion of exposure retained, a smooth flow of algorithmic developments, and a wealth of valuable and far-reaching applications have been delivered. Deep learning by itself is a well-established paradigm, however when cast in a context of a certain problem, the domain knowledge contributes to its enhanced usage. This is a crux of well thought-out design process and here the book offers this important enhancement to the general design strategy.

All in all, the book is a well-rounded volume which will be of genuine interest to the academic community as well as practitioners with a vital interest in Machine Learning. The Editors should be congratulated on bringing this timely and useful research material to the readers.

Prof. Dr. Witold Pedrycz
Fellow IEEE

University of Alberta
Alberta, Canada

Systems Research Institute Polish Academy of Sciences
Warsaw, Poland

Preface

Artificial Intelligence [1], in general, and Machine Learning [2], in particular, are scientific areas of very active research worldwide. However, while many researchers continuously report on new findings and innovative applications, skeptics warn us of potentially existential threats to humankind [3, 4]. The debate among zealots and skeptics of Artificial Intelligence is ongoing and only in the future we may see this issue resolved. For the moment, it seems that perhaps Nick Bostrom's statement constitutes the most reliable view: "...*artificial intelligence (AI) is not an ongoing or imminent global catastrophic risk. Nor is it as uncontroversially a serious cause for concern. However, from a long term perspective, the development of general artificial intelligence exceeding that of the human brain can be seen as one of the main challenges to the future of humanity (arguably, even as the main challenge)*" [5].

For the moment, it is also certain that a new era is arising in human civilization, which has been called "the 4th Industrial Revolution" [6, 7], and Artificial Intelligence is one of the key technologies driving it.

Within the broad discipline of Artificial Intelligence, a sub-field of Machine Learning, called *Deep Learning,* stands out due to its worldwide pace of growth both in new theoretical results and in new technological application areas (for example, see [8] for an-easy-to-follow first read on Deep Learning). While growing at a formidable rate, Deep Learning is also achieving high scores in successful technological applications and promises major impact in science, technology, and the society.

The book at hand aims at exposing its reader to some of the most significant recent advances in deep learning-based technological applications and consists of an editorial note and an additional fifteen (15) chapters. All chapters in the book were invited from authors who work in the corresponding chapter theme and are recognized for their significant research contributions. In more detail, the chapters in the book are organized into six parts, namely (1) *Deep Learning in Sensing,* (2) *Deep Learning in Social Media and IOT,* (3) *Deep Learning in the Medical Field,* (4) *Deep Learning in Systems Control,* (5) *Deep Learning in Feature Vector Processing,* and (6) *Evaluation of Algorithm Performance.*

This research book is directed toward professors, researchers, scientists, engineers, and students in computer science-related disciplines. It is also directed toward readers who come from other disciplines and are interested in becoming versed in some of the most recent deep learning-based technological applications. An extensive list of bibliographic references at the end of each chapter guides the readers to probe deeper into their application areas of interest. We hope that all of them will find it useful and inspiring in their works and researches.

We are grateful to the authors and reviewers for their excellent contributions and visionary ideas. We are also thankful to Springer for agreeing to publish this book in its *Learning and Analytics in Intelligent Systems* series. Last, but not least, we are grateful to the Springer staff for their excellent work in producing this book.

Piraeus, Greece   George A. Tsihrintzis
Sydney, Australia   Lakhmi C. Jain

References

1. E. Rich, K. Knight, S. B. Nair, *Artificial Intelligence*, 3rd edn. (Tata McGraw-Hill Publishing Company, 2010)
2. J. Watt, R. Borhani, A. K. Katsaggelos, *Machine Learning Refined—Foundations, Algorithms and Applications*, 2nd edn. (Cambridge University Press, 2020)
3. J. Barrat, *Our Final Invention—Artificial Intelligence and the End of the Human Era*, Thomas Dunn Books, 2013
4. N. Bostrom, *Superintelligence—Paths, Dangers, Startegies* (Oxford University Press, 2014)
5. N. Bostrom, M. M. Ćirković (eds.), *Global Catastrophic Risks*, (Oxford University Press, 2008), p. 17
6. J. Toonders, Data Is the New Oil of the Digital Economy, *Wired* (https://www.wired.com/insights/2014/07/data-new-oil-digital-economy/)
7. K. Schwabd, The Fourth Industrial Revolution—What It Means and How to Respond, *Foreign Affairs* (2015), https://www.foreignaffairs.com/articles/2015-12-12/fourth-industrial-revolution Accessed 12 December 2015
8. J. Patterson, A. Gibson, *Deep Learning—A Practitioner's Approach*, O' Reilly, 2017

Contents

Chapter 1
Machine Learning Paradigms: Introduction to Deep Learning-Based Technological Applications

George A. Tsihrintzis and Lakhmi C. Jain

Abstract At the dawn of the 4th Industrial Revolution, the field of *Deep Learning* (a sub-field of *Artificial Intelligence* and *Machine Learning*) is growing continuously and rapidly, developing both theoretically and towards applications in increasingly many and diverse disciplines. The book at hand aims at exposing its readers to some of the most significant recent advances in Deep Learning-based technological applications. As such, the book is directed towards professors, researchers, scientists, engineers and students in computer science-related disciplines. It is also directed towards readers who come from other disciplines and are interested in becoming versed in some of the most recent technological progress in this active research field. An extensive list of bibliographic references at the end of each chapter guides the readers to probe deeper into their areas of interest.

1.1 Editorial Note

At the dawn of the 4th Industrial Revolution [1, 2], Artificial Intelligence [3], in general, and Machine Learning [4], in particular, are fields of very active and intense research worldwide [5–11]. More specifically, Machine Learning was introduced in 1959 in a nominal paper by Arthur Samuel [12] and has grown into a multi-disciplinary approach to Artificial Intelligence which aims at incorporating learning abilities into machines. In other words, Machine Learning aims at developing mechanisms, methodologies, procedures and algorithms that allow machines to become better and more efficient at performing specific tasks, either on their own or with

G. A. Tsihrintzis (✉)
Department of Informatics, University of Piraeus, Piraeus 185 34, Greece
e-mail: geoatsi@unipi.gr

L. C. Jain
Liverpool Hope University, Liverpool, UK
e-mail: jainlakhmi@gmail.com

University of Technology Sydney, NSW, Australia

G. A. Tsihrintzis and L. C. Jain (eds.), *Machine Learning Paradigms*,
Learning and Analytics in Intelligent Systems 18,
https://doi.org/10.1007/978-3-030-49724-8_1

the help of a supervisor/instructor. Within the umbrella of Machine Learning, the sub-field of *Deep Learning* (for example, see [13] for an-easy-to-follow first read on Deep Learning) stands out due to its worldwide pace of growth both in new theoretical results and in new technological application areas. While growing at a formidable rate, Deep Learning is also achieving high scores in successful technological applications and promises major impact in science, technology and the society.

The advances of Deep Learning have already had a significant effect on many aspects of everyday life, the workplace and human relationships and they are expected to have an even wider and deeper effect in the foreseeable future. Specifically, some of the technological application areas, where the use of Deep Learning-based approaches has met with success, include the following:

1. *Deep Learning in Sensing.*
2. *Deep Learning in Social Media and IOT.*
3. *Deep Learning in the Medical Field.*
4. *Deep Learning in Systems Control.*
5. *Deep Learning in Feature Vector Processing.*
6. *Evaluation of Algorithm Performance.*

The book at hand aims at updating the relevant research communities, including professors, researchers, scientists, engineers and students in computer science-related disciplines, as well as the general reader from other disciplines, on the most recent advances in Deep Learning-based technological applications.

More specifically, the book at hand consists of an editorial chapter (This chapter) and an additional fifteen (15) chapters. All chapters in the book were invited from authors who work in the corresponding chapter theme and are recognized for their significant research contributions. In more detail, the chapters in the book are organized into six parts, as follows:

The *first part* of the book consists of three chapters devoted to *Advances in Deep Learning in Sensing.*

Specifically, Chap. 2, by Naeha Sharif, Uzair Nadeem, Mohammed Bennamoun, Wei Liu, and Syed Afaq Ali Shah is on *"Vision to Language: Methods, Metrics and Datasets."* The authors discuss the fundamentals of generating natural language description of images as well as the prominent and the state-of-the-art methods, their limitations, various challenges in image captioning and future directions to push this technology further for practical real-world applications.

Chapter 3, by Arvind W. Kiwelekar, Geetanjali S. Mahamunkar, Laxman D. Netak and Valmik B. Nikam is on *"Deep Learning Techniques for Geospatial Data Analysis."* The authors presents a survey on the current state of the applications of deep learning techniques for analyzing geospatial data, with emphasis on remote sensing, GPS and RFID data.

Chapter 4, by Chairi Kiourt, George Pavlidis and Stella Markantonatou, is on *"Deep Learning Approaches in Food Recognition."* The authors outline a design from scratch, transfer learning and platform-based approaches for automatic image-based food recognition.

The *second part* of the book consists of two chapters devoted to *Advances in Deep Learning in Social Media and IOT*.

Specifically, Chap. 5, by Akrivi Krouska, Christos Troussas, and Maria Virvou, is on *"Deep Learning for Twitter Sentiment Analysis: the Effect of pre-trained Word Embedding."* The authors have made data analysis with huge numbers of tweets taken as big data and thereby classifying their polarity using a deep learning approach with four notable pre-trained word vectors, namely Google's Word2Vec, Stanford's Crawl GloVe, Stanford's Twitter GloVe, and Facebook's FastText.

Chapter 6, by Luke Holbrook and Miltiadis Alamaniotis, is entitled *"A Good Defense is a Strong DNN: Defending the IoT with Deep Neural Networks."* The authors provide a new deep learning-based comprehensive solution for securing networks connecting IoT devices and discuss a corresponding new protocol.

The *third part* of the book consists of two chapters devoted to *Advances in Deep Learning in the Medical Field*.

Specifically, Chap. 7, by Shymaa Abou Arkoub, Amir Hajjam El Hassani, Fabrice Lauri, Mohammad Hajjar, Bassam Daya, Sophie Hecquet and Sébastien Aubry, is on *"Survey on Deep Learning Techniques for Medical Imaging Application Area."* The authors highlight the primary deep learning techniques relevant to the medical imaging area and provide fundamental knowledge of deep learning methods, while specifying current limitations and future research directions.

Chapter 8, by Krzysztof Kotowski, Katarzyna Stapor and Jeremi Ochab, is entitled *"Deep Learning Methods in Electroencephalography."* The authors discuss the state-of-the-art architectures and approaches to classification, segmentation, and enhancement of EEG recordings in applications in brain-computer interfaces, medical diagnostics and emotion recognition.

The *fourth part* of the book consists of two chapters devoted to *Advances in Deep Learning in Systems Control*.

Specifically, Chap. 9, by Roxana-Elena Tudoroiu, Mohammed Zaheeruddin, Sorin Mihai Radu, Dumitru Dan Burdescu, Nicolae Tudoroiu, is on *"The Implementation and the Design of a Hybrid-digital PI Control Strategy Based on MISO Adaptive Neural Network Fuzzy Inference System Models–A MIMO Centrifugal Chiller Case Study."* The authors develop accurate multi-input single-output adaptive neuro-fuzzy inference system models, suitable for modelling the nonlinear dynamics of any process or control system and present new contributions to improve significantly in terms of accuracy, simplicity and real time implementation the control of single-input single-output autoregressive moving average models with exogenous input.

Chapter 10, by Recep Özalp, Nuri Köksal Varol, Burak Tasci and Ayşegül Ucar, is entitled *"A Review of Deep Reinforcement Learning Algorithms and Comparative*

Results on Inverted Pendulum System." The authors present the use of deep reinforcement learning algorithms to control the cart-pole balancing problem, with potential application in robotics and autonomous vehicles.

The *fifth part* of the book consists of three chapters devoted to *Advances in Deep Learning in Feature Vector Processing.*

Specifically, Chap. 11, by Kostas Liagkouras and Kostas Metaxiotis, is on "*Stock Market Forecasting by using Support Vector Machines.*" The authors investigate stock prices forecasting by using a support vector machine with some well-known stock technical indicators and macroeconomic variables as input.

Chapter 12, by Brian Hilburn, is on "*An Experimental Exploration of Machine Deep Learning for Drone Conflict Prediction.*" The author experimentally explores via low-fidelity offline simulations the potential benefits of Deep Learning for drone conflict prediction.

Chapter 13, by Georgios Kostopoulos, Maria Tsiakmaki, Sotiris Kotsiantis and Omiros Ragos, is on "*Deep Dense Neural Network for Early Prediction of Failure-Prone Students.*" The authors evaluate the efficiency of deep dense neural networks with a view to early predicting failure-prone students in distance higher education.

Finally, the *sixth part* of the book consists of three chapters devoted to *Evaluation of Algorithm Performance.*

Specifically, Chap. 14, by Gregory Koronakos and Dionisios Sotiropoulos, is on "*Non-parametric Performance Measurement with Artificial Neural Networks.*" The authors develop a novel approach for large scale performance assessments, which integrates Data Envelopment Analysis (DEA) with Artificial Neural Networks (ANNs) to accelerate the evaluation process and reduce the computational effort.

Chapter 15, by Alexandros Tzanetos and Georgios Dounias, is on "*A Comprehensive Survey on the Applications of Swarm Intelligence and Bio-inspired Evolutionary Strategies.*" The authors survey Swarm Intelligence methods as well as methods that are not inspired by swarms, flocks or groups, but still derive their inspiration from animal behavior to be used in optimization problems which can be mathematically formulated and are hard to be solved with simple or naïve heuristic methods.

Chapter 16, by Alessia Amelio and Ivo Rumenov Draganov, is on "*Detecting Magnetic Field Levels Emitted by Tablet Computers* via *Clustering Algorithms.*" The authors study the problem of detecting ranges of extremely low frequency magnetic fields emitted from tablet computers and analyse and cluster the measurement results using seven well-known algorithms.

In this book, we have presented some significant advances in Deep Learning-based technological application areas. The book is directed towards professors, researchers, scientists, engineers and students in computer science-related disciplines. It is also directed towards readers who come from other disciplines and are interested in becoming versed in some of the most recent advances in these active technological areas. We hope that all of them will find the book useful and inspiring in their works and researches.

This book has also come as a sixth volume, following five previous volumes devoted to aspects of various Machine Learning Paradigms [5–7, 9, 11]. As societal demand continues to pose challenging problems, which require ever more efficient tools, methodologies, systems and computer science-based technologies to be devised to address them, the readers may expect that additional related volumes will appear in the future.

References

1. J. Toonders, Data is the new oil of the digital economy, *Wired,* (2014) https://www.wired.com/insights/2014/07/data-new-oil-digital-economy/
2. K. Schwabd, The fourth industrial revolution-what it means and how to respond, *Foreign Affairs.* December 12 (2015). https://www.foreignaffairs.com/articles/2015-12-12/fourth-industrial-revolution
3. E. Rich, K. Knight, S.B. Nair, *Artificial Intelligence*, 3rd edn. Tata McGraw-Hill Publishing Company (2010)
4. J. Watt, R. Borhani, A.K. Katsaggelos, *Machine Learning Refined–Foundations, Algorithms and Applications*, 2nd edn. Cambridge University Press (2020)
5. A.S. Lampropoulos, G.A. Tsihrintzis, *Machine Learning Paradigms–Applications In Recommender Systems,* vol. 92, in *Intelligent Systems Reference Library Book Series.* Springer (2015)
6. D.N. Sotiropoulos, G.A. Tsihrintzis, *Machine Learning Paradigms–Artificial Immune Systems and their Application in Software Personalization,* vol. 118, in *Intelligent Systems Reference Library Book Series.* Springer (2017)
7. G.A. Tsihrintzis, D.N. Sotiropoulos, L.C. Jain (Eds.), *Machine Learning Paradigms–Advances in Data Analytics*, vol. 149, in *Intelligent Systems Reference Library Book Series.* Springer (2018)
8. A.E.Hassanien (Ed.), *Machine Learning Paradigms: Theory and Application*, vol. 801 in *Studies in Computational Intelligence Book Series.* Springer (2019)
9. G.A. Tsihrintzis, M. Virvou, E. Sakkopoulos, L.C. Jain (Eds.), *Machine Learning Paradigms-Applications of Learning and Analytics in Intelligent Systems*, vol. 1 in *Learning and Analytics in Intelligent Systems Book Series.* Springer (2019)
10. J.K. Mandal, S. Mukhopadhyay, P. Dutta, K. Dasgupta (Eds.), *Algorithms in Machine Learning Paradigms*, vol.870 in *Studies in Computational Intelligence Book Series.* Springer (2020)
11. M. Virvou, E. Alepis, G.A. Tsihrintzis, L.C. Jain (Eds.), *Machine Learning Paradigms–Advances in Learning Analytics*, vol. 158 in *Intelligent Systems Reference Library Book Series.* Springer (2020)
12. A. Samuel, Some studies in machine learning using the game of checkers, IBM J. **3**(3), 210–229 (1959)
13. J.Patterson, A. Gibson, *Deep Learning–A Practitioner's Approach*, O' Reilly (2017)

Part I
Deep Learning in Sensing

Chapter 2
Vision to Language: Methods, Metrics and Datasets

Naeha Sharif, Uzair Nadeem, Syed Afaq Ali Shah, Mohammed Bennamoun, and Wei Liu

We may hope that machines will eventually compete with men in all purely intellectual fields.

Alan Turing, Computing Machinery and Intelligence (1950).

Abstract Alan Turing's pioneering vision of machines in the 1950s, that are capable of thinking like humans is still what Artificial Intelligence (AI) and Deep Learning research aspires to manifest, 70 years on. With replicating or modeling human intelligence as the ultimate goal, AI's Holy Grail is to create systems that can perceive and reason about the world like humans and perform tasks such as visual interpretation/processing, speech recognition, decision-making and language understanding. In this quest, two of the dominant subfields of AI, Computer Vision and Natural Language Processing, attempt to create systems that can fully understand the visual world and achieve human-like language processing, respectively. To be able to interpret and describe visual content in natural language is one of the most distinguished capabilities of a human. While humans find it rather easy to accomplish, it is very hard for a machine to mimic this complex process. The past decade has seen significant research effort on the computational tasks that involve translation between and fusion of the two modalities of human communication, namely Vision and Language. Moreover, the unprecedented success of deep learning has further propelled research on tasks that link images to sentences, instead of just tags (as done in Object Recognition and Classification). This chapter discusses the fundamentals of generating natural language description of images as well as the prominent and the state-of-the-art methods, their limitations, various challenges in image captioning and future

N. Sharif · U. Nadeem (✉) · M. Bennamoun · W. Liu
The University of Western Australia, Perth, Australia
e-mail: uzair.nadeem@research.uwa.edu.au

S. A. A. Shah
Murdoch University, Murdoch, Australia

© The Editor(s) (if applicable) and The Author(s), under exclusive license
to Springer Nature Switzerland AG 2020
G. A. Tsihrintzis and L. C. Jain (eds.), *Machine Learning Paradigms*,
Learning and Analytics in Intelligent Systems 18,
https://doi.org/10.1007/978-3-030-49724-8_2

9

directions to push this technology further for practical real-world applications. It also serves as a reference to a comprehensive list of data resources for training deep captioning models and metrics that are currently in use for model evaluation.

Keywords Vision · Language · Image captioning · Deep models · Evaluation · Datasets · Metrics

2.1 Introduction

Artificial Intelligence's (AI) ultimate aim is to build systems that can perform human-like intellectual tasks [1, 2]. Various cognitive functions such as visual perception, planning, memory, language or learning are indispensable parts of human intelligence. Visual and language understanding are the two key cognitive abilities that enable humans to describe their surroundings, communicate with others, and plan a course of actions. To create artificial systems that can mimic these human traits requires insights and tools from two dominant sub-fields of AI: Computer Vision (CV) and Natural Language Processing (NLP). In particular, vision-to-language tasks such as image captioning [3], visual question answering [4], visual story telling [5] or video description [6] integrate computer vision and language processing as shown in Fig. 2.1. Amongst these, the fundamental task is image captioning, which underpins the research of other aforementioned visual understanding tasks. As such, it is the focal point of this chapter.

Computer Vision aims to create models that possess holistic visual scene understanding i.e., a model that can perform tasks similar to the human visual system. Such tasks can be categorised into low-level, mid-level and high-level vision tasks as shown in Fig. 2.2. Building blocks of holistic scene understanding such as segmentation, classification, and recognition have been the main focus of the computer vision community over the past few decades. Natural language processing, traditionally aims to understand the linguistic underpinning of meaning, using words as the basic building blocks. With the recent success of deep learning models [7, 8], NLP research community has made headway in designing powerful machine translation, question answering and dialogue systems. Likewise, with recent advances in core vision tasks, CV researchers are now more interested to compose human-like sentences for images, instead of just generating a list of labels. Figure 2.2 shows a similar trend line.

For a long time, language and vision related tasks have been studied independently of one another. However, in the last decade, the research community has shown a great enthusiasm towards the tasks that integrate both of these modalities. Amongst the variety of such multi-modal tasks, we would primarily focus on image captioning in this chapter, which has been under research for more than a decade. Other relevant tasks such as visual question answering, visual storytelling, video captioning and visual dialogue are relativity new. Therefore, our objective in this chapter is to discuss

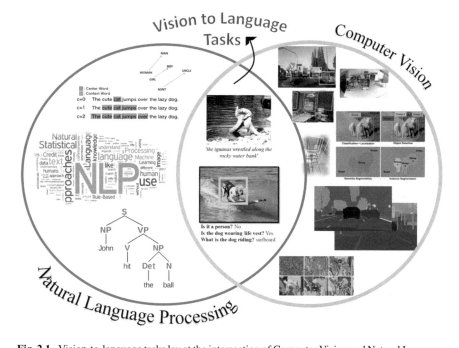

Fig. 2.1 Vision-to-language tasks lay at the intersection of Computer Vision and Natural Language Processing

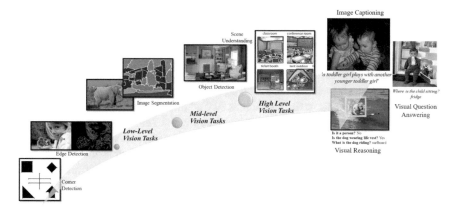

Fig. 2.2 Various low-to-high level (left to right) Computer Vision tasks

one of the pioneer vision-to-language tasks in depth and to briefly touch upon other spin off's, which are in their primary stages.

Describing the visual world in natural language is one of the most distinguished capabilities of a human being. While humans find it rather easy to construct sentences from images, this task is not as straightforward for machines. Language and vision are two different modalities and translation between them involves a variety of complex processes. It not only involves holistic visual understanding but also language modeling to construct human-like sentences. In the past few decades, rapid progress has been made towards the individual tasks such as recognition, segmentation, and classification. An array of these vision tasks can provide us with a list of unstructured labels (e.g., object labels and attributes, scene properties and human-object interactions) that alone cannot be used as a description. A good image description has to be comprehensive yet succinct (talk about the salient objects in an image) and grammatically correct. Moreover, the description of an image can reflect any aspect of an image, such as the objects and their attributes, characteristics of a scene (e.g., indoor/outdoor), interaction between people and objects, or it can even talk about the objects that are not present in the image (e.g., it can talk about children waiting for the school bus although the bus is not in the image since it has not arrived).

It is also pertinent to shed some light on different categories of image description. As argued in the literature [9, 10], image descriptions can be divided into three main classes: *Conceptual, Non-visual and Perceptual* descriptions. *Conceptual descriptions* are the ones that offer a concrete account of the visual content, i.e., scenes, entities, their attributes, relationships, and activities. These are the ones that are most relevant to the image understanding that we seek to attain for automatic image captioning. *Non-visual descriptions* capture information that cannot be explicitly inferred from the image content alone, e.g., about the situation, time and location. These are the type of descriptions that people tend to provide when sharing images on social media. Lastly, *perceptual descriptions* provide low-level visual aspects of images e.g., whether it is a painting or a sketch, or what the dominating colors or patterns are. In this chapter, we focus primarily on studying the elementary category of descriptions i.e., the content-based conceptual descriptions.

Generating image descriptions is an intricate but significant task, as progress in this research area has substantial practical implications. Trillions of images are available on the web,[1] and the estimated number is constantly growing every second. Photo sharing and social websites such as Flickr[2] and Facebook,[3] online news cites like Yahoo, BBC and other web pages keep adding their share of visual data to this gigantic assemblage. Finding pictures in this ever-growing collection is a significant problem. Search engines such as Google [4] rely heavily on the text around the pictures for image retrieval task. Automatically associating accurate descriptions with the respective images can improve content-based retrieval and enable the search engines to handle longer and targeted queries. Image descriptions can also help organize our photo collections, aid domestic robots to communicate their visual observations with the users, and most importantly, assist the visually impaired to better understand their surroundings.

---

[1] https://focus.mylio.com/tech-today/heres-how-many-digital-photos-will-be-taken-in-2017-repost-oct.

[2] https://flickr.com/.

[3] https://www.facebook.com/.

[4] www.google.com.

This chapter is organized as follows: in Sect. 2.2, we discuss the various challenges related to image captioning. In Sect. 2.3, we review various captioning models and their taxonomy. Apart from reviewing the state-of-the-art deep captioning models, through this work, we attempt to organize the existing models into a comprehensive taxonomy. We then discuss various methods for assessing captioning models in Sect. 2.4, followed by an arguably exhaustive list of existing datasets for image captioning. In Sect. 2.5, we also review the existing datasets and classify them into various categories, based on their properties and applications for which they can be used. In Sect. 2.6 we will look into various practical applications of visual captioning. Some of the prominent vision-to-language tasks are briefly covered in Sect. 2.7 and the final section presents the conclusion with an outlook to some future directions.

2.2   Challenges in Image Captioning

The ability to aptly describe an image in natural language feels so natural to us, we often forget how difficult it is for a machine to accomplish the same task. In standard settings, an image is fed to a computer in the form of an array of numbers indicating the intensity of light at any position, as shown in Fig. 2.3. To generate descriptions for an image, this simple array of numbers must be transformed into high-level semantic concepts or syntactically correct sequence(s) of words i.e., sentence(s), which are semantically meaningful in depicting interactions, events and activities. This is a very complex task and though a rapid progress has been made in this research area, the problem at hand is far from being solved. In what follows, we will be discussing few of the major challenges in this line of research.

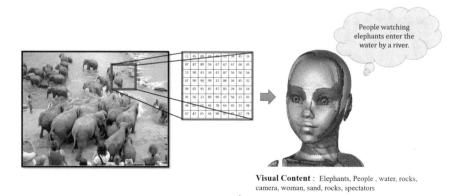

Visual Content : Elephants, People , water, rocks, camera, woman, sand, rocks, spectators

Fig. 2.3 Image captioning is a challenging task, which involves translation of an array of numbers (image) to a well-formed sequence of words

2.2.1 Understanding and Predicting 'Importance' in Images

State-of-the-art visual detectors can recognize hundreds and thousands of object categories and treat them with equal importance. However, when it comes to describing an image, humans do not give equal importance to all objects, only a few "salient" ones are mentioned [12]. Figure 2.4 shows that although the image contains more than 20 objects, only as few as 5–6 are stated in the description by people. Therefore, in order to derive more human-centric descriptions, we need to know what people care the most about in images. Are all of the recognized objects equally important? or is the recognition of all objects important? What are the important factors that relate to human perception? How can the perceived importance affect image description? Answers to all these questions are challenging to obtain but are the key to generating more natural-sounding, human-like captions.

2.2.2 Visual Correctness of Words

One of the most important aspects of caption quality is the accuracy with which the visual content is described. This requires models to have a deep visual understanding. Moreover, it is noticed that humans lay more importance on the correctness of visual content in the descriptions compared to specificity (mentioning only the salient content) [13]. Rohrbarch et al. [14] highlighted that image captioning models tend to "hallucinate" objects, i.e., the objects that are not present in the image are mentioned in the caption. Object hallucination is a major concern, which reflects poor visual understanding as well as over-reliance on language priors [14]. Captioning models tend to memorize or rely on the combination of words that frequently appear together in the corpus. Miltenberg et al. [15] presented an error analysis of descriptions generated by an attention-based captioning model to reveal that 80% of the descriptions contain at least a single mistake and 26% are completely wrong.

Fig. 2.4 Not all content is salient. Descriptions written by people for an image in MSCOCO dataset [11] are shown on right. Not all detected objects are mentioned in the descriptions, some objects (elephant, man, woman) seem to hold more importance than others (umbrella, cycle, trees etc.)

Predicting activities, events, attributes, salient objects and spatial information accurately from a single image is a challenge and is very crucial to the success of image captioning systems.

2.2.3   Automatic Evaluation Metrics

The heterogeneity of input and output data of image captioning models makes evaluation very challenging. The availability of reliable and effective evaluation measures is vital for the advancement of image captioning research. Metrics like BLEU [16], ROUGE [17], and METEOR [18] were originally designed to evaluate the output of machine translation or text summarization engines. Their suitability for the evaluation of image descriptions is widely questioned [10]. A number of hand-crafted measures such as CIDEr [19], SPICE [20] and WMD [21] have been proposed for captioning, but they have various drawbacks [15]. Miltenberg et al. [15] stressed that text-based metrics fail to highlight the strengths and weaknesses of existing models. Recently, learning-based metrics [23] have been proposed for caption evaluation, which rely on data to learn evaluation. These metrics have improved upon the performance of existing measures in terms of system-level correlation with human judgements. However, they are not robust to sentence perturbations [24] and still fail to achieve high correlation with human judgements at the caption-level.

2.2.4   Image Specificity

Existing automatic image evaluation metrics such as METEOR, ROGUE, BLEU, and CIDEr compare a set of human-provided reference descriptions with the generated

'People lined up for a train'   'A typical fruit and vegetable market'
'People are waiting at a terminal'   'Watermelons look fresh'
'Long line at the train station'   "people are buying fruit"
People lined up at a train terminal   'Salesman is weighing the purchase'

Fig. 2.5 The captions a for 'specific' image (left) are semantically consistent. Whereas, those for an 'ambiguous' image (right) are more diverse

descriptions of an image. The evaluation procedure does not take into account the fact that images can either be ambiguous or specific (Fig. 2.5). Ambiguous images can have several different plausible descriptions associated with them i.e., they can be described in many different ways. Whereas, for content-specific images, only a narrow range of descriptions are considered reasonable. This causes a problem in the evaluation process. Ambiguous images often lead to a penalty of a low score, since the generated descriptions have a higher chance of not matching the reference descriptions, compared to the specific images. Moreover, human level of agreement needs to be taken into account during evaluation. This challenge needs to be addressed in order to improve the performance of captioning systems in real-world applications.

2.2.5 Natural-Sounding Descriptions

Composing human-like, natural-sounding descriptions for images is the prime objective of any automatic image captioning model. Generating a list of labels for the objects and entities that are present in an image seems fairly convenient, instead of well-formed natural sounding sentences. However, we do not want our future AI systems to just provide name tags, but to communicate with their users in the most natural way. There are various characteristics which make natural language unique and challenging to replicate. Diversity, personalization, and humour are a few of the human language traits, that are further discussed below:

2.2.5.1 Personalization

We are most aware of the fact that every individual person tends to describe an image in their own way, which involves the role of perception, knowledge, vocabulary and other external factors. These descriptions might even vary depending on age and gender. Incorporating that into a machine is a fairly complicated task. Generating description of images becomes more challenging if it is done with a user-specific perception. For instance, a librarian would require a different kind of description compared to an art critic for the same photograph. Personalization is what makes descriptions more natural and diverse. Tackling this issue is essential for the future AI systems, but is deemed perplexing.

2.2.5.2 Diversity

Humans are capable of describing what they see from unique and diverse perspectives. Figure 2.6 shows how various people tend to describe an image in different ways. This feature is crucial in designing future AI systems that can communicate with their users in the most natural way. However, designing systems that are capable of generating diverse, as well as unique descriptions, for a particular scene is an open

Descriptions

Person1: Glass is half empty

Person2: Glass is half full

Person3: Is there water in the glass?

Person4 : This is an elegant glass

Person5 : Water inside the glass looks clean

Person 6: Water appears to be swirling inside

Person7 : Glass is transparent

Fig. 2.6 An example of diverse descriptions generated by humans for the given image

challenge. Current systems which employ neural network language models generate sentences that are very similar to the ones in the training sets [25, 26]. Devlin et al. highlighted in their work [25] that their best model is only able to generate 47% of unique descriptions. It is a challenge to devise a framework to annotate images with descriptions that show uniqueness and diversity in terms of expression and is the prime focus of the ongoing research [27] in this field.

2.2.5.3 Understanding and Predicting Humour in Images

Humour is an important characteristic of human expression. Sense of humor is found to correlate positively with various aspects of social competence [28], creativity and intelligence [29]. We come across a lot of visual content which is humorous, and we as humans are apt at understanding it. However, how can a machine be able to understand and describe that? What content in an image causes it to be funny and how to compose a humorous description? Or in general, how to detect and express sentiments? Understanding and describing visual humour remains as an unsolved research problem and is fraught with challenges.

2.3 Image Captioning Models and Their Taxonomy

In the last decade, a plethora of literature on image captioning has been published. The existing literature on image captioning can be divided into two main categories: (1) Generation-based models and (2) Example-lookup based models (Fig. 2.7).

The first type of models (generation based) generate a novel image description from the predicted image content [30–32]. Generation based approaches predict the semantic meaning of the visual content in the form of a vector representation, which is fed to the language models to generate an image description. These approaches differ in terms of the training framework (end-to-end trainable vs. compositional). The end-to-end trainable frameworks can be further categorised into encoder-decoder versus reinforcement learning based paradigms. In the former framework, the encoder mod-

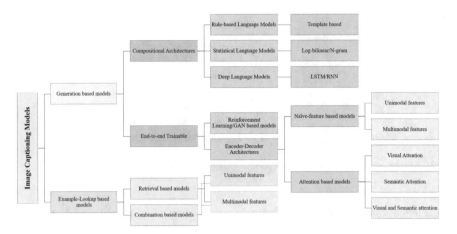

Fig. 2.7 Taxonomy of Image Captioning models

ule consists of vision algorithms, which encode the visual information to be fed to the decoder. The decoder is typically a language model, which transforms/decodes the input information into a well-formed sequence of words, i.e., a caption. Encoder-decoder based frameworks are rigorously researched upon, and the proposed models in this category differ in terms of the encoding i.e., the features (naïve vs. attention-based features) that are fed mainly as an input to the decoder. With some exceptions, most of these models use RNN/GRU/LSTM as language decoders. The compositional architectures, on the other hand are not end-to-end trainable. The visual content determination and language generation modules are trained separately, and then fused to perform captioning. These models are mainly language driven and differ in terms of the type of language models, in some cases, the overall framework.

Example-lookup based algorithms make use of existing human-authored captions to describe images [33, 34]. For any input image example-lookup based models search for the most similar images in a database of captioned images. They either transfer the description of the most similar images to the input image [35] or synthesize a novel description by using the description of a set of the most similar images [36]. As argued by Hodosh et al. [10], framing the image description problem as a natural language generation task takes the focus away from the underlying problem of image understanding. Therefore, example-lookup based algorithms avoid building sentences from scratch and make use of available descriptions in the image-caption database, thus, mainly focusing on image understanding. These models can further be divided into subgroups of unimodal or multimodal-feature space models. The former exploits the similarities in the visual-space to retrieve descriptions, whereas the later uses a multimodal-space (common space for image and text retrieval) to transfer descriptions.

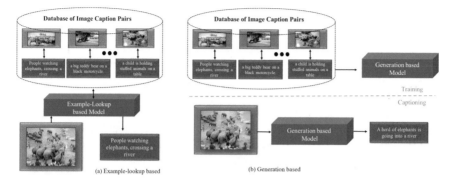

(a) Example-lookup based   (b) Generation based

Fig. 2.8 An overview of example-lookup and generation based captioning models. The former model retrieves the most appropriate caption from a database image-caption pairs, while the later first trains on a dataset and then generates novel captions

2.3.1 Example Lookup-Based Models

These algorithms deal with image description as a retrieval problem and try to find the closest match in a database of image-caption pairs and use that as an output, as shown in Fig. 2.8. The descriptions associated with the retrieved images can either be mapped directly (retrieval-based methods) or can be used to synthesize novel descriptions (combination-based methods).

2.3.1.1 Retrieval-Based Methods

Retrieval-based methods rely on finding the best match in a database but are restricted by the size of their training database. The retrieval can be performed in unimodal space or mulitmodal (semantic) space and based on that, these approaches can be divided into the categories described below:

– **Unimodal-feature based Methods**

   Unimodal (visual/linguistic) methods, find the closest match to source modality (image) in a database, in the space of the source (image)—for examples given an image (visual modality) to be captioned, the unimodal feature-based methods would search for the best match in the visual feature space of images. In one of the earliest work in 2011, Ordonez et al. [35], proposed a model which uses global features such as GIST [37] and Tiny images [38] to retrieve a set of similar images for an input source image. The retrieved set of images are re-ranked using estimates of image content (objects, actions, stuff, attributes, and scenes) to output the most adequate caption. Patterson et.al [39] further extended the baseline proposed in [35] and showed improved performance by using a different set of scene attributes-based global features. Devlin et al. [25] proposed a method in which activations of a Convolutional Neural Network are used as global image features and the

K-nearest neighbor retrieval method is used to extract a set of images that are similar to the query image in the visual space. An F-score of n-gram overlap is computed between a set of candidate captions to choose the final output description. The caption that has the highest mean n-gram overlap with the other candidates is selected as the output.

Unimodal-feature based methods, often require an extra post-processing step, such as re-ranking the retrieved captions [25, 35, 40], since similarity in the unimodal feature space might not always imply a good output caption. Nevertheless, one of the advantages of such approaches is that they only require unimodal features (features of a single modality) for performing retrieval [41].

- **Multimodal-feature based Methods**
  An alternative approach is to use multi-modal semantic space to find similitude during retrieval. Farhadi et.al. [42], proposed a model which assumed an intermediate 'meaning space' between images and captions. Both images and captions in a database, are mapped to the space of meanings, where the meaning of each modality is represented as a triplet of ⟨objects, actions, scene⟩ . The captions that are most similar in meaning to the query image are retrieved as candidates. This approach is symmetric, i.e., it is possible to search for the best caption, given an image, and vice versa. Hodosh et al. [10] use KCCA to learn a semantic space. Socher et al. [43] learn coordinated visual (CNN outputs) and textual representations (compositional sentence vectors) using neural networks. Karpathy et al. [44] further extend the work in [43] and propose a model to embed visual fragments (objects in images) and linguistic fragments (dependency tree relationships) into a multimodal-feature space.

  Mulitmodal feature-based approaches perform better compared to their unimodal counterparts, as they use a more meaningful semantic space for retrieval, which is representative of both visual and linguistic modalities. Furthermore, the multimodal space allows symmetry, i.e., image retrieval and image annotation [41], i.e., for a given image, set of the most similar captions can be retrieved, or given a caption, the corresponding images can be fetched. However, constructing a multimodal-space is not as straightforward and is data dependent.

2.3.1.2 Combination-Based Methods

These methods take the retrieval-based models one step ahead. Instead of just retrieving candidate captions from a database, combination-based models make use of intricate rules to generate output captions, based on a number of retrieved descriptions [41]. Kuznetsova et al. [36] devised a combination-based[5] model to retrieve and combine human-composed phrases to describe images that are visually alike. Kiros et al. [45] proposed an Encoder-Decoder pipeline that learns joint multimodal

---

[5]Models make use of intricate rules to generate output captions, based on a number of retrieved descriptions/captions [41].

embedding space for ranking and generating captions. The proposed model not only retrieves the most suitable descriptions for a given image but also generates novel descriptions. The encoder uses a deep convolutional network [46] and a long short term memory (LSTM) [47] to project image and sentence features to a joint embedding space. The decoder uses a structure-content neural language model (SC-NLM) to generate new captions from scratch. While the Encoder handles the ranking of captions and images, the decoder tackles the caption generation part. In [48], Mao et al. proposed a multimodal Recurrent Neural Network (MRNN) for sentence retrieval and generation. Their proposed architecture is comprised of a vision model (containing Deep CNN), language model (containing embedding and recurrent layers) and a multi-modal part.

Karpathy et al. [34], proposed a model which generates captions for various regions in an image. A deep neural network model was used to learn the alignment between sentence segments and image regions. Image and sentence representation are computed using Region Convolutional Neural Network (RCNN) and Bidirectional Recurrent Neural Network respectively. The visual and textual representations are projected to a multimodal space to learn the alignment between the two modalities. A multimodal recurrent neural network uses the inferred alignments to generate novel dense descriptions of images. In [49], Chen et.al, proposed a model that generates novel image descriptions and also reconstructs visual features given an image description. In this work, an RNN is used for the language modeling task. Libret et al. [50] learn a common space of image and phrase representations. For a given image, the most appropriate phrases are retrieved and combined via a log-bilinear language model to generate a novel caption.

The performance of example lookup-based approaches relies heavily on the database size and quality. Though a larger database might result in an improved performance, but it might have an impact on the inference speed. Moreover, it is also unreasonable to expect the dataset to contain relevant candidate captions for all possible input images.

2.3.2 Generation-based Models

Treating image description as a retrieval task enables comparative analysis between existing methods. However, retrieval based approaches lack the ability to accurately describe an unseen combination of objects and scenes in the input images. Approaches which directly generate descriptions from an image adopt a general pipeline. The overall pipeline can be divided into two parts, (1) visual content detection and (2) language generation (Fig. 2.9).

Visual content detection, relates to predicting/determining the visual information to be conveyed through the caption. The objective of this phase is to convert the image into an intermediate representation, which adequately reflects the meaning of an image. In most of the conventional works [51, 52], various computer vision algorithms were used to detect objects, determine their characteristics and relationships.

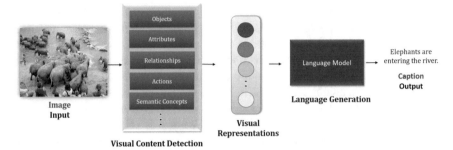

Fig. 2.9 An overview of the generation based image captioning framework

State-of-the-art captioning systems rely on high-level features acquired from a deep network (Convolutional Neural Networks) as a compact image representation. Some other recent works augment the high-level CNN features with semantic information such as vector of confidences of the visual concepts/ attributes etc., to improve upon the performance [32, 53].

Image representation can be in the form of tuples which reflect various aspects of an image (detected objects, their visual attributes and spatial relationships, scene type etc.) [54–56], high-level features from a CNN (Convolutional Neural Network) [22, 26, 57, 58] or a combination of both [31].

Language generation is the second phase which involves the generation of a natural language description. The visual and/or semantic representations of the first phase (visual content determination) are fed to a language model to form grammatically correct sentences using natural language generation techniques (n-gram, template based, RNN etc.). The language models can be pre-trained (on relevant language corpus) or trained from scratch on captions.

2.3.2.1 Compositional Architectures

Image captioning models that have a compositional architecture are not end-to-end trainable, and comprise of independent visual detection and language generation blocks. These models focus on optimizing each building block (visual and linguistic) independently. Though the majority of the compositional models rely on CNNs to extract image representations, they differ in terms of the language generation models, as well as the overall framework. In this section, we classify the compositional architectures based on the models that they use for language generation i.e., (1) Rule-based language models (2) Statistical Language models and (3) Recurrent Neural Network Language models.

– **Rule-based language model-based captioning**
   Amongst the rule-based language models, template-based systems are the most commonly used for captioning. Template-based generation systems directly map

labels, tags/detections (non-linguistic inputs) to a linguistic structure (template) [59]. The template contains gaps/slots, which are filled by relevant information to output well-formed sentence. For example '*[ ] is [ ] at the [ ]*' represents a template, where the gaps '[ ]' can be filled by detecting relevant information from the input image, to output a proper caption i.e., '*[train] is [arriving] at the [station]*'. Though the output sentences are well-formed, template-based sentences are considered inferior to other counterparts in terms of quality and variation [60]. Yang et al. and Kulkurni et al. in [52] and [51] respectively, build image descriptions from scratch; filling up the predefined sentence templates with detected instances of an object, attributes, action, scene and spatial relationship. Yang et al. [52] use trained classifiers to predict objects and scenes, whereas a large generic corpus (English Gigaword) is leveraged upon to predict verbs (actions) and prepositions (spatial relationship). A Hidden Markov Model (HMM) is then used to compose captions. Kulkerni et al. [51] predict objects and their attributes (e.g., fluffy dog, green grass etc.) and the spatial relationships in the image. The predicted triple ⟨objects, attributes, spatial relationship⟩ are then used for filling predefined templates, e.g., "This is a photograph of ⟨count⟩, object(s)." Elliot et al. in their work [61] use Visual Dependency Representations (VDR) to capture the relationship between objects in the images. VDR corresponding to the input image is then traversed to fill caption templates.

- **Statistical Language model-based captioning**
  Statistical language models evaluate the probability of a particular word w_t in a sentence, given a context w_{t-n+1}^{t-1}. N-gram, log-bilinear and maximum entropy language models are some of the models that come under this category and are used for captioning. Li et al. [62] and Kulkarni et al. [51] compose image captions, using visual detections and n-gram language models. In [30] Yatskar et., build a more comprehensive language generation model from densely labeled images incorporating human annotations for objects, parts, attributes, actions, and scenes. Given rich image annotations, sentences are generated using a generative language model, followed by a re-ranker. Fang et.al, in [55] elude the need of hand-labeled data to detect a list of objects, attributes, relationships etc. by using weakly super-vised word detectors that are trained on a dataset of images and their corresponding descriptions. The detected words are fed to a Maximum Entropy Language Model (MELM) to generate sentences. In the final stage, a set of generated sentences are re-ranked according to a predefined criteria.
  Tran et al. [63] address the challenge of describing images in the wild. Similar to [55], Tran et al. select MELM for their work .Visual detectors of the proposed model in [63] are trained on a broad range of visual concepts, as well as celebrities and landmarks. Along with a deep multi-modal similarity model (DMSM) and a confidence model, their proposed method achieved the state-of-the-art results for the novel captioning task.

- **Deep Language model-based captioning**
  The majority of the recent image captioning models use Recurrent Neural Network (RNN) based deep language models. RNNs are the state-of-the-art deep language

models, which gained popularity after a successful performance in machine translation [64, 65]. Hendircks et al. [66] proposed a *Deep Compositional Captioner* to address novel object captioning. Their proposed framework consists of a three main blocks (1) lexical classifier, (2) language model and a (3) caption model. Lexical classifier is trained to detect visual semantic concepts in the images, whereas, an LSTM[6] is used for language modelling. The caption model integrates both a lexical classifier and language model to learn a joint captioning model. Ma et al. [68] propose a two-stage framework for generating image captions, by feeding LSTM with structural words. A multi-task based method is used to generate a tetrad of ⟨objects, attributes, activities, scenes⟩, which is decoded into a caption using an LSTM. Wang et al. [69] address the challenge of generating diverse and accurate factual image descriptions. They propose a two-stage, coarse-to-fine model to generate captions. The first stage involves the generation of a skeleton sentence that contains the main objects and relationships. In the second stage the skeleton sentence is refined by predicting and merging the attributes for each skeletal word.

2.3.2.2 End-to-end Trainable Architectures

The two main types of models that we would discuss under this category are (1) encoder-decoder and (2) reinforcement-learning based models.

- **Encoder-Decoder Models**

  A more recent line of work [22, 26, 53, 57] uses end-to-end deep learning approaches to generate captions for images. An encoder-decoder based architecture is based on a single joint model, which takes an image as an input and is trained end-to-end to maximise the probability of generating a sequence of words i.e., caption. The main inspiration behind these encoder-decoder type models, comes from Machine Translation (MT). Sustkever et al. [65] proposed a sequence-to-sequence learning with neural networks for translating a source to a target sentence. Their proposed method consists of a multilayered Long Short-Term Memory (LSTM), which encodes the source sequence into vector/embedding, and another deep LSTM which decodes the vector to a target sequence.

  An encoder takes an input (image/sentence) and transforms it into a feature map. This feature vector represents the input information. The decoder, converts the input feature vector/embedding to a target output. Inspired by the state-of-the-art performance of sequence-to-sequence models in MT, Vinyals et al. [26] proposed a model which uses a Convolutional Neural Network (CNN) as an encoder to generate a compact representation of an image. High-level image features (form layer Fc7) extracted from the CNN are fed to a Recurrent Neural Network (RNN) [70], which acts as a decoder. Various models have been proposed in the literature that follow the CNN-RNN baseline presented in [26] and have shown improvements.

---

[6]LSTM networks are a special kind of RNN models that are capable of learning long-term dependencies [67].

We further divide the work that comes under this category into (1) Naïve-feature based and (2) Attention-based models. Some of the prominent works in each category are discussed below.

- **Naïve feature-based models** extract vector encoding(s) from the encoder. These encodings are a representative of the visual information and can be the CNN activations, high-level concepts, or a combination of hand-picked features. Traditional encoder-decoder based models require careful feature engineering to ensure that the decoder is fed with the right amount of useful visual information, which can be transformed into caption. Various other works [26, 71, 72], successfully adapted the encoder-decoder architecture, in which visual features (fixed-length vector representations) encoded by a CNN are fed to a RNN, which act as a decoder. Wu et al. [73], replaced the image feature vector (CNN activations) with a high-level semantic representation of the image content. Where, the proposed semantic representation is a set of attribute likelihoods for an image. Yao et al. in [31] used both the image representations extracted from CNN and a list of detected attributes for given images as an input to the RNN. Both of these works [31, 73] reported an improved performance.

  Mathews et al., [74] address the challenging task of generating captioning with sentiments, by using a novel switching RNN in their framework. The switching RNN consists of two CNN-RNN pipelines running in parallel, where one pipeline focuses on generating factual captions, the other learns to generate words with sentiments. The two pipelines are optimally combined and trained to generate emotional image captions.

  Gan et al. [75] develop a Semantic Compositional Network (SCN) for captioning. Their proposed framework not only feeds the LSTMs with a visual feature vector (extracted from a CNN) but also uses a SCN to weigh the LSTM weight matrices via the vector of probabilities of semantic concepts. You et al. [76] propose two different models to inject sentiments to images. The proposed systems are built on top of the traditional encoder-decoder model [26], where the LSTM model contains a sentiment cell. The updated sentiment state and current hidden state of the LSTM are employed to predict the next word in the sequence.

  Instead of using RNNs for language modelling, some methods have also used CNN as language decoders [77–79]. The CNN-CNN approach has shown comparable/improved performance over some of the CNN-RNN based encoder-decoder schemes.

- **Attention-based models**

  Attention-based models draw inspiration from human visual attention [82], which allows humans to focus on the most salient information of the visual input. Similarly, attention mechanism in artificial neural models allows them to differentiate parts of the source input. In the case of image captioning the decoder is trained to attend to the salient information at each step of the output caption generation. Attention-based models in image captioning can be divided into three categories, based on the type of information (visual [22], semantic [80], visual-semantic) which they attend to (Fig. 2.10).

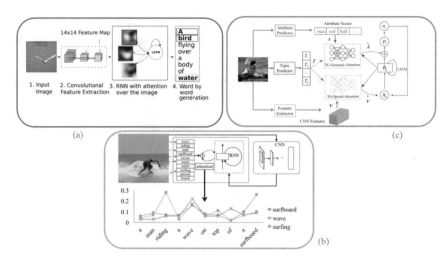

Fig. 2.10 An overview of three different types of attention models in the literature. (**a**) Visual attention based [22], (**b**) semantic attention based [80] and (**c**) visual-semantic attention based [81]

(a) ***Visual attention-based models***: In [22], Xu et al. train an end-to-end system (based on a CNN-RNN pipeline) which learns to attend to salient regions in the images while generating corresponding words in the output sequence. The attention mechanism proposed in [22] showed promising results on benchmark datasets. In another state-of-the-art approach [58], Lu et al. introduced an adaptive attention framework, which learns to attend to the visual input only when required. The learned model in [58] decides whether to attend to the image and where in order to extract meaningful information for caption generation.

Yao et al., [83] use Graph Convolutional Networks (GCN) to model semantic and spatial relationships between objects in an image. The learnt region-level features are then fed to attention-LSTM decoder for caption generation. Anderson et al., [32] propose a top-down bottom-up attention model for image captioning. The bottom-up attention model consists of Faster-RCNN [84] and a CNN, which generates the image bounding box features for the object class and also predicts the object attributes. The proposed framework also consists of a two-layer LSTM, where the first layer is characterised as the top-down attention model and attends to the image features generated by the bottom-up attention model. The second layer of the LSTM acts as the language model, which generated captions based on the attended image feature and the output of attention LSTM.

(b) *Semantic attention-based models*: You et al., [80] proposed a model based on semantic attention. The proposed model uses a CNN, which generates a vector of visual concepts and an LSTM, which attends to the semantically important concepts while generating the output sentence. The initial state of the language model (LSTM) captures the global visual information in the form of a feature vector, which is also extracted from the CNN. However, the model does not attend to the visual features/ image regions during caption generation. Zhang et al. [85] also propose a semantic attention-based model for describing remote sensing images and show promising results.

(c) *Visual-semantic attention-based models*: In a recent work, Zhu et al. [81] propose a novel attention mechanism. Their proposed model extracts three different types of feature vectors, visual, attributes and topic. A topic guided visual attention is used to select the salient image regions, whereas a topic guided semantic attention is applied on the predicted attribute vector. The proposed model shows improvement over other baseline models, which use either visual [22] or semantic attention [80].

- **Adversarial Networks and Reinforcement-learning based models**
  Existing encoder-decoder based models have shown promising results, but they lack the ability to generate variable and diverse descriptions. Diversity in expression is one of the human characteristics that the captioning models aim for. As discussed in the literature [86], captioning models which are trained to maximize the likelihood of the training samples, tend to regurgitate the training set examples. Thus, impacting on the diversity and variability of the output. To address this concern Dai et al. [86] proposed the first conditional GAN (Generative Adversarial Network) [87] based framework for image captioning, consisting of a *generator* and an *evaluator*. Both *generator* and *evaluator* have a neural network-based architecture (CNN-LSTM), which operate in a contrastive manner. The *generator* generates diverse and relevant captions, while the *evaluator* evaluates the generated descriptions for semantic relevance to the image and naturalness. Shetty et al. [88] also propose an adversarial training framework to diversify the generated captions. However their work mainly differs from [86] in terms of the discriminator (evaluator) design and the use of Gumbel Softmax [89] to handle back-propagation through a discrete sequence-of-words.
  In another contemporary work, Ren et al. [90], introduced a decision-making framework, comprising of a *policy* and *value* network. The *policy* and *value* networks are trained using an actor-critic reinforcement learning model. The *policy network*, which consists of a CNN and RNN, serves as a *local guidance*. It predicts the next output word according to the current state of the language model (RNN). The *value* network, which consists of a CNN, RNN and MLP (Multi layer perceptron), provides a lookahead guidance for all plausible extensions of the current state. Both networks are trained together using a novel visual-semantic embedding reward.

2.4 Assessment of Image Captioning Models

Evaluation of image captioning models is a daunting task, which is vital for the optimization and advancement of the captioning systems. A straight forward way to assess a captioning model is to evaluate its output captions for quality. However, it takes proficiency in language as well as intellectual ability to judge whether the output caption (usually known as the candidate caption) is truly representative of the image. Moreover, a multitude of correct captions corresponding to the input image, add to the complexity of evaluation, which may also depend on the intended application for captioning. For example, the tone or sentiment of a caption might be more important in some applications, but trivial in others. In this section, we will discuss the three different methods that can be used to evaluate captioning models:

1. Human evaluation.
2. Automatic evaluation metrics.
3. Distractor task(s) based fine-grained evaluation.

2.4.1 Human Evaluation

Human judgements are considered the gold-standard for caption evaluation. Humans find it very easy to compare images with captions, and the intricate biological processes behind this intellectual ability have been under research for decades [91–93]. Human understanding of the real world is far superior than machines, which allows humans to easily pin point important errors in the captions. For example, a caption 'a kangaroo is flying in the desert' is syntactically correct but the word 'flying' impairs its semantic quality, because kangaroos 'cannot' fly. Though the error is limited to a single word, but is very significant. Humans can easily note such errors just using common sense, even without looking at the image. Common-sense reasoning is an active research area in the knowledge representation community, for the past several decades [94–97].

The quality of a caption can be judged based on various linguistic characteristics such as fluency, adequacy and descriptiveness. In the literature, different systematic methods [10, 56] have been used to gather human expert scores for the caption quality. Amongst the various quality aspects, fluency and adequacy are the two most important ones [98]. *Adequacy* refers to the semantic correctness of the captions, disregarding the sentence grammar/structure, whereas, *fluency* indicates how grammatically well-formed a sentence is.

Bringing humans into the evaluation loop almost always creates room for variation in judgement. Therefore, to help reduce variance in the collected assessments, usually more than one subject is used to assess each caption. Some of the methods which were used in the literature to collect human judgements for captions include:

- **Forced-Choice Evaluation**: Human subjects are asked to choose the best caption amongst a pool of candidates, for a given image. The choice can be based on relevance of the caption to image content [56], its grammar, overall quality [99, 100] or meaningful incorporation of objects in the captions [101]. Another method found in the literature [102], is similar to a Turing test, in which human judges are asked to distinguish between human and machine generated captions for given images.
- **Likert-Scale Evaluation** : Manual assessments are also gathered using likert-scales[7] for various caption quality aspects, such as readability, relevance, grammaticality, semantic correctness, order of the mentioned objects, thoroughness, human-likeness, creativity or whether the salient objects are mentioned in the caption etc., [51, 52, 62, 63, 103].

Human judgements hold a fundamental importance in the evaluation of captioning systems. However, they are subjective, expensive and time consuming to acquire. In this fast paced era, we need cost-effective and efficient solutions, and this is where automatic metrics, come in handy. Automatic metrics, make the evaluation process faster, less resource-intensive, and easier to replicate.

2.4.2 Automatic Evaluation Metrics

Automatic evaluation measures are a vital part of the image captioning research cycle, and play a crucial role in the optimization of system parameters and performance comparison between captioning models. Automatic evaluation metrics aim to replicate the human judgement process, such that there is a high correlation between their scores and human judgements. Unlike humans, it is yet not possible for machines to directly compare images with captions (text), as they are two different modalities. Therefore, the evaluation is carried out by comparing reference caption(s), which is considered a true representative of the image content, against the candidate caption. The reference captions, are written by humans, whereas, the candidate captions are machine generated. Automatic measures capture the similarity between the candidate and reference captions, using different linguistic methods and generate a quality score, the higher the better.

One of the earliest metrics used for image captioning BLEU [16] was designed to measure the n-gram[8] . The higher the overlap, the better the quality of the generated text. BLEU was originally designed to judge machine translations. Various other n-gram based metrics such as METEOR and ROUGE from other domains (Machine Translation/Summarisation) were adopted for caption evaluation. Mainly

---

[7]Likert-scale is an ordered rating scale used for collecting user response for questions. The range of the scale usually represents the intensity of the response, for example 'excellent, good, normal, bad, worst' can represent the likert-items, where the range is from 'excellent' and 'worst'.

[8]"gram" corresponds to a word.

because of the similarity of the evaluation process in the other domains, where a reference translation/summary is compared with a hypothesis (machine generated translation/summary). Performance of such n-gram based measures is highly dependant on the number of references. The greater the n-gram space, the better the probability to hit the correct words.

CIDEr [19] was the first n-gram based measure, designed specifically for image captioning and showed a higher correlation to human judgments, compared to other commonly used measures. However, it was soon realized by the research community that n-gram overlap might not be the optimal indicator of caption quality. It is quite possible that captions that may contain the same words (lexicons) might convey a totally different meaning. For example. 'a cat ate a mouse' and 'a mouse ate a cat', both of these sentences have the same words, but convey a very different semantic meaning. On the other hand, captions that might consist of different words, can be semantically alike. Therefore, based on the limitations of lexical measures, a semantic metric SPICE [20] was proposed and since then a number of measures have been developed, to improve caption evaluation.

The quality of automatic evaluation metrics is usually assessed by measuring the correlation of their scores against human assessments. This correlation can be measured by comparing scores of captions, known as *caption-level* correlation or by comparing the system scores known as *system-level* correlation. The 'system-level score' corresponds to the average score of the captions, which are generated by a system. The automatic metrics need to faithfully reflect the caption quality, Otherwise, it would have a detrimental effect on the progress of this research area. As the competition between systems could take a wrong turn, by encouraging the wrong objective of achieving a 'higher score' instead of a 'better caption'.

We divide the captioning metrics into the two categories:

- Hand-designed metrics.
- Data-driven metrics.

Figure 2.11 shows an overall taxonomy of the image captioning metrics.

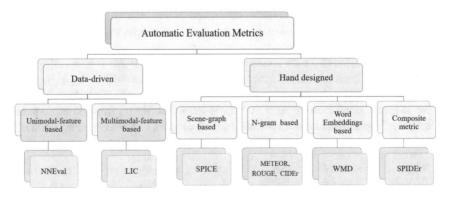

Fig. 2.11 Taxonomy of existing metrics for generation based captioning systems

2.4.2.1   Hand-Designed Metrics

Hand-designed measures are a set of handcrafted criterias/features that measure the similarity between texts. The similarity between captions is not evaluated directly using the caption in its original form. Instead, the captions are first converted into another representation, such as n-grams, scene-graphs or bag-of-word embeddings. Such representations are then used to measure the similarity between the candidate and reference captions. The similarity measure, can either be based on recall, precision, F-measure, etc.,. For instance, BLEU is based on the n-gram overlap precision between the candidate and reference caption. As discussed in [98], automatic metrics capture the correspondence between captions at different linguistic levels:

Table 2.1 Properties of various automatic captioning metrics

Name	Caption representation	Similarity measure	Linguistic-levels	Score range	Domain
BLEU [16]	N-grams	N-gram precision	Lexical and syntactic	0–1	Machine translation
METEOR [18]	Unigrams	F-mean based on unigram matches	Lexical and semantic	0–1	Machine translation
ROGUE-L [17]	N-gram sequences	F-score based on longest common sequence	Lexical and syntactic	0–1	Machine summarisation
WMD [104]	Word embeddings	Word movers distance	Lexical and semantic	0–1	Document similarity
CIDEr [19]	Tf-idf weighted N-grams	Sum of N-gram Cosine similarity	Lexical and syntactic	0–10	Image captioning
SPICE [20]	Semantic scene graph	F-score based on scene graph similarities	Semantic	0–1	Image captioning
SPIDEr [105]	–	–	Lexical, semantic and syntactic	0–1	Image captioning
LIC [23]	Sentence embeddings	Learned	Unknown	0–1	Image captioning
NNEval [24]	Vector of metric scores	Learned	Lexical, semantic and syntactic	0–1	Image captioning
LCEval [108]	Vector of metric scores	Learned	Lexical, semantic and syntactic	0–1	Image captioning

lexical, semantic and syntactic. For example, uni-gram measures such as $BLEU_1$ as they only assesses the lexical similarity between captions. Whereas, SPICE captures the semantic similarity between candidate and reference captions. Other measures like METEOR and WMD primarily capture the lexical correctness of the candidate caption but also some level of the semantic correspondance via synonym matching and word embeddings. Table 2.1 gives an overview of various properties of existing hand-designed and data-driven methods. Various existing measures are summarized below:

- BLEU [16] is a precision based metric, which counts the n-gram matches between the candidate and reference texts. BLEU score is computed via the geometric averaging of modified n-gram precisions scores, multiplied by a brevity penalty factor to penalize short sentences.
- METEOR [18] is a recall oriented metric, which matches unigrams between the candidate and reference captions. It also utilises synonymy and stemming of words for a better alignment between texts. Whereas, the metric score is defined as the harmonic mean of unigram precision and n-gram recall, where recall is weighted higher than precision.
- ROUGE-L [17] measures the similarity between captions by computing the Longest Common Subsequence (LCS). Where the longest common subsequence (LCS) of two captions refers to as common sequence of words with maximum length. The LCS-based F-score is termed as Rouge-L.
- CIDEr [19] assesses the agreement between the candidate and reference captions using n-gram correspondence. The commonly used n-grams are down-weighed by computing the *Term Frequency Inverse Document frequency* weighting. The average cosine similarity between the n-grams of the candidate and the reference caption is referred to as $CIDEr_n$ score. The mean of $CIDEr_n$ scores is known as CIDEr, where $n = 1, 2, 3, 4$.
- SPICE [20] evaluates the semantic correspondence between captions. The candidate and reference texts are converted to a semantic representation known as a *scene graph*. A scene-graph encodes the objects, attributes and relationships that are present in the captions. A set of logical tuples corresponding to the scene graphs are formed to compute an F-score. Where, the F-score is based on the conjunction of tuples.
- WMD [104] estimates the distance between two sentences using word embeddings. The minimum amount of distance that the embedded words of one caption need to cover to reach the embedded words of the other caption is termed as 'Word Movers Distance'. Each caption is represented as a weighted point cloud of word embeddings $d \in R_N$, whereas the distance between two words i and j is set as the Euclidean distance between their corresponding word2vec embeddings [106].
- Lexical-Gap [107] is a recent metric that measures the semantic richness of a description as a measure of quality. The metric is based on the lexical diversity of the entire output of the captioning system against the training corpus. A lexical diversity ratio is computed between the reference description corpus and the

machine generated corpus. This measures helps to improve the correlation of the existing measures by taking into account the semantic diversity of captions.

- Metrics for Retrieval-based Captioning Systems [10]

  - Recall (R@k): k refers to the position. In retrieval based systems, the goal is to retrieve the correct caption from a set. Recall at position k refers to the percentage of test queries for which the image captioning model retrieves the correct caption amongst the top k results. Where in the literature the usual values of k are 1, 5 and 10 (R@1, R@5, R@10).
  - The median rank score: estimates the median position of the correct caption in a ranked list of candidate captions. The lower value of this score represents a high performance, whereas, a high score represents a bad performance.

2.4.2.2 Data-Driven Metrics

The linguistic features to measure the correspondence between texts can be learned via a data-driven approach, in which a learning algorithm is used. Some of the recently proposed learning-based measures are discussed below:

- NNEval [24] is one of the first learning based captioning measure, which is trained to perform binary classification between human and machine captions. NNEval first converts the captions into another representation, which is a vector of metric scores. The vector is then fed to a neural network which classifies the input caption as machine or human generated. The probability of classification is used as a metric score.
- LCEval [108] is an extension of NNEval, which aims to capture lexical, syntactic and semantic correspondence between captions for quality evaluation. LCEval gives an insight into the parameters which impact a learned metric's performance, and investigates the relationship between different linguistic features and the caption-level correlation of the learned metrics.
- LIC is [23] another learned metric that is trained to distinguish between human and machine generated captions. It not only uses captions for training, but also the source, i.e., Image. Image features are extracted from a CNN, whereas, the candidate and reference captions are encoded into a sentence embedding via LSTM. All the features are combined via compact bilinear pooling and are fed to a softmax classifier. The probability of a caption belonging to the human class is taken as the metric score.

Learning-based measures have shown improvement in terms of caption and system-level correlation with human judgements. However there is still room for improvement. As captioning systems continue to improve, more powerful linguistic features need to be incorporated into such measures, along with the use of diverse training examples.

2.4.3 Distraction Task(s) Based Methods

Captioning systems need to have both visual and linguistic understanding to generate adequate descriptions. Apart from analyzing the output captions, researchers have developed another systematic way to assess the captioning models. Remember, that the ultimate goal of computer vision is the deep visual understanding, and captions are considered to be the representative of that. Humans can easily identify the decoy captions, or the ones that are most suited for an image. Given any image, humans can compose captions for it, and given a caption, humans can imagine images. Given, a set of captions, humans have the ability to choose the most appropriate/adequate one. There is a need to assess the fact that whether captioning systems are just memorizing the existing captions or whether they have actually developed some linguistic and visual sense [109].

The existing automatic measures are coarse, [110] and better methods for fine-grained evaluation of captioning systems are required. In order to judge the depth of visual and linguistic understanding of captioning systems, various datasets have been proposed in the literature [109–111]. Such datasets contain images paired with correct and perturbed (incorrect) captions. The captioning models are required to distinguish between the correct and incorrect captions, and/or detect the incorrect word or/and correct the wrong word. One such example is the **FOIL dataset** (Find One Mismatch between Image and Language Caption) [111], which contains one-foil caption (caption containing an incorrect word) per image. Shekhar et al. [111], proposed three tasks to evaluate the captioning systems using FOIL dataset; (1) classification of the foil caption, (2) detection and (3) correction of the foil word. Ding et al. [109] also released a **machine comprehension dataset** to exploit the extent to which the captioning models are able to understand both the visual and linguistic modalities. The task is to identify the caption that best suits the image from a set of decoys, i.e, the accuracy of selecting the correct target. The accuracy reflects the visual understanding of the models. Hodosh et al. [110] designed a set of **binary-choice tasks** to exploit the strengths and weaknesses of the image captioning systems. Each task consists of images paired with correct and incorrect captions. The captioning system has to choose the correct caption over the incorrect one. The authors built this dataset as an extension of Flickr30K [122]. For each task, correct sentences are perturbed in a different way. For example, the incorrect sentence might just be a part of the original caption, or have a different scene or place mention and may convey an entirely different semantic meaning. The authors conclude that the image captioning task is far from being solved, and highlighted various deficiencies of the existing models.

The primary purpose of these tasks is to assess the visual and linguistic under-standing of machines. Humans can quite conveniently detect semantic mistakes in the sentences, or choose the best suited caption for an image. These tasks tend to expect the same from the captioning models. Though this is a good step towards the fine-grained assessment of models, but is still far from perfect and it will not be unfair to say that, there is still no substitute for human quality assessments.

2.5 Datasets for Image Captioning

Humans learn in a variety of ways, via different sensory modalities. For example, what we see through our eyes is registered in our brains; what we read, develops not only our language facilities, but various other skills. Perhaps, the importance of data (via sensory organs) for learning is undeniable, even for humans. The more a child is exposed to different environments, the more he/she develops their intellectual skills, reasoning and language abilities etc. Similarly, AI algorithms also foster through data that might be images, labels, text, sound etc. The significance of data is undeniable.

Datasets are vital for the advancement of research in any field and also developing a synergy between research communities. Large datasets, such as ImageNet [112] have played a crucial role in advancing the computer vision research and have expedited the performance of various learning algorithms, resulting in the deep learning boom. Benchmark datasets are used to conduct various challenges, such as MSCOCO to solve existing problems and to reach new horizons. For computer vision, the data consists of images and labels, for natural language processing its annotated text, and for a domain like image captioning the dataset needs to contain both images and their corresponding captions.

Although a vast amount of image data is available online and is ever increasing, the availability of images paired with captions and suitable descriptions is relatively scarce. Therefore, the construction of an image captioning dataset is not only chal- lenging but also time consuming and resource intensive. Image Captioning datasets are collected systematically, following a set of guidelines, to provide the flavour of diverse examples to the learning algorithms. As it says '*a picture is worth a thousand words*', for each given image, there can be numerous plausible descriptions. Every person might describe an image in their own way. Factual, humorous, brief, negative are some of the styles that one might adopt. Therefore, what kind of descriptions should be collected for each image?

For tasks such as image classification and object recognition, a single label is often agreed upon. But in terms of image description, three different people might describe an image in three different ways or more. The manner in which each person describes depends on their linguistic ability, experience, observation, and style etc.,. Thus each dataset for image captioning has its uniqueness. Collected using different human subjects, some datasets focus more on including a variety of images, while others are driven by the need of inducing sentiments in their descriptions. Some of the datasets might be application specific, such as medicine, news, and sports. The availability of a variety of such datasets is essential to develop sophisticated models. Some of the general properties of a good dataset are that: it should be reusable, publicly available to foster research and development and it should be representative of a wide range of real world scenarios.

The textual content for the captioning datasets is usually crowd sourced via some popular crowd sourcing website, such as AMT [113], where human workers are asked to follow certain guidelines, adhere to certain linguistic requirements. The images are often sourced from popular social websites such as Flickr[9] or other publicly available image resource. Quality control is also very important. After the images and their respective captions are collected, they undergo a post pre-processing step to qualify for use in the training/testing of the captioning models.

Various datasets have been developed for the advancement of this field of research with a specific intention, catering to a specific limitation or a research gap. The images in some datasets focus on particular aspects, such as activities, while others focus more on linguistics, for example how the image should be described. There is no single dataset that caters for all the needs (such as diversity, sentiments, and succinctness) but each has its own significance. Some of the datasets are based on crowd sourced captions, others are based on user-generated captions, the images in some dataset might be domain specific, like cartoons, charts, abstract scenes, x-rays, etc., whereas, others might have generic images harvested from the Internet or Flickr. Visual data is not that straightforward to obtain due to copyright issues. Commercial datasets are typically larger in size, but are not available to the public and research community for free. What people put on public platforms e.g., Flickr might not be the most representative of the real world scenarios. Real world is far from just random clicks. Moreover, there are differences in the language content as well. Some datasets have more descriptive captions, others might have brief, or captions with sentiments. Some might be multilingual and others might just have captions in one language. Datasets also vary in terms of their size, i.e., number of image-caption pairs. The details of various image captioning datasets are given in Table 2.2. Based on the different features of the existing datasets we categorize them as follows:

- Generic Captioning datasets

  - Unilingual
  - Multilingual.

- Stylised Captioning Datasets
- Domain Specific Captioning Datasets.

2.5.1 Generic Captioning Datasets

These datasets contain images that correspond to generic activities or scenes and differ in terms of the number of images and their source (internet or crowdsourced), as well as the language (English, Chinese, Japanese, and Dutch etc.,) of paired captions. However, their captions are generally factual, i.e., they correspond to the visual content (Fig. 2.12). Amongst these datasets some contain only unilingual (such as English) captions, whereas others have multilingual captions. We therefore, further categorize them into *Unilingual* and *Multilingual* datasets. The details of these datasets are discussed below:

---

[9]https://www.flickr.com/.

Table 2.2 Various image captioning datasets, organized in a chronological order

Dataset	Images	Image source	Captions	Languages	Release	Features
IAPR TC-12	20,000	Internet and viventura[10]	70,460	English/German/Spanish	2006	Multi-lingual generic captions
PASCAL1K	1,000	PASCAL VOC [114]	5,000	English	2010	Generic captions
SBU 1M	1,000,000	Flickr[11]	1,000,000	English	2011	Generic captions
BBC news dataset	3,361	BBC News	3,361	English	2011	Domain specific captions
Abstract	10,020	Manmade via AMT	60,120	English	2013	Generic captions
VLT2K	2,424	PASCAL VOC 2011 [115]	7,272	English	2013	Generic captions
Flickr8K	8,100	Flickr	40,500	English	2013	Generic captions
Flickr30K	31,783	Flickr	1,58,915	English	2014	Generic captions
MSCOCO	1,64,062	Flickr	8,20,310	English	2015	Generic captions
Déjà images	4,000,000	Flickr	1,80,000	English	2015	Generic captions
Japanese pascal	1,000	PASCAL1K [42]	5,000	English/Japanese	2015	Multi-lingual generic captions
Cartoon captions	50	New Yorker[12]	2,50,000	English	2015	Domain specific captions
MIC testset	31,014	MSCOCO [116]	1,86,084	English/German /French	2016	Multi-lingual generic captions
KBK-1M	1,603,396	Dutch newspapers	1,603,396	Dutch	2016	Domain specific captions
Bilingual caption	1,000	MSCOCO	1,000	English/German	2016	Multi-lingual generic captions
Multi30K	1,000	Flickr30K	1,000	English/German	2016	Multi-lingual generic captions
Flickr8K-CN	8,000	Flickr8K	45,000	English/Chinese	2016	Multi-lingual generic captions
SentiCap	2,359	MSCOCO	8,869	English	2016	Stylized captions
Visual genome	1,08,077	MSCOCO and YFCC100M [117]	5,403,840	English	2017	Generic captions

(continued)

Table 2.2 (continued)

Dataset	Images	Image source	Captions	Languages	Release	Features
Stock3M	3,217,654	Stock website	3,217,654	English	2017	Generic captions
STAIR	26,500	MSCOCO	1,31,740	Japanese/English	2017	Multi-lingual generic captions
AIC-ICC	2,40,000	Internet	1,200,000	English/Chinese	2017	Mutli-lingual generic captions
Flick30K-CN	1,000	Flickr30K	5,000	English/Chinese	2017	Multi-lingual generic captions
RSICD	10,921	Remote sensing images	54,605	English	2017	Domain specific captions
FlickrStyle10K	10,000	Flickr	20,000	English	2017	Stylized captions
COCO-CN	20,342	MSCOCO	27,218	English/Chinese	2018	Multi-lingual generic captions
Conceptual captions	3,330,000	Internet	3,330,000	English	2018	Generic captions
SMACDA	1,000	PASCAL50S [19]	50	English	2019	Generic captions, eye-tracking data
Nocaps	15,100	Open images [118]	1,66,100	English	2019	Domain specific captions
Personality-captions	2,01,858	YFCC100M	2,41,858	English	2019	Stylized captions
Good news	4,66,000	Internet	4,66,000	English	2019	Domain specific captions
Fig caps	1,09,512	Machine generated [119]	2,19,024	English	2019	Domain specific captions

[10]https://www.viventura.de/.
[11]https://www.flickr.com/.
[12]https://www.newyorker.com/tag/newspapers.

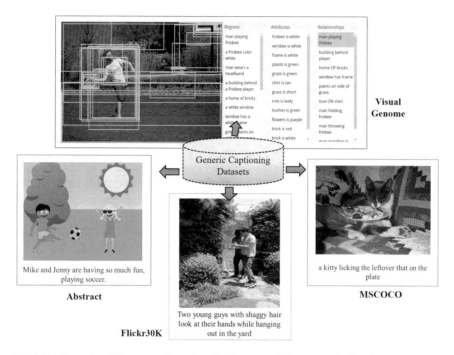

Fig. 2.12 Examples of image-caption pairs of various generic image captioning datasets

2.5.1.1 Unilingual Datasets

– *Pascal1K* [42] is a popular dataset and its images were selected from the PASCAL
  Visual Object Classes (VOC) 2008 development and training set [42]. To obtain a
  description of each image, Amazon Mechanical Turk (AMT) [43] was used, where
  each image was described by five different people. The captions in PASCAL1K
  are simple and mainly focus on people and animals performing some actions.
– *SBU1M* [35] is a Web-Scale collection of one million images with their associated
  captions. It was harvested from Flicker via queries for objects, stuff, attributes,
  actions, and scenes. The authors initially retrieved a larger number of images
  (more than 1 million) with the original human-authored captions and then filtered
  those down to only 1 million photos. The remaining 1 million images have captions
  that are relevant and visually descriptive of the image content.
– *Abstract* [120] dataset is composed of 10,020 abstract images created from a collection of 80 pieces of clipart. The images were created through AMT, depicting
  the scenes of two children (a boy and a girl) playing outside. The AMT workers
  followed some specific guidelines in the process of creating these images. Descriptions for these abstract images were generated by a different set of workers, instead
  of the ones who created the images.

- *The Visual and Linguistic Tree-bank (VLTK)* [121] contains images from PAS-
  CAL VOC 2011 action recognition dataset. Each image is accompanied by three
  descriptions which were obtained using AMT. Each description consists of two
  sentences, the first talks about the action taking place in the image, persons performing it and the region of interest. Whereas, the second sentence verbalizes
  the other most important regions in the image that are not directly involved in the
  action. A subset of 341 Images in this dataset is annotated with polygons around the
  regions that are mentioned in the descriptions, along with the visual dependency
  representations of the annotated regions.
- The images for *Flickr8K* [10] were chosen to represent a wide variety of scenes
  and situations, because the authors were interested in conceptual captions that
  describe events, scenes, and entities. Having more captions per image caters for
  the need of diversity.
- *Flickr30K* [122] is an extension of Flickr8K and consists of 31,783 photos acquired
  from Flickr. The images in this dataset are the photographs of various day-to-day
  activities, scenes or events. Each image is annotated with 5 descriptions obtained
  through AMT service following the same guidelines as Flickr. Flickr30K provides
  detailed grounding of phrases in the image which can help find correspondences
  between image regions and parts of the captions, thus helping in the grounded
  language understanding.
- *MSCOCO* [116] is one of the largest image recognition, segmentation, and captioning dataset. It is a widely used dataset for image description research and it
  also gave rise to the MSCOCO captioning challenge. It makes use of the images
  collected by Microsoft COCO through queries of various scene types and object
  categories on Flickr. Currently, it consists of 123,287 images, each of which is
  associated with five captions that were generated through AMT service.
- *Déjà images* [123] dataset is a large image-caption corpus containing four million
  images along with 1,80,000 unique user-generated captions compiled from Flickr.
  This dataset was collected by initially downloading 760 million images by querying
  Flickr for a set of 693 nouns. These were then filtered down to keep only those
  image-sentence pairs that have at least one verb, preposition, or adjective. From the
  remaining captions, only those that were repetitive were retained. In Deja Image
  Captions dataset a single caption corresponds to numerous images.
- *Visual Genome* [124] consists of 108k images, where each image is densely annotated for objects, attributes and pairwise relationships between objects. Visual
  genome offers human-generated descriptions of multiple image regions. In contrast to other image-caption datasets, whose captions typically describe the salient
  regions, visual genome provides scene level, as well as various low-level descriptions with up to 50 descriptions per image. The region descriptions of images in
  this dataset were crowd sourced through AMT.
- *Stock3M* [69] contains user uploaded images on stock website. The uploaded
  images are also accompanied by user written captions. The images have a great
  variety and the captions are short but challenging.

2.5.1.2 Multilingual Datasets

- *IAPR TC-12* [125] is one of the earliest datasets, containing 20,000 photographs with descriptions expressed in multiple languages (English, German and Spanish). The majority of images for this dataset were obtained from a travel company 'Viventura'[10] , while the rest of them were retrieved via various search engines such as Yahoo, Google and Bing. This dataset includes pictures of people, animals, a range of sports and actions, cities, landscapes and other aspects of contemporary life. Images are annotated with descriptions comprised of one to five sentences. The first sentence(s) describe(s) the most salient aspects, whereas the latter describes the surroundings or settings of an image.
- *Japanese Pascal* [126] dataset consists of images paired with English and Japanese text. This dataset was created to facilitate image mediated cross-lingual document retrieval. The images and English captions were sourced from UIUC Pascal Sentence Dataset and each English sentence was translated by a Japanese professional translators to obtain Japanese captions.
- *MIC testset* [127] has 1000 images, and is a multilingual version of a portion of MSCOCO data. Authors collected French and German translations for each English caption via crowd-sourcing. The translations were then verified by three native speakers.
- *Bilingual Captions* [128] is a German-English parallel dataset, which was constructed using the MSCOCO image corpus by randomly selecting 1000 images from the training corpus. Out of five English captions per image, one of the randomly chosen caption was translated into German by a native German.
- *Multi30K* [129] extends the Flickr30K dataset by pairing 1000 test images with German translations. Some of the German captions are translations by professional German translators and some are sourced independently without the use of English captions. The quality of capions in this dataset was assessed both manually and also via automated checks.
- *Flickr8K-CN* [130] contains Chinese captions for unlabeled images, and is a bilingual extension of Flick8k dataset. The authors used a local crowd-sourcing service for translating the English descriptions of Flick8k to Chinese.
- *STAIR* [131] is a large scale Japanese Image caption dataset based on images from MSCOCO. Authors annotated 164,042 MSCOCO images with five Japanese captions each, amounting to about 820,310 captions in total. The annotations are provided by part-time workers and crowd-sourcing workers. It took half a year and 2100 workers to complete this dataset.
- *SMACDA* [132] is an interesting dataset that has eye-tracking data paired with images and their descriptions. The images were sourced from PASCAL50S [19]. Five subjects participated in the data collection and human fixations as well as verbal descriptions were recorded for each image. The oral descriptions were converted to textual ones for all images.

---

[10]https://www.viventura.de/.

- *AIC-ICC (AI Challenger-Image Chinese Captioning)* dataset [133]: The images in this datasets were collected from search engines by following certain guidelines to ensure that each selected image can be described in a single sentence, and it should depict multiple objects to form complex scenes. Each image is annotated with five captions written by five different Chinese native speakers.
- *Flick30K-CN* [134] is another dataset for cross-lingual image captioning. It was constructed by extending Flickr30K training and validation sets. Each image in Flickr30K was annotated by Chinese captions using Baidu[11] translation. Whereas, the test set images are manually annotated by five Chinese students.
- *COCO-CN* [135] extends MSCOCO with Chinese captions and tags. The images were annotated using an interface and via a predefined methodology. For any given image, the most appropriate Chinese caption is retrieved from a pool of Chinese captions obtained via machine translation on MSCOCO captions. Then the annotators are asked to write a caption, given the retrieved caption and image, using their own viewpoint.
- *Conceptual Captions* [136] is a large dataset consisting of about 3.3 million image-caption pairs. The images and paired captions were harvested from the web to represent a diversity in styles. An automatic pipeline was created to filter out captions and images that do not fulfill the quality criteria.

2.5.2 Stylised Captioning Datasets

Collecting stylized captions is more challenging, compared to collecting factual (generic) captions. The captions in stylised captioning datasets contain various sentiments or styles (Fig. 2.13). Furthermore, sentiments in the captions might not be visual i.e., the are not explicitly reflected by the visual content. Algorithms trained on such datasets can learn human style or can generate humours or personalized captions.

- *SentiCap* [74] is one of the pioneering stylized captioning dataset. The images in this dataset were obtained from MSCOCO. Howerver, the captions with positive or negative sentiments corresponding to each image were crowd-sourced. For each image, 3 captions have positive sentiments and 3 have negative sentiments. For quality validation, the captions were tested for descriptiveness and sentiment correctness.
- *FlickrStyle10K* [137] is built on the Flickr30k dataset. The annotations were collected using AMT. The annotators were shown some factual captions and were asked to inject styles to make it e.g., more romantic, or funny. The dataset contains 10,000 images in total. 9,000 images are accompanied by a humorous and romantic caption each. Whereas, the remaining 1000 images have five humorous and romantic captions each.

---

[11] http://research.baidu.com/Index.

Personality Captions

Sweet	That is a lovely sandwich.
Dramatic	This sandwich looks so delicious! My goodness!
Anxious	I'm afraid this might make me sick if I eat it.
Sympathetic	I feel so bad for that carrot, about to be consumed.
Arrogant	I make better food than this
Optimistic	It will taste positively wonderful!
Money-minded	I would totally pay \$100 for this plate.

a group of nice people holding umbrellas in the rain

a lonely train pulling into a train station

"A group of people stand on the beach, enjoying the beauty of nature"
"A group of people looking for pokemon go"

SentiCap   **FlickrStyle10K**

Fig. 2.13 Image-caption pairs of various stylised captioning datasets

– *Personality-Captions* [100] is one of the most recent stylized dataset. The images were sourced from YFCC100M dataset, whereas, the captions were collected using crowd-workers. This dataset contains captions corresponding to around 215 personality traits such as sweet, dramatic and optimistic.

2.5.3 Domain Specific Captioning Datasets

The images in this dataset correspond to a particular domain or application, for example medical x-rays, news images, news cartoons or remote sensing images etc. Few of the categories are discussed below:

2.5.3.1 News Image Captioning Datasets

BBC News [54] dataset is one of the earliest and biggest news dataset consisting of images, captions and news articles on a wide range of topics. The captions were free to obtain but are a bit noisy, since they were not created with the vision and language tasks in mind.

KBK-1M [138] contains a vast variety of images extracted from Dutch Newspapers that were printed between 1992–1994 paired with captions, articles and locations of known images. Captions in this dataset are mostly non-factual i.e., they refer to things that might not be present in the images, for example political viewpoints and opinions.

Good News [139] is the most recent news image captioning dataset consisting of 4,66,000 image caption pairs. The authors used Newyork Times API to retrieve articles from 2010–2018. This is the largest currently available data for news captioning. The captions are longer and more descriptive compared to the generic captions in MSCOCO.

2.5.3.2   Cartoon Captioning Dataset

Cartoon captions [140] contains a large set of cartoons and their captions that can help the design of more engaging agents. The cartoons are taken from the New Yorker caption contest, where the contestants are asked to write a funny caption for news cartoons. The dataset has around 50 cartoons, each paired with at least 5000 human generated captions.

2.5.3.3   Remote Sensing Captioning Dataset

RSICD-Remote Sensing Image Captioning dataset [141] has more than 10,000 images. The visual objects in remote sensing images are quite different from the ones in the generic images. There is no sense of direction in such images, and may have rotation, and translation variations. The captions for the images in this dataset were written by volunteers with annotation experience and relevant knowledge. Each image is accompanied by five captions.

2.5.3.4   Figure Captioning Dataset

FigCap [142] is a dataset for figure captioning. The figures are synthetically generated and are paired with captions based on the fundamental data. This dataset has five types of figures; horizontal bar chart, vertical bar chart, pie chart, line plot and dotted line plot. The descriptions paired with images are both high level descriptions and detailed. However, both figures and captions in this dataset are synthetic.

2.5.3.5   Novel Captioning Dataset

Nocaps [143] is a recent dataset for novel object captioning that contains 15,100 images sourced from the Open Images validation and test sets. Generating captions for rare objects is a challenging and interesting task. This dataset contains images that on average have 4 unique object classes and this dataset is not overly dominated by commonly occurring classes such as person, car etc. For each image 11 captions are collected via AMT amongst them 10 serve as human references, whereas one is used for oracle testing.

2.6 Applications of Visual Captioning

Over the years, visual captioning has evolved from a purely research topic to one with practical applications and commercial implications. It can play a major role in various fields such as assistive technology, media, robotics, sports and medical.

2.6.1 Medical Image Captioning

Recent advances in computer vision and natural language processing have propelled the progress in the automatic diagnosis of diseases and the description of medical scans. Now-a-days automatic description generation of medical scans and tests has recently attracted the attention of the medical community. Doctors and specialists have to write reports of their findings, diagnosis and treatment. However, manual report writing can be time consuming, subjective and prone to errors.

Jing et al., [144] explored a framework to predict classification tags for the diagnosis from X-rays and to generate a description of the results. Their model used a convolutional neural network (CNN) to extract visual features of input image (i.e. X-ray) and fed those to a multi-label classification (MLC) network to predict the corresponding diagnosis tags. Each tag was then represented by a word-embedding vector in the tag vocabulary. These embedded vectors corresponded to the semantic features of the image under consideration. Next, the visual features from the CNN and the semantic features from the embedded vectors were combined with a co-attention model to generate a context-vector. This context-vector was then decoded with two LSTM based networks, connected in series, to generate descriptions for the medical report. As medical reports often contain multiple sentences, an interesting fact of the generated reports was that, instead of just a caption consisting of a single sentence, long reports were generated in a hierarchical manner. The first LSTM network generated a list of topic-vectors at a high level which were subsequently fed to a second LSTM network. The second network then generated a sentence for each of the topic-vectors. In this way, the generated descriptions composed of a sentence for each of the high level topics identified by the first LSTM. In addition to X-rays, this method can potentially be applied for other types of medical scans as well.

One of the main criticisms on the use of deep learning in the medical field is that deep networks are perceived as "black-boxes", and do not provide a proper justification of the different features or the variables in their decision to the satisfaction of the medical experts. To overcome this problem, Zhang et al. [145] proposed a mechanism for an interpretable framework for the automatic diagnosis of diseases and generation of reports from medical images in clinical pathology. It consists of three components: an image model, a language model and an attention mechanism [22]. The image model uses a CNN to extract features from the input images and to classify the input image. The language model is based on Long Short Term Memory (LSTM) network, generates diagnosis reports using image features. The attention mechanism [22] enables the framework to provide a justification of the results by highlighting the image regions on the basis of which it took the decision. The qualitative analysis of the comparison between the key image regions predicted by the system and the important regions identified by experts showed good correlation.

2.6.2 Life-Logging

Another potential application of visual captioning is to summarize and arrange the pictures and videos captured via smart phones or other portable cameras. Image captioning systems usually work on single images, while video captioning requires a continuous video. However, most of the pictures we take during a day or an event are not continuous, but related to each other. For a system to be able to describe or summarize a bunch of pictures, it should be able to find various complex relationships among them. Moreover, these relationships and context also heavily rely on the person taking the picture or the person in context, i.e., the captioning system should be personalized according to the person for which 'life-logging' has to be generated. With the text description of daily life photos, it becomes easier to retrieve media related to a particular event from a large collection. Fan et al., [146] introduced a similar system, consisting of a combination of CNN and LSTM networks. Their systems can be used to automatically generate textual description of a day's events from images just like diary writing. Their framework first generates multiple captions for each image as a first person narrative. Then the captions of different images are combined into a description of the sequence of images. Although there is still a lot of room for improvement in this application, future research may result in automatic life-loggers for individuals.

2.6.3 Commentary for Sports' Videos

Description generation from videos can also be used to provide commentary on sports videos. This is different from traditional video description models since commentary is much longer than a caption or a simple sentence, which not only requires to focus on minor details, but it also needs to differentiate between different teams and players in the game.

Yu et al., [147] used a combination of CNNs and LSTMs to generate long descriptions from videos of Basketball games. Their algorithm first segments the players of the two teams and the ball, which is followed by the posture detection of the players. Another neural network was is to detect the actions of the players in the video clips.

Then, the segmentation, posture and action information, along with the optical flow for motion information, is fused with an LSTM Network. A bi-directional LSTM network is then used to generate descriptions from the fused information. In this way, their framework is able to generate more detailed and an accurate description for short basketball clips, compared to other traditional frameworks of video captioning.

2.6.4   Captioning for Newspapers

An interesting topic in natural language processing is the summarization of text. This can be combined with image captioning to create a combined model when both textual and visual information is available, e.g., in newspapers and magazines. From the perspective of visual captioning, this problem is closer to the category of visual question answering, as it requires to analyze both modalities, a textual description and a visual image, to produce a caption for the given data.

Feng et al., [54] proposed a two-step system to generate captions for news articles. A Bag Of Words approach is used to select important keywords from the text article. For the associated image, SIFT [148] features are extracted and a probabilistic model is used to detect the different objects in the image, which are then converted to Bag of Visual Words [149]. This allows for both the textual and the visual information to be represented in the same manner so that it can be processed in the form of dependent variables. In the second step, a description is generated from the gathered features and keywords. Their system is able to generate decent captions for the images in news articles, although it can potentially be improved further with the use of state-of-the-art deep networks instead of probabilistic models.

2.6.5   Captioning for Assistive Technology

One of the major applications of visual captioning is its use for assistive technology. Visual captioning can be used to automatically describe images, videos, social media feed and even common day-to-day activities to the visually challenged and blind people. It can enable the blind people to carry out their daily tasks more freely and without the help of other humans.

Now-a-days, social media has become a part-and-parcel of our lives. A lot of content on social media is in the form of pictures and videos. The visually impaired people feel left out, compared to their fellows in this aspect of life, as they can only enjoy and communicate on social media with textual messages. With the help of visual captioning, it is possible to describe the visual content on social media in a textual form. Wu et al., [150] from Facebook research introduced a system to automatically generate descriptions from the social media images. They use an object detection and scene recognition algorithm to describe an image to its users.

The system detects faces (with some expressions), objects and themes and describes the detected objects with more than 80% confidence to the users.

Another practical application of this technology is available in the form of Microsoft's "Seeing AI" App[12] (for Apple devices) and other similar apps for smart phone users. These apps use the devices' cameras to detect faces, objects and scenes in real time. They can also predict various facial expressions, gender and age. They can also be used to describe the scenes visible through the smart phones' cameras and can relate various actions and relationships between objects in the scene to some extent. Moreover, they can read text and recognize bar-codes to identify different products. This can be broadened to relieve people from focusing on a screen, e.g., as the Google Home recipes and Adobe book readers do. Basically, the integration with a voice-enabled personal assistant is another important application area.

2.6.6 Other Applications

In addition to the applications discussed above, visual captioning has many more applications e.g., media retrieval and robotics. With the large amount of visual data being generated, its organization and retrieval has become a difficult problem. Visual captioning can be used to provide textual descriptions for images and videos, to store them in an organized fashion. Any individual can then retrieve videos or pictures from a large database using a meaningful textual description or use some automatic text-based retrieval methods [151].

Visual captioning and question answering is also very useful for humanoid and assistive robots. Humanoid robots need to decipher information from visual streams and reason about it in real time, in order to have a useful and meaningful interaction with humans. Similarly, it will help the assistive robots to analyze and describe a situation to the people they are helping, in order to aid them to carry out their day-to-day activities.

There has been a great advancement in the area of visual captioning systems in the last decade, especially due to the rapid advancement of deep learning. With more improvement over time, many of the current challenges in the field of visual captioning will be overcome and devices with the ability of visual description will become a common part of our daily life.

2.7 Extensions of Image Captioning to Other Vision-to-Language Tasks

Image captioning, its taxonomy, various challenges, applications and evaluation measures have been discussed in detail in the previous sections. In this section, we briefly

---

[12]https://www.microsoft.com/en-us/ai/seeing-ai.

discuss some applications which are closely related to image captioning such as video captioning, visual dialogue, visual question and answering and visual story telling. We compare these tasks with image captioning and outline their similarities or differences. Although these applications fall in different domains, the captioning output may still be used to help inform the output of these tasks. The aforementioned applications are rapidly evolving and their relevant literature is becoming rich day-by-day, for the sake of completeness we briefly discuss the most recent or the most representative state-of-the-art techniques that are related to these tasks.

2.7.1   Visual Question Answering

Visual Question Answering (VQA) is a research area, which focuses on the design of autonomous systems to answer questions posed in natural language, related to the image content. In contrast to image captioning system, which takes an image as input and produces its caption, VQA system takes an 'image and a natural language question' about that image as input and the system then generates a suitable answer

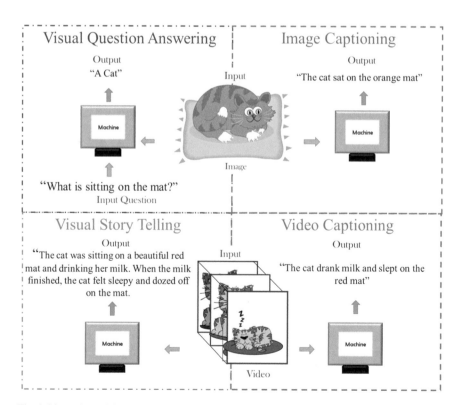

Fig. 2.14 Various vision-to-language tasks related to image captioning

in natural language, as the output [4]. For example, as shown in Fig. 2.14, the input to VQA is an image of a cat and the natural language question is, 'What is sitting on the mat?'. A VQA system produces 'A cat' as its natural language answer or output.

For VQA systems to accurately provide an answer for a given image, they systems require a vast set of capabilities, such as object detection (e.g., detect cat and mat in Fig. 2.14), object recognition (e.g., recognise cat and mat in Fig. 2.14), activity recognition (e.g., cat is sitting), question understanding, knowledge base reasoning (e.g., is there a mat in the image?) and question understanding (question parsing/interpretation). Visual Question Answering is therefore more challenging than image captioning. This can be inferred from the fact that the question to be answered by a VQA system is not determined until run time and a VQA system requires additional information, which is not necessarily present in the input image. The required information may be specific knowledge about the image content or common sense. For example, given an image, such as Fig. 2.14, showing 'a cat on the mat', if one asks a question 'why is the cat sitting on the mat?', to provide an answer to this question, the VQA system must not only detect the scene e.g., 'room', but must also know that 'the mat is often used as a resting place'.

Noh et al., [152] proposed a transfer learning technique for VQA task, which learns visual concepts without questions and then transfers the learned concepts to VQA models. A neural network, coined as task conditional visual classifier, is trained to learn visual concepts. The classifier, whose task is defined by a task feature (inferred from the question), is used as an answering unit. An unsupervised task discovery technique based on linguistic knowledge sources is also proposed to train the task conditional visual classifier. Their experimental results show that the proposed framework generalizes to VQA task with answers, which are out of vocabulary. Further explanations about task feature and unsupervised task discovery is beyond the scope of this chapter and readers are referred to [152] for details.

2.7.2 *Visual Storytelling*

Storytelling is an old human activity which provides a way to engage, share advice, educate, instil morals and preserve culture [5]. Inspired by this, AI research has started to focus on this task to bring about human like understanding and intelligence into this emerging domain. The aforementioned VQA task, including image captioning, focus on the direct and literal description of an input image, which itself contains static moments and is devoid of context. These tasks are therefore far from the capabilities that are required for naturalistic interactions.

In contrast, visual storytelling relies on videos or a sequence of images, as shown in Fig. 2.14, which portray the occurrence of events. From vision perspective, a trained visual storytelling system can reason about a visual moment in a sequence of images. From the language point of view, moving from a literal description (such as the case for image captioning) to narrative enables the visual storytelling system to produce more conversational and illustrative language. For example, the description 'The cat sat on the orange mat' (image captioning) in Fig. 2.14 captures image content, which is literal. On the other hand, the description 'The cat was sitting on a beautiful red mat and drinking her milk. When the milk finished, the cat felt sleepy and dozed off on the mat' (Visual Storytelling in Fig. 2.14) provides further inference about drinking milk and feeling sleeping.

The field of visual story telling field has significantly progressed in the recent years and a number of research works have recently emerged. For completeness, we briefly discuss just one recent visual storytelling method. Huang et al., [5] developed the first dataset, called SIND, for visual storytelling and proposed a story generation model. To train their story generation model, a sequence-to-sequence recurrent neural net was adopted. The use of this network architecture is justified by the fact that it extends a single-image captioning to multiple images. A recurrent neural network runs over the fc7 vector to encode an image sequence (i.e., to extract useful visual features). The story decoder model then learns to produce one word at a time of the story by using softmax loss. Their baseline model has been shown to generate good visual stories.

2.7.3   Video Captioning

Video captioning received inspirations from the task of image captioning. However, in contrast to image captioning where the input is a static image, the input to a video captioning system is a video (Fig. 2.14) or a sequence of images. Video captioning therefore aims to produce a caption in natural language to describe the visual content of a video. As this task involves the processing of a sequence of events, which is in the form of a video, video captioning is more challenging than image captioning. Video captioning requires an understanding of scenes by capturing the details of the events in the input video and their occurrences. A video captioning system therefore focuses not only on the dynamic object of interest, but also on its surroundings and it takes into consideration the background, actions, objects, human-object interactions, occurrence of events, and the order in which those events occur. For example, as illustrated in Fig. 2.14, the video caption system captures the object of interest (i.e., cat), its actions and sequence of events (e.g., drank milk and slept) and produces a caption, 'The cat drank milk and slept on the red mat'.

While video captioning is a trivial task for most people, it is very challenging for machines. Most existing video captioning techniques rely on an encoder-decoder framework. The encoder part learns distinctive visual features and the decoder part generates captions. The limitation of the encoder-decoder framework is that it can only process one source video at a time. Thus, the focus of the framework is only restricted to one word and its corresponding visual features. As a result, existing video captioning methods fail to exploit the relationship between words and their various visual contexts across input videos. To overcome this problem, Pei et al., [153] proposed the Memory-Attended Recurrent Network (MARN) for video captioning. In MARN, they adopted the encoder-decoder framework, but introduced an attention based recurrent decoder and a memory module (called attended memory decoder) to store the descriptive information for each word in the vocabulary. This architecture captured the full correspondence between a word and its visual contexts, thus providing a detailed understanding for each word. MARN has been shown to improve and enhance video captioning quality on Microsoft Research-Video to Text (MSRVTT) and Microsoft Research Video Description Corpus (MSVD) datasets.

2.7.4 Visual Dialogue

Despite the rapid progress of image captioning, video captioning and visual question answering tasks, it is evident that these tasks are still not capable of performing a two-way communication with the users. For instance, a trained image captioning system can converse with humans by producing image captions for a given input image and without any input from the human in the form of a dialog. Similarly, VQA task represents a single round of dialog with human (e.g., 'What is sitting on the mat?', the VQA system produces 'A cat') and does not take into account any previous questions asked by the user or answers provided by the system. In addition, VQA systems do not facilitate any follow-up questions from humans.

To enable active machine-human interaction and to take the intersection of vision and language to the next level, the concept of visual dialog was introduced. Visual dialog is an emerging research theme that lies at the intersection between natural language processing, dialogue management, and computer vision. In contrast to image captioning, visual dialog enables a two-way communication by holding a meaningful dialog with users in natural language about visual content. Given an image, history of dialogues (i.e., previous questions and answers), and a new question about the image, the visual dialog system attempts to answer the question. Because of its active human-machine interaction nature, visual dialog finds vast applications in assisting visually impaired users to understand their surroundings, data analysis and robotic navigation.

Guo et al., [154] proposed a two-stage image-question-answer synergistic network for the visual dialog task. In the first stage, called the primary stage, representative vectors of the image, dialog history, and initial question are learnt. To achieve this, objects and their features are detected in the input image using Faster-RCNNN. These features are then encoded using a convolutional neural network. As the question and dialog history contain text data, these are encoded using LSTM. All the candidate answers which are generated in the primary stage are then scored based on their relevance to the image and question pair. In the second stage, dubbed the synergistic stage, answers in synergy with image and question are ranked based on their probability of correctness. The proposed technique has been shown to achieve the state-of-the-art visual dialog performance on the Visual Dialog v1.0 dataset.

2.7.5 Visual Grounding

Visual grounding can be considered as the reverse task of image captioning and is now receiving a lot of attention from the research community. Unlike other vision-to-text tasks (discussed above), a visual grounding system takes a natural language query in the form of a word or sentence as input and aims to locate the most relevant object or region of interest in the input image. To achieve this, a visual grounding system considers the main focus in a query, it finally attempts to understand an image and then locates the object of interest in the image. Visual grounding is an important technique for human-machine interaction and can enable a machine to attain human like understanding of the real-world.

Deng et al., [155] proposed Accumulated Attention (A-ATT) method for visual grounding. Given an image containing objects and a query, their A-ATT method learns a hypothesis to map the query to the target object. A-ATT achieves this by estimating attention on three modalities including sentence, object and feature maps in three multiple stages, where each stage exploits the attention of the other modalities to achieve the desired mapping. A-ATT extracts distinctive and informative visual and language features and has been shown to achieve superior performance.

2.8 Conclusion and Future Works

Vision-to-language tasks have garnered a lot of attention over the past decade. Amongst various vision-to-language tasks, the prime focus of this chapter was 'Image Captioning', which is a fundamental and pioneering problem in this area. Image captioning underpins the research of various visual understanding tasks, such as visual grounding, visual question answering and visual story telling. The goal of captioning models is to mimic the humans trait of describing the visual input (images) in natural language. However, this is not a straight forward task, because for machine generated captions to be human-like, they need to have various quality traits. Two of the most important characteristics are adequacy/relevance and fluency. The caption should be relevant to the image, semantically correct and syntactically well-formed. Apart from these, a caption can have other superlative qualities such as descriptiveness, brevity, humor, style, which in our opinion are optional and domain specific.

Over the past decade, significant progress has been made towards developing powerful captioning models, evaluation metrics and captioning datasets. The state-of-the-art image captioning models are based on deep encoder-decoder based architecture. However, such encoder-decoder based baselines generate captions which lack novelty, creativity and diversity in expression. A recent trend towards reinforcement learning and GAN based methods has shown improvement in terms of generating diverse, human-like expressions. Generating stylised captions is another interesting direction, which is in its primary stages.

The internet has a plethora of unpaired linguistic and visual data, but the availability of images paired with captions/descriptions is relatively scarce. Models that can fuse linguistic knowledge and CV learning, without relying much on the paired data (image-caption) can be very useful. Unsupervised/semi-supervised learning based captioning is an emerging trend in this domain. Novel object captioning is another promising direction. As far as humans are concerned, they can deal easily with novel objects either by using common sense, relying on linguistic sense, or simply by saying that they don't know the object. Humans can describe an unknown object or scene via its attributes or its resemblance to something similar, instead of giving a wrong answer. State-of-the-art models lack this capability, and this research gap needs to be addressed.

The assessment of models and their outputs hold paramount importance for the advancement of any field. We discussed various assessment methods for captioning, amongst which the automatic evaluation metrics are the most popular. However, existing captioning metrics are far from perfect. Most of the progress in captioning is misleading due to the flaws of the existing measures. A captioning model that generates inaccurate or senseless descriptions, can be a hazard and has no real world application. The descriptions should be well-grounded in the image. It is not only important to analyse the visual understanding of the models, but also how well they can generalize beyond the training sets. As pointed in the literature [156], captioning models tend to exploit biases in dataset to generate plausible outputs. Therefore, it is important to understand what the captioning models are learning, identify the shortfalls of datasets and their biases, and whether models are just memorizing the captions or truly understanding the visual content. This is where effective evaluation methodologies come into play. To understand why a model fails/succeeds and what are its capacities, shortcomings and strengths.

Various other vision-to-language tasks discussed in this chapter, such as VQA, visual grounding and visual dialogue have been studied independently in the literature. The majority of these tasks are the representative of the visual and linguistic understanding of a machine model. However, any model that has a sound visual and linguistic understanding should be capable of performing relevant tasks, with little modification. For example a strong captioning model can serve as a baseline for VQA. Common baseline models should be developed that can be helpful for various vision-to-language tasks, instead of just a specific one. Moreover, integrating different vision to language tasks is the key to create futuristic models.

References

1. M. Alan, Turing, computing machinery and intelligence. Mind **59**(236), 433–460 (1950)
2. A.M. Turing, Computing machinery and intelligence, in Parsing the Turing Test (Springer, 2009), pp. 23–65
3. R. Bernardi, R. Cakici, D. Elliott, A. Erdem, E. Erdem, N. Ikizler-Cinbis, F. Keller, A. Muscat, B. Plank et al., Automatic description generation from images: A survey of models, datasets, and evaluation measures. J. Artif. Intell. Res. (JAIR) **55**, 409–442 (2016)

4. S. Antol, A. Agrawal, J. Lu, M. Mitchell, D. Batra, C. Lawrence Zitnick, D. Parikh, Vqa: Visual question answering, in *Proceedings of the IEEE International Conference on Computer Vision* (2015), pp. 2425–2433
5. T.-H.K. Huang, F. Ferraro, N. Mostafazadeh, I. Misra, A. Agrawal, J. Devlin, R. Girshick, X. He, P. Kohli, D. Batra, et al., Visual storytelling, in *Proceedings of the 2016 Conference of the North American Chapter of the Association for Computational Linguistics: Human Language Technologies* (2016), pp. 1233–1239
6. S. Venugopalan, M. Rohrbach, J. Donahue, R. Mooney, T. Darrell, K. Saenko, Sequence to sequence-video to text, in *Proceedings Of The Ieee International Conference On Computer Vision* (2015), pp. 4534–4542
7. S. Khan, H. Rahmani, S.A.A. Shah, M. Bennamoun, A guide to convolutional neural networks for computer vision. Synth. Lect. Comput. Vis. **8**(1), 1–207 (2018)
8. U. Nadeem, S. A. A. Shah, F. Sohel, R. Togneri, M. Bennamoun, Deep learning for scene understanding, in *Handbook of Deep Learning Applications*. (Springer, 2019), pp. 21–51
9. A. Jaimes, S.-F. Chang, Conceptual framework for indexing visual information at multiple levels, in Internet Imaging, Vol. 3964, *International Society for Optics and Photonics* (1999), pp. 2–15
10. M. Hodosh, P. Young, J. Hockenmaier, Framing image description as a ranking task: Data, models and evaluation metrics. J. Artif. Intell. Res. **47**, 853–899 (2013)
11. T.-Y. Lin, M. Maire, S. Belongie, J. Hays, P. Perona, D. Ramanan, P. Dollár, C. L. Zitnick, Microsoft coco: Common objects in context, in *European conference on computer vision*. (Springer, 2014), pp. 740–755
12. A.C. Berg, T.L. Berg, H. Daume, J. Dodge, A. Goyal, X. Han, A. Mensch, M. Mitchell, A. Sood, K. Stratos, et al., Understanding and predicting importance in images, in *2012 IEEE Conference on Computer Vision and Pattern Recognition*, (IEEE, 2012), pp. 3562–3569
13. A. Rohrbach, A. Torabi, M. Rohrbach, N. Tandon, P. Chris, L. Hugo, C. Aaron, B. Schiele, Movie description. https://arxiv.org/pdf/1605.03705.pdf
14. A. Rohrbach, L.A. Hendricks, K. Burns, T. Darrell, K. Saenko, Object hallucination in image captioning. arXiv:1809.02156
15. E. van Miltenburg, D. Elliott, Room for improvement in automatic image description: an error analysis. arXiv:1704.04198
16. K. Papineni, S. Roukos, T. Ward, W.-J. Zhu, Bleu: a method for automatic evaluation of machine translation, in *Proceedings of the 40th Annual Meeting on Association For Computational Linguistics, Association For Computational Linguistics* (2002), pp. 311–318
17. C.-Y. Lin, Rouge: a package for automatic evaluation of summaries, Text Summarization Branches Out
18. S. Banerjee, A. Lavie, Meteor: An automatic metric for mt evaluation with improved correlation with human judgments, in: *Proceedings of the ACL Workshop On Intrinsic and Extrinsic Evaluation Measures For Machine Translation And/or Summarization* (2005), pp. 65–72
19. R. Vedantam, C. Lawrence Zitnick, D. Parikh, Cider: Consensus-based image description evaluation, in *Proceedings Of The IEEE Conference on Computer Vision And Pattern Recognition* (2015), pp. 4566–4575
20. P. Anderson, B. Fernando, M. Johnson, S. Gould, Spice: Semantic propositional image caption evaluation, in *European Conference on Computer Vision*. (Springer, 2016) pp. 382–398
21. M. Kilickaya, A. Erdem, N. Ikizler-Cinbis, E. Erdem, Re-evaluating automatic metrics for image captioning. arXiv:1612.07600
22. K. Xu, J. Ba, R. Kiros, K. Cho, A. Courville, R. Salakhudinov, R. Zemel, Y. Bengio, Show, attend and tell: Neural image caption generation with visual attention, in *International Conference on Machine Learning* (2015), pp. 2048–2057
23. Y. Cui, G. Yang, A. Veit, X. Huang, S. Belongie, Learning to evaluate image captioning, in *Proceedings of the IEEE Conference on Computer Vision and Pattern Recognition* (2018), pp. 5804–5812
24. N. Sharif, L. White, M. Bennamoun, S. Afaq Ali Shah, Nneval: Neural network based evaluation metric for image captioning, in *Proceedings of the European Conference on Computer Vision (ECCV)* (2018), pp. 37–53

25. J. Devlin, H. Cheng, H. Fang, S. Gupta, L. Deng, X. He, G. Zweig, M. Mitchell, Language models for image captioning: The quirks and what works. arXiv:1505.01809
26. O. Vinyals, A. Toshev, S. Bengio, D. Erhan, Show and tell: A neural image caption generator, in *2015 IEEE Conference On Computer Vision and Pattern Recognition (CVPR)*, (IEEE, 2015), pp. 3156–3164
27. A. Deshpande, J. Aneja, L. Wang, A.G. Schwing, D. Forsyth, Fast, diverse and accurate image captioning guided by part-of-speech, in *Proceedings of the IEEE Conference on Computer Vision and Pattern Recognition* (2019), pp. 10695–10704
28. J.A. Yip, R.A. Martin, Sense of humor, emotional intelligence, and social competence. J. Res. Pers. **40**(6), 1202–1208 (2006)
29. W.E. Hauck, J.W. Thomas, The relationship of humor to intelligence, creativity, and intentional and incidental learning. J. Exp. Educ. **40**(4), 52–55 (1972)
30. M. Yatskar, M. Galley, L. Vanderwende, L. Zettlemoyer, See no evil, say no evil: Description generation from densely labeled images, in *Proceedings of the Third Joint Conference on Lexical and Computational Semantics (* SEM 2014)* (2014), pp. 110–120
31. T. Yao, Y. Pan, Y. Li, Z. Qiu, T. Mei, Boosting image captioning with attributes, in *Proceedings of the IEEE International Conference on Computer Vision* (2017), pp. 4894–4902
32. P. Anderson, X. He, C. Buehler, D. Teney, M. Johnson, S. Gould, L. Zhang, Bottom-up and top-down attention for image captioning and visual question answering, in *Proceedings of the IEEE Conference on Computer Vision and Pattern Recognition* (2018), pp. 6077–6086
33. M. Hodosh, J. Hockenmaier, Sentence-based image description with scalable, explicit models, in *Proceedings of the IEEE Conference on Computer Vision and Pattern Recognition Workshops* (2013), pp. 294–300
34. A. Karpathy, L. Fei-Fei, Deep visual-semantic alignments for generating image descriptions, in *Proceedings of the IEEE conference on computer vision and pattern recognition* (2015), pp. 3128–3137
35. V. Ordonez, G. Kulkarni, T. L. Berg, Im2text: Describing images using 1 million captioned photographs, in *Advances in neural information processing systems* (2011), pp. 1143–1151
36. P. Kuznetsova, V. Ordonez, A. C. Berg, T. L. Berg, Y. Choi, Collective generation of natural image descriptions, in *Proceedings of the 50th Annual Meeting of the Association for Computational Linguistics: Long Papers-Vol. 1, Association for Computational Linguistics* (2012), pp. 359–368
37. A. Oliva, A. Torralba, Modeling the shape of the scene: a holistic representation of the spatial envelope. Int. J. Comput. Vis. **42**(3), 145–175 (2001)
38. A. Torralba, R. Fergus, W.T. Freeman, 80 million tiny images: a large data set for nonparametric object and scene recognition. IEEE Trans. Pattern Anal. Mach. Intell. **30**(11), 1958–1970 (2008)
39. G. Patterson, C. Xu, H. Su, J. Hays, The sun attribute database: beyond categories for deeper scene understanding. Int. J. Comput. Vis. **108**(1–2), 59–81 (2014)
40. S. Yagcioglu, E. Erdem, A. Erdem, R. Cakici, A distributed representation based query expansion approach for image captioning, in *Proceedings of the 53rd Annual Meeting of the Association for Computational Linguistics and the 7th International Joint Conference on Natural Language Processing* (Vol. 2: Short Papers) (2015), pp. 106–111
41. T. Baltrušaitis, C. Ahuja, L.-P. Morency, Multimodal machine learning: a survey and taxonomy. IEEE Trans. Pattern Anal. Mach. Intell. **41**(2), 423–443 (2018)
42. A. Farhadi, M. Hejrati, M.A. Sadeghi, P. Young, C. Rashtchian, J. Hockenmaier, D. Forsyth, Every picture tells a story: Generating sentences from images, in: *European conference on computer vision.* (Springer, 2010), pp. 15–29
43. R. Socher, A. Karpathy, Q.V. Le, C.D. Manning, A.Y. Ng, Grounded compositional semantics for finding and describing images with sentences. Trans. Assoc. Comput. Linguist. **2**, 207–218 (2014)
44. A. Karpathy, A. Joulin, L. F. Fei-Fei, Deep fragment embeddings for bidirectional image sentence mapping, in *Advances in Neural Information Processing Systems* (2014), pp. 1889–1897

45. R. Kiros, R. Salakhutdinov, R.S. Zemel, Unifying visual-semantic embeddings with multi-modal neural language models. arXiv:1411.2539
46. A. Krizhevsky, I. Sutskever, G. E. Hinton, Imagenet classification with deep convolutional neural networks, in *Advances in Neural Information Processing Systems* (2012), pp. 1097–1105
47. M. Sundermeyer, R. Schlüter, H. Ney, Lstm neural networks for language modeling, in *Thirteenth Annual Conference of the International Speech Communication Association* (2012)
48. J. Mao, X. Wei, Y. Yang, J. Wang, Z. Huang, A. L. Yuille, Learning like a child: Fast novel visual concept learning from sentence descriptions of images, in *Proceedings of the IEEE International Conference On Computer Vision* (2015), pp. 2533–2541
49. X. Chen, C. Lawrence Zitnick, Mind's eye: A recurrent visual representation for image caption generation, in *Proceedings of the IEEE Conference On Computer Vision And Pattern Recognition* (2015), pp. 2422–2431
50. R. Lebret, P. O. Pinheiro, R. Collobert, Phrase-based image captioning. arXiv:1502.03671
51. G. Kulkarni, V. Premraj, S. Dhar, S. Li, Y. Choi, A.C. Berg, T.L. Berg, Baby talk: Understanding and generating image descriptions, in *Proceedings of the 24th CVPR*. (Citeseer, 2011)
52. Y. Yang, C.L. Teo, H. Daumé III, Y. Aloimonos, Corpus-guided sentence generation of natural images, in *Proceedings of the Conference on Empirical Methods in Natural Language Processing, Association for Computational Linguistics* (2011), pp. 444–454
53. T. Yao, Y. Pan, Y. Li, Z. Qiu, T. Mei, Boosting image captioning with attributes. OpenReview **2**(5), 8 (2016)
54. Y. Feng, M. Lapata, Automatic caption generation for news images. IEEE Trans. Pattern Anal. Mach. Intell. **35**(4), 797–812 (2012)
55. H. Fang, S. Gupta, F. Iandola, R. Srivastava, L. Deng, P. Dollár, J. Gao, X. He, M. Mitchell, J. Platt, et al., From captions to visual concepts and back
56. G. Kulkarni, V. Premraj, V. Ordonez, S. Dhar, S. Li, Y. Choi, A.C. Berg, T.L. Berg, Babytalk: Understanding and generating simple image descriptions. IEEE Trans. Pattern Anal. Mach. Intell **35**(12), 2891–2903 (2013)
57. J. Johnson, A. Karpathy, L. Fei-Fei, Densecap: Fully convolutional localization networks for dense captioning, in *Proceedings of the IEEE Conference on Computer Vision and Pattern Recognition* (2016), pp. 4565–4574
58. J. Lu, C. Xiong, D. Parikh, R. Socher, Knowing when to look: Adaptive attention via a visual sentinel for image captioning, in *Proceedings of the IEEE Conference on Computer Vision and Pattern Recognition (CVPR)*, (Vol. 6) (2017)
59. K.V. Deemter, M. Theune, E. Krahmer, Real versus template-based natural language generation: A false opposition? Comput. Linguisti. **31**(1), 15–24 (2005)
60. E. Reiter, R. Dale, Building applied natural language generation systems. Nat. Lang. Eng. **3**(1), 57–87 (1997)
61. D. Elliott, F. Keller, Image description using visual dependency representations, in *Proceedings of the 2013 Conference on Empirical Methods in Natural Language Processing* (2013), pp. 1292–1302
62. S. Li, G. Kulkarni, T.L. Berg, A.C. Berg, Y. Choi, Composing simple image descriptions using web-scale n-grams, in *Proceedings of the Fifteenth Conference on Computational Natural Language Learning, Association for Computational Linguistics* (2011), pp. 220–228
63. K. Tran, X. He, L. Zhang, J. Sun, C. Carapcea, C. Thrasher, C. Buehler, C. Sienkiewicz, Rich image captioning in the wild, in *Proceedings of the IEEE Conference on Computer Vision and Pattern Recognition Workshops* (2016), pp. 49–56
64. A. Aker, R. Gaizauskas, Generating image descriptions using dependency relational patterns, in *Proceedings of the 48th Annual Meeting of The Association For Computational Linguistics, Association For Computational Linguistics* (2010), pp. 1250–1258
65. I. Sutskever, O. Vinyals, Q. V. Le, Sequence to sequence learning with neural networks, in: *Advances in Neural Information Processing Systems* (2014), pp. 3104–3112

66. L. Anne Hendricks, S. Venugopalan, M. Rohrbach, R. Mooney, K. Saenko, T. Darrell, Deep compositional captioning: Describing novel object categories without paired training data, in *Proceedings of the IEEE Conference on Computer Vision and Pattern Recognition* (2016), pp. 1–10
67. S. Hochreiter, J. Schmidhuber, Long short-term memory. Neural Comput. **9**(8), 1735–1780 (1997)
68. S. Ma, Y. Han, Describing images by feeding lstm with structural words, in *2016 IEEE International Conference on Multimedia and Expo (ICME).* (IEEE, 2016), pp. 1–6
69. Y. Wang, Z. Lin, X. Shen, S. Cohen, G.W. Cottrell, Skeleton key: Image captioning by skeleton-attribute decomposition, in *Proceedings of the IEEE Conference on Computer Vision and Pattern Recognition* (2017), pp. 7272–7281
70. T. Mikolov, M. Karafiát, L. Burget, J. Černocký, S. Khudanpur, Recurrent neural network based language model, in *Eleventh Annual Conference Of The International Speech Communication Association* (2010)
71. J. Mao, W. Xu, Y. Yang, J. Wang, Z. Huang, A. Yuille, Deep captioning with multimodal recurrent neural networks (m-rnn). arXiv:1412.6632
72. J. Donahue, L. Anne Hendricks, S. Guadarrama, M. Rohrbach, S. Venugopalan, K. Saenko, T. Darrell, Long-term recurrent convolutional networks for visual recognition and description, in *Proceedings of the IEEE Conference On Computer Vision And Pattern Recognition* (2015), pp. 2625–2634
73. Q. Wu, C. Shen, L. Liu, A. Dick, A. Van Den Hengel, What value do explicit high level concepts have in vision to language problems?, in *Proceedings of the IEEE Conference On Computer Vision And Pattern Recognition* (2016), pp. 203–212
74. A.P. Mathews, L. Xie, X. He, Senticap: Generating image descriptions with sentiments, in *Thirtieth AAAI Conference on Artificial Intelligence* (2016)
75. Z. Gan, C. Gan, X. He, Y. Pu, K. Tran, J. Gao, L. Carin, L. Deng, Semantic compositional networks for visual captioning, in *Proceedings of the IEEE Conference On Computer Vision And Pattern Recognition* (2017), pp. 5630–5639
76. Q. You, H. Jin, J. Luo, Image captioning at will: a versatile scheme for effectively injecting sentiments into image descriptions. arXiv:1801.10121
77. J. Gu, G. Wang, J. Cai, T. Chen, An empirical study of language cnn for image captioning, in *Proceedings of the IEEE International Conference on Computer Vision* (2017), pp. 1222–1231
78. J. Aneja, A. Deshpande, A.G. Schwing, Convolutional image captioning, in *Proceedings of the IEEE Conference on Computer Vision and Pattern Recognition* (2018), pp. 5561–5570
79. Q. Wang, A.B. Chan, Cnn+ cnn: convolutional decoders for image captioning. arXiv:1805.09019
80. Q. You, H. Jin, Z. Wang, C. Fang, J. Luo, Image captioning with semantic attention, in *Proceedings of the IEEE Conference on Computer Vision and Pattern Recognition* (2016) pp. 4651–4659
81. Z. Zhu, Z. Xue, Z. Yuan, Topic-guided attention for image captioning, in *2018 25th IEEE International Conference on Image Processing (ICIP)* (IEEE, 2018), pp. 2615–2619
82. M. Corbetta, G.L. Shulman, Control of goal-directed and stimulus-driven attention in the brain. Nat. Rev. Neurosci. **3**(3), 201 (2002)
83. T. Yao, Y. Pan, Y. Li, T. Mei, Exploring visual relationship for image captioning, in *Proceedings of the European Conference on Computer Vision (ECCV)* (2018), pp. 684–699
84. R. Faster, Towards real-time object detection with region proposal networks shaoqing ren [j], Kaiming He, Ross Girshick, and Jian Sun
85. X. Zhang, X. Wang, X. Tang, H. Zhou, C. Li, Description generation for remote sensing images using attribute attention mechanism. Remote Sens. **11**(6), 612 (2019)
86. B. Dai, S. Fidler, R. Urtasun, D. Lin, Towards diverse and natural image descriptions via a conditional gan, in *Proceedings of the IEEE International Conference on Computer Vision* (2017), pp. 2970–2979
87. M. Mirza, S. Osindero, Conditional generative adversarial nets. arXiv:1411.1784

88. R. Shetty, M. Rohrbach, L. Anne Hendricks, M. Fritz, B. Schiele, Speaking the same language: Matching machine to human captions by adversarial training, in *Proceedings of the IEEE International Conference on Computer Vision*, (2017), pp. 4135–4144
89. E. Jang, S. Gu, B. Poole, Categorical reparameterization with gumbel-softmax. arXiv:1611.01144
90. Z. Ren, X. Wang, N. Zhang, X. Lv, L.-J. Li, Deep reinforcement learning-based image captioning with embedding reward, in *Proceedings of the IEEE Conference on Computer Vision and Pattern Recognition* (2017), pp. 290–298
91. H.H. Clark, W.G. Chase, On the process of comparing sentences against pictures. Cogn. Psychol. **3**(3), 472–517 (1972)
92. W.T. Fitch, Empirical approaches to the study of language evolution. Psychon. Bull. Rev. **24**(1), 3–33 (2017)
93. W.T. Mitchell, *Iconology: image, text, ideology* (University of Chicago Press, 2013)
94. G. Lakemeyer, B. Nebel, Foundations of knowledge representation and reasoning, in *Foundations of Knowledge Representation and Reasoning*, (Springer, 1994), pp. 1–12
95. L.-J. Zang, C. Cao, Y.-N. Cao, Y.-M. Wu, C. Cun-Gen, A survey of commonsense knowledge acquisition. J. Comput. Sci. Technol **28**(4), 689–719 (2013)
96. K. Dinakar, B. Jones, C. Havasi, H. Lieberman, R. Picard, Common sense reasoning for detection, prevention, and mitigation of cyberbullying. ACM Trans. Interact. Intell. Syst. (TiiS) **2**(3), 18 (2012)
97. E. Davis, *Representations of Commonsense Knowledge* (Morgan Kaufmann, 2014)
98. N. Sharif, L. White, M. Bennamoun, S.A.A. Shah, Learning-based composite metrics for improved caption evaluation, in *Proceedings of ACL 2018, Student Research Workshop* (2018), pp. 14–20
99. P. Kuznetsova, V. Ordonez, T.L. Berg, Y. Choi, Treetalk: Composition and compression of trees for image descriptions. Tran. Assoc. Comput. Linguist. **2**, 351–362 (2014)
100. K. Shuster, S. Humeau, H. Hu, A. Bordes, J. Weston, Engaging image captioning via personality, in Proceedings *of the IEEE Conference on Computer Vision and Pattern Recognition* (2019), pp. 12516–12526
101. S. Venugopalan, L. Anne Hendricks, M. Rohrbach, R. Mooney, T. Darrell, K. Saenko, Captioning images with diverse objects, in *Proceedings of the IEEE Conference on Computer Vision and Pattern Recognition* (2017), pp. 5753–5761
102. K. Fu, J. Jin, R. Cui, F. Sha, C. Zhang, Aligning where to see and what to tell: image captioning with region-based attention and scene-specific contexts. IEEE Trans. Pattern Anal. Mach. Intell. **39**(12), 2321–2334 (2016)
103. M. Mitchell, X. Han, J. Dodge, A. Mensch, A. Goyal, A. Berg, K. Yamaguchi, T. Berg, K. Stratos, H. Daumé III, Midge: Generating image descriptions from computer vision detections, in *Proceedings of the 13th Conference of the European Chapter of the Association for Computational Linguistics, Association for Computational Linguistics* (2012), pp. 747–756
104. M. Kusner, Y. Sun, N. Kolkin, K. Weinberger, From word embeddings to document distances, in *International Conference on Machine Learning* (2015), pp. 957–966
105. S. Liu, Z. Zhu, N. Ye, S. Guadarrama, K. Murphy, Improved image captioning via policy gradient optimization of spider. arXiv:1612.00370
106. T. Mikolov, I. Sutskever, K. Chen, G. S. Corrado, J. Dean, Distributed representations of words and phrases and their compositionality, in *Advances In Neural Information Processing Systems* (2013), pp. 3111–3119
107. A. Kershaw, M. Bober, The lexical gap: An improved measure of automated image description quality, in *Proceedings of the 13th International Conference on Computational Semantics-Student Papers* (2019), pp. 15–23
108. N. Sharif, L. White, M. Bennamoun, W. Liu, S.A.A. Shah, Lceval: Learned composite metric for caption evaluation. Int. J. Comput. Vis. **127**(10), 1586–1610 (2019)
109. N. Ding, S. Goodman, F. Sha, R. Soricut, Understanding image and text simultaneously: a dual vision-language machine comprehension task. arXiv:1612.07833

110. M. Hodosh, J. Hockenmaier, Focused evaluation for image description with binary forced-choice tasks, in *Proceedings of the 5th Workshop on Vision and Language* (2016), pp. 19–28
111. R. Shekhar, S. Pezzelle, Y. Klimovich, A. Herbelot, M. Nabi, E. Sangineto, R. Bernardi, Foil it! find one mismatch between image and language caption. arXiv:1705.01359
112. J. Deng, W. Dong, R. Socher, L.-J. Li, K. Li, L. Fei-Fei, Imagenet: A large-scale hierarchical image database, in *IEEE Conference on Computer Vision And Pattern Recognition*. (IEEE ,2009), pp. 248–255
113. M. Buhrmester, T. Kwang, S.D. Gosling, Amazon's mechanical turk: a new source of inexpensive, yet high-quality, data? Perspect. Psychol. Sci. **6**(1), 3–5 (2011)
114. M. Everingham, L. Van Gool, C.K. Williams, J. Winn, A. Zisserman, The pascal visual object classes (voc) challenge. Int. J. Comput. Vis **88**(2), 303–338 (2010)
115. M. Everingham, L. Van Gool, C. K. I. Williams, J. Winn, A. Zisserman, The PASCAL Visual Object Classes Challenge (VOC2011) Results (2011). http://www.pascal-network. org/challenges/VOC/voc2011/workshop/index.html
116. X. Chen, H. Fang, T.-Y. Lin, R. Vedantam, S. Gupta, P. Dollár, C. L. Zitnick, Microsoft coco captions: Data collection and evaluation server. arXiv:1504.00325
117. B. Thomee, D. A. Shamma, G. Friedland, B. Elizalde, K. Ni, D. Poland, D. Borth, L.-J. Li, Yfcc100m: The new data in multimedia research. arXiv:1503.01817
118. A. Kuznetsova, H. Rom, N. Alldrin, J. Uijlings, I. Krasin, J. Pont-Tuset, S. Kamali, S. Popov, M. Malloci, T. Duerig, et al., The open images dataset v4: Unified image classification, object detection, and visual relationship detection at scale. arXiv:1811.00982
119. S. E. Kahou, V. Michalski, A. Atkinson, Á. Kádár, A. Trischler, Y. Bengio, Figureqa: An annotated figure dataset for visual reasoning. arXiv:1710.07300
120. C. L. Zitnick, D. Parikh, L. Vanderwende, Learning the visual interpretation of sentences, in *Proceedings of the IEEE International Conference on Computer Vision* (2013), pp. 1681–1688
121. D. Elliott, F. Keller, A treebank of visual and linguistic data
122. B. A. Plummer, L. Wang, C. M. Cervantes, J. C. Caicedo, J. Hockenmaier, S. Lazebnik, Flickr30k entities: Collecting region-to-phrase correspondences for richer image-to-sentence models, in *2015 IEEE International Conference on Computer Vision (ICCV)*. (IEEE, 2015), pp. 2641–2649
123. J. Chen, P. Kuznetsova, D. Warren, Y. Choi, Déja image-captions: A corpus of expressive descriptions in repetition, in *Proceedings of the 2015 Conference of the North American Chapter of the Association for Computational Linguistics: Human Language Technologies* (2015), pp. 504–514
124. R. Krishna, Y. Zhu, O. Groth, J. Johnson, K. Hata, J. Kravitz, S. Chen, Y. Kalantidis, L.-J. Li, D.A. Shamma et al., Visual genome: Connecting language and vision using crowdsourced dense image annotations. Int. J. Comput. Vis. **123**(1), 32–73 (2017)
125. M. Grubinger, P. Clough, H. Müller, T. Deselaers, The iapr tc-12 benchmark: A new evaluation resource for visual information systems, in *International workshop ontoImage*, (Vol. 2) (2006)
126. R. Funaki, H. Nakayama, Image-mediated learning for zero-shot cross-lingual document retrieval, in *Proceedings of the 2015 Conference on Empirical Methods in Natural Language Processing* (2015), pp. 585–590
127. J. Rajendran, M. M. Khapra, S. Chandar, B. Ravindran, Bridge correlational neural networks for multilingual multimodal representation learning. arXiv:1510.03519
128. J. Hitschler, S. Schamoni, S. Riezler, Multimodal pivots for image caption translation. arXiv:1601.03916
129. D. Elliott, S. Frank, K. Sima'an, L. Specia, Multi30k: Multilingual english-german image descriptions. arXiv:1605.00459
130. X. Li, W. Lan, J. Dong, H. Liu, Adding chinese captions to images, in *Proceedings of the 2016 ACM on International Conference on Multimedia Retrieval, ACM* (2016), pp. 271–275
131. Y. Yoshikawa, Y. Shigeto, A. Takeuchi, Stair captions: Constructing a large-scale japanese image caption dataset. arXiv:1705.00823
132. S. He, H. R. Tavakoli, A. Borji, N. Pugeault, A synchronized multi-modal attention-caption dataset and analysis. arXiv:1903.02499

133. J. Wu, H. Zheng, B. Zhao, Y. Li, B. Yan, R. Liang, W. Wang, S. Zhou, G. Lin, Y. Fu, et al., Ai challenger: a large-scale dataset for going deeper in image understanding. arXiv:1711.06475
134. W. Lan, X. Li, J. Dong, Fluency-guided cross-lingual image captioning, in *Proceedings of the 25th ACM International Conference on Multimedia, ACM* (2017), pp. 1549–1557
135. X. Li, C. Xu, X. Wang, W. Lan, Z. Jia, G. Yang, J. Xu, Coco-cn for cross-lingual image tagging, captioning and retrieval, *IEEE Transactions on Multimedia*
136. P. Sharma, N. Ding, S. Goodman, R. Soricut, Conceptual captions: A cleaned, hypernymed, image alt-text dataset for automatic image captioning, in *Proceedings of the 56th Annual Meeting of the Association for Computational Linguistics* , (Vol. 1: Long Papers) (2018), pp. 2556–2565
137. C. Gan, Z. Gan, X. He, J. Gao, L. Deng, Stylenet: Generating attractive visual captions with styles, in *Proceedings of the IEEE Conference on Computer Vision and Pattern Recognition* (2017), pp. 3137–3146
138. D. Elliott, M. Kleppe, 1 million captioned dutch newspaper images
139. A.F. Biten, L. Gomez, M. Rusinol, D. Karatzas, Good news, everyone! context driven entity-aware captioning for news images, in *Proceedings of the IEEE Conference on Computer Vision and Pattern Recognition* (2019), pp. 12466–12475
140. D. Radev, A. Stent, J. Tetreault, A. Pappu, A. Iliakopoulou, A. Chanfreau, P. de Juan, J. Vallmitjana, A. Jaimes, R. Jha, et al., Humor in collective discourse: Unsupervised funniness detection in the new yorker cartoon caption contest. arXiv:1506.08126
141. X. Lu, B. Wang, X. Zheng, X. Li, Exploring models and data for remote sensing image caption generation. IEEE Tran. Geosci. Remote Sen. **56**(4), 2183–2195 (2017)
142. C. Chen, R. Zhang, E. Koh, S. Kim, S. Cohen, T. Yu, R. Rossi, R. Bunescu, Figure captioning with reasoning and sequence-level training. arXiv:1906.02850
143. H. Agrawal, K. Desai, X. Chen, R. Jain, D. Batra, D. Parikh, S. Lee, P. Anderson, nocaps: novel object captioning at scale. arXiv:1812.08658
144. B. Jing, P. Xie, E. Xing, On the automatic generation of medical imaging reports, in *Proceedings of the 56th Annual Meeting of the Association for Computational Linguistics* (Vol. 1: Long Papers), (2018), pp. 2577–2586
145. Z. Zhang, Y. Xie, F. Xing, M. McGough, L. Yang, Mdnet: A semantically and visually interpretable medical image diagnosis network, in *Proceedings of the IEEE Conference on Computer Vision And Pattern Recognition* (2017), pp. 6428–6436
146. C. Fan, Z. Zhang, D.J. Crandall, Deepdiary: lifelogging image captioning and summarization. J. Vis. Commun. Image Rep. **55**, 40–55 (2018)
147. H. Yu, S. Cheng, B. Ni, M. Wang, J. Zhang, X. Yang, Fine-grained video captioning for sports narrative, in *Proceedings of the IEEE Conference on Computer Vision and Pattern Recognition* (2018), pp. 6006–6015
148. D.G. Lowe, Distinctive image features from scale-invariant keypoints. Int. J.Comput. Vis. **60**(2), 91–110 (2004)
149. L. Fei-Fei, Recognizing and learning object categories, CVPR Short Course (2007)
150. S. Wu, J. Wieland, O. Farivar, J. Schiller, Automatic alt-text: Computer-generated image descriptions for blind users on a social network service, in *Proceedings of the 2017 ACM Conference on Computer Supported Cooperative Work and Social Computing, ACM* (2017), pp. 1180–1192
151. Y. Alemu, J.-b. Koh, M. Ikram, D.-K. Kim, Image retrieval in multimedia databases: A survey, in *2009 Fifth International Conference on Intelligent Information Hiding and Multimedia Signal Processing*, (IEEE, 2009), pp. 681–689
152. H. Noh, T. Kim, J. Mun, B. Han, Transfer learning via unsupervised task discovery for visual question answering, in *Proceedings of the IEEE Conference on Computer Vision and Pattern Recognition* (2019), pp. 8385–8394
153. W. Pei, J. Zhang, X. Wang, L. Ke, X. Shen, Y.-W. Tai, Memory-attended recurrent network for video captioning, in *Proceedings of the IEEE Conference on Computer Vision and Pattern Recognition* (2019), pp. 8347–8356

154. D. Guo, C. Xu, D. Tao, Image-question-answer synergistic network for visual dialog, in *Proceedings of the IEEE Conference on Computer Vision and Pattern Recognition* (2019), pp. 10434–10443
155. C. Deng, Q. Wu, Q. Wu, F. Hu, F. Lyu, M. Tan, Visual grounding via accumulated attention, in *Proceedings of the IEEE Conference on Computer Vision and Pattern Recognition* (2018), pp. 7746–7755
156. L.A. Hendricks, K. Burns, K. Saenko, T. Darrell, A. Rohrbach, Women also: Overcoming bias in captioning models, in *European Conference on Computer Vision*, (Springer, 2018), pp. 793–811

Chapter 3
Deep Learning Techniques for Geospatial Data Analysis

Arvind W. Kiwelekar, Geetanjali S. Mahamunkar, Laxman D. Netak, and Valmik B Nikam

Abstract Consumer electronic devices such as mobile handsets, goods tagged with RFID labels, location and position sensors are continuously generating a vast amount of location enriched data called geospatial data. Conventionally such geospatial data is used for military applications. In recent times, many useful civilian applications have been designed and deployed around such geospatial data. For example, a recommendation system to suggest restaurants or places of attraction to a tourist visiting a particular locality. At the same time, civic bodies are harnessing geospatial data generated through remote sensing devices to provide better services to citizens such as traffic monitoring, pothole identification, and weather reporting. Typically such applications are leveraged upon non-hierarchical machine learning techniques such as Naive-Bayes Classifiers, Support Vector Machines, and decision trees. Recent advances in the field of deep-learning showed that Neural Network-based techniques outperform conventional techniques and provide effective solutions for many geospatial data analysis tasks such as object recognition, image classification, and scene understanding. The chapter presents a survey on the current state of the applications of deep learning techniques for analyzing geospatial data. The chapter is organized as below: (i) A brief overview of deep learning algorithms. (ii)Geospatial Analysis: a Data Science Perspective (iii) Deep-learning techniques for Remote Sensing

A. W. Kiwelekar (✉) · G. S. Mahamunkar · L. D. Netak
Department of Computer Engineering, Dr. Babasaheb Ambedkar Technological University, Lonere 402103, Raigad, India
e-mail: awk@dbatu.ac.in

G. S. Mahamunkar
e-mail: gsmahamunkar@gmail.com

L. D. Netak
e-mail: ldnetak@dbatu.ac.in

V. B. Nikam
Department of Information and Technology, Veermata Jijabai Technical Institute, Mumbai 4000019, India
e-mail: vbnikam@it.vjti.ac.in

© The Editor(s) (if applicable) and The Author(s), under exclusive license to Springer Nature Switzerland AG 2020
G. A. Tsihrintzis and L. C. Jain (eds.), *Machine Learning Paradigms*, Learning and Analytics in Intelligent Systems 18, https://doi.org/10.1007/978-3-030-49724-8_3

data analytics tasks (iv) Deep-learning techniques for GPS data analytics (iv) Deep-learning techniques for RFID data analytics.

3.1 Introduction

Deep learning has emerged as a preferred technique to build intelligent products and services in various application domains. The resurgence of deep learning in recent times is attributed to three key factors. The first one is the availability high-performance GPUs necessary to execute computation-intensive deep learning algorithms. Second, the availability of such hardware at an affordable price. Also, the third and most important key factor responsible for the success of deep learning algorithms is the availability of open datasets in various application domains such as ImageNet [1] required to train the deep learning algorithms extensively [2].

The conventional application domains in which deep learning techniques have been applied effectively are speech recognition, image processing, language modelling and understanding, natural language processing and information retrieval [3]. All these application domains include processing and retrieving of useful information from raw multimedia data.

The success of deep learning techniques in these fields triggered its application in other fields such as Biomedicine [4], Drug Discovery [5], and Geographical Information System [6].

The chapter presents a state of the art review on applications of deep learning techniques for geospatial data analysis, one of the fields which is increasingly applying deep learning techniques to understand our planet earth.

3.2 Deep Learning: A Brief Overview

The field of Deep Learning is a sub-field of Machine Learning which studies the techniques for establishing a relationship between input feature variables and one or more output variables. Many tasks, such as classification and prediction, can be represented as a mapping between input feature variables and output variable(s).

For example, the price of a house in a city is a function of input feature variables such as the number of rooms, built-up area, the locality in a city, and other such parameters. The goal of the machine learning algorithms is to learn a mapping function from the given input data set, which is referred to as a training data set.

For example, the variables $x_1, x_2, \ldots x_n$ which are input feature variables and the variable y which is an output variable can be represented as

$$y = f(x_1, x_2, \ldots x_n)$$

The input variables are also referred to as independent variables, features, and predictors. The function f is referred to as a *model* or *hypothesis function*.

The training data set may or may not include the values of y, i.e., output variable. When a machine learning algorithm learns the function f from both features and output variable, the algorithm is referred to as a *supervised* algorithm. An *unsupervised learning* algorithm learns the function f from input features only without knowing the values of the output variable. Both kinds of the algorithm have been widely in use to develop intelligent product and services.

There exist many machine learning algorithms (e.g., Linear Regression, Logistic Regression, Support Vector Machine [7]) which are practical to learn simple tasks such as predicting house prices based on input features (e.g. the number of bedrooms, built-up area, locality). The goal of these learning algorithms is to reduce the error in predicting the value of the output variable. This minimizing goal is captured by a function called *cost function*. The *stochastic gradient descent* is one of the optimization techniques that is commonly used to achieve the minimization goal.

The conventional machine learning algorithms have been found useful and effective in learning simple tasks such as predicting house prices and classifying the tumour as malignant or benign. In such situations, the relationship between input features and the output variable is a simple linear function involving few predefined features.

However, they fail to perform effectively in situations where it is difficult to identify the features required for prediction—for example, recognizing the plant, rivers, animals in a given image. In such situations, the number of input features required for accurate prediction is large, and the relationship between input features and the output variable is also complex and non-linear one.

The set of algorithms that belongs to Deep Learning outperforms as compared to conventional machine learning methods in such situations. This section briefly describes deep learning techniques that have been found useful for geospatial data analysis. For more detailed and elaborate discussion on these techniques, one can refer [2].

3.2.1 Deep Learning Architectures

The deep learning techniques attempt to reproduce or emulate the working of the human brain in an artificial context. A network of neurons is the structural and functional unit of the human brain. Likewise, the Artificial Neural Network (ANN) is the fundamental element underlying most of the deep learning techniques.

3.2.2 Deep Neural Networks

A simple ANN consists of three essential elements: (i) Input Layer (ii) Hidden Layer
and (iii) Output Layer. An input layer consists of values of input features, and an
output layer consists of values of output variables. A hidden layer is referred to as
hidden because values held by the neurons in a hidden layer are not visible during
information processing.

A layer consists of more than one information processing nodes called *neurons*.
An artificial neuron is an information processing node taking n-inputs and producing
k-outputs. It is essentially a mapping function whose output is defined in two steps.
The first step is the sum of weighted multiplication of all of its input. Mathematically,
it can be represented as:

$$Y_i = g\left(\sum_j W_{ij} * a_j \right),$$

where Y_i is output of ith node, W_{ij} is the weight of jth input on ith node and a_j is
the value of jth input. This operation implements a matrix multiplication operation
which is a linear function. In the second step, the output of the first step is fed to
a non-linear function called an *activation function*. A neural network may use any
one of the functions (i) Sigmoid function (ii) Hyperbolic tangent function or (iii)
Rectified Linear Unit (ReLU).

The modern *deep neural networks* consist of more than one hidden layer, as shown
in Fig. 3.1 and ReLU as an activation function.

Training of the DNN is an iterative process which usually implements a stochastic
gradient algorithm to find the parameters or weights of input features. The weights
of the parameters are randomly initialized or used from a pre-trained model. An error
in the prediction is calculated at the output layer. At a hidden layer, the gradient of
error which is a partial derivative of the error with respect to the existing values of

Fig. 3.1 Deep neural
network

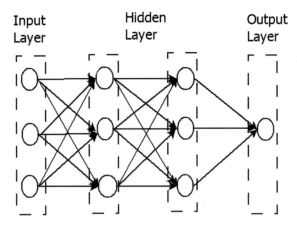

weights is fed back to update the values of weights at the input layer during next iteration. The process is known as back-propagation of gradients.

This seemingly simple strategy of learning parameters or weights of a model works effectively to detect features required for classification and prediction in many image processing and speech recognition tasks, provided that hardware required to do matrix multiplication and a large data-set is available. Thus eliminating the need for manual feature engineering required for non-hierarchical classification and prediction mechanism.

The deep neural networks have been successfully used to detect objects such as handwritten digits, and pedestrians [8]. In general, DNNs have been found efficient in handling 2-dimensional data that can be represented through a simple matrix.

3.2.3 Convolutional Neural Network (CNN)

As seen in the previous section, the input of a hidden layer of DNN is connected to all the outputs of the previous layer making computations and handling the number of connections unmanageable in case of high-dimensional matrices. Such situations arise when image sizes are of high resolutions (1024 × 1024). Also, the DNNs perform well when the data-set is 2-dimensional. But the majority of multi-modal data sets, for example, coloured images, videos, speech, and text are of 3-dimensional nature.

The *Convolutional Neural Networks* (CNN) are employed when the data set is three dimensional in nature and matrix size is very large. For example, a high-resolution coloured image has three channels of pixels (i.e., Red, Blue, Green) of size 1024 × 1024.

The architecture of CNN, as shown in Fig. 3.2 can be divided into multiple stages. These stages perform pre-processing activities such as identifying low- level features and reducing the number of features, followed by the main task of classification and prediction as done by a fully connected neural network.

The pre-processing activities are done by *convolution layers* and *pooling layers*.

The *convolution layer* performs step-wise convolution operation on the given input data size and a filter bank to create a feature map. Here, a filter is a matrix of learnable weights representing a pattern or a motif. The purpose of the step is to identify

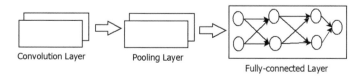

Convolution Layer   Pooling Layer

Fully-connected Layer

Fig. 3.2 Convolutional neural network

low-level features—for example, edges in an image. The underlying assumption of this layer is that low-level features correlate with a pixel configuration [8].

The *pooling layer* implements an aggregation operation such as either addition or maximization or average with an intention to share the weights among multiple connections. Thus reducing the number of features required for classification and/or prediction.

These pre-processing stages drastically reduce the number of features in a fully connected part of CNN.

The CNNs have been found useful in many image processing and speech recognition tasks. Because the working of CNN is based on the assumption that many high-level features are the compositions of low-level features. For example, a group of edges constitute a pattern or a motif, a group of motif constitute a part of an image, and a group of parts constitute an object [8].

3.2.4   Recurrent Neural Networks (RNN)

The third type of neural network is Recurrent Neural Networks (RNN). It has found applications in many fields such as speech recognition and machine language translation.

The RNN learns the dependency relationship among the sequence of input data. Unlike DNN and CNN, which learn relationships among feature variables and output variables, the RNN learns the relationship between data items fed to the network at different instances. To do this, RNN maintains a state vector or memory, which is implemented by connecting the hidden layer of current input to the hidden layer of previous input, as shown in Fig. 3.3. Hence, the output of RNN is not only the function of current input data but also the input data observed so far. As a result, the RNN may give two different outputs for the same input at two separate instances (Fig. 3.4).

A variant of RNN called RNN with LSTM (Long Short Term Memory) uses a memory unit to remember long term dependencies between data items in a sequence. Such networks have been found useful in question answering systems and symbolic reasoning, which draw a conclusion from a series of premises.

3.2.5   Auto-Encoders (AE)

The autoencoder is an *unsupervised* deep neural network architecture. Unlike supervised deep neural networks (e.g., DNN, CNN, RNN) which establish a mapping between input and output variables, autoencoders identify patterns in input data intending to transform the input data items. Reducing the dimensions of input vector is one example of transformation activity. In such variants, autoencoders transform an n-dimensional input vector into k-dimensional output vector such that $k < n$.

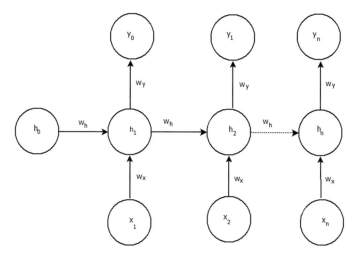

Fig. 3.3 Recurrent neural network

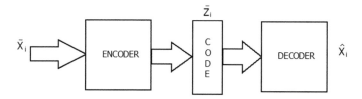

Fig. 3.4 Auto-encoders

These networks learn two different hypothesis function called *encode* and *decode* as shown in Fig. 3.4. The purpose of the function *encode* is to transform input vector x in some other representation z which is defined as

$$z = encode(x)$$

where z, and x are vectors with different representations. Similarly, the purpose of the function *decode* is to restore the vector z to its original form:

$$x = decode(z)$$

The functions *encode* and *decode* may be simple linear functions or a complex non-linear function. In case the data is highly non-linear an architecture called deep autoencoder with hidden layers is employed.

Autoencoders are typically employed for dimensionality reduction and pre-training a deep neural network. It has been observed that the performance of CNN or RNN improves when they are pre-trained with autoencoders [9].

3.3 Geospatial Analysis: A Data Science Perspective

The purpose of geospatial data analysis is to understand the planet Earth by collecting, interpreting and visualizing the data about objects, properties and events happening on, above and below the surface of Earth. Tools that we use to manage this information influence our knowledge of the Earth. So this section briefly reviews the technologies that enable geospatial data analysis.

3.3.1 Enabling Technologies for Geospatial Data Collection

The tools that are being used to collect location-specific information include (i) Remote Sensing (ii) Unmanned Aerial Vehicles (UAV) (iii) Global Positioning System (GPS) and (iv) Radio Frequency Identifiers (RFID). This section briefly explains these techniques.

1. *Remote Sensing*

   Remote sensing is a widely used technique to observe and acquire information about Earth without having any physical contacts. The principle of electromagnetic radiation is used to detect and observe objects on Earth. With Sun as the primary source of electromagnetic radiations, sensors on remote sensing satellites detect and record energy level of radiations reflected from the objects on Earth.

   Remote sensing can be either active or passive. In *passive* remote sensing, the reflected energy level is used to detect objects on Earth while in *active* remote sensing the time delay between emission and delay is used to detect the location, speed, and direction of the object.

   The images collected by a remote sensor are characterized primarily by attributes such as *spectral resolution* and *spatial resolution*. Spatial resolution describes the amount of physical area on Earth corresponding to a pixel in a raster image. Typically it corresponds to 1–1000 m area. Spectral resolution corresponds to the number of electromagnetic bands used in a pixel which typically corresponds to 3 bands (e.g., Red, Green and Blue) to seven visible colours. In hyper-spectral imaging produced by some remote sensing mechanism, 100–1000 bands correspond to a pixel.

   Data acquired through remote sensing have been useful in many geospatial analysis activities such as precision agriculture [10], in hydrology [11], for monitoring soil moisture [12] to name a few.

2. *Drones and Unmanned Aerial Vehicles (UAV)*

   Drones and UAVs are emerging as a cost effective alternative to conventional satellite-based method of remote sensing for Earth surface imaging [13]. Like satellite based remote sensing, UAVs are equipped with multi-spectral cameras, infrared cameras, and thermal cameras. A spatial accuracy of 0.5–2.5 m has been reported when UAVs are used for remote sensing [14]. Remote sensing

with UAVs have found applications for precision agriculture [14] and civilian security applications [15].
3. *Global Positioning Systems (GPS)*
   The devices equipped with Global Positioning Systems (GPS) can receive signals from GPS satellites and can calculate its accurate position. Smartphones, Cars and dedicated handheld devices are examples of GPS. These devices are specifically used for navigation purpose showing directions to a destination on maps, and monitoring and tracking of movements of objects of interest. Such devices can identify their locations with an accuracy of 5 m–30 cm. The location data collected from GPS devices have been used to identify driving styles [16], to predict traffic conditions [17] and transportation management.
4. *Radio Frequency Identification* (RFID)
   It is a low-cost technological alternative to GPS used for asset management. It is typically preferred when assets to be managed move around a shorter range. Unlike GPS device, RFID tags are transmitter of radio waves which are received by RFID tracker to identify the location of RFID tagged device. Recently data generated by RFID devices have been found useful to recognise human activities [18], and to predict order completion time [19].

3.3.2 Geospatial Data Models

The geospatial data models represent information about earth surface and locations of objects of reference. The data model used to represent this information depends on the mode used to collect the data. The raster and vector data models are used, when the mode of data collection used is remote sensing and UAV, while, GPS and RFID data models are used when mode of data collection GPS and RFID respectively.

The three data models that are prevalent in the field of remote sensing and UAV are raster, vector and Triangular Irregular Networks (TIN). In raster data model the earth surface is represented through points, lines and polygons. In vector representation, the earth surface is represented through cell matrices that store numeric values. In TIN data model the earth surface is represented as non-overlapping contiguous triangles.

The GPS data contains the location information of GPS enabled device in terms of longitude, latitude, timestamps and the satellites used to locate the device.

The RFID data contains information about Identification number of the RFID tag, location and timestamp.

3.3.3 Geospatial Data Management

Geographic Information System (GIS) integrates various aspects of geospatial analysis into one unified tool. It combines multiple geospatial analysis activities such as capturing of data, storage of data, querying data, presenting and visualizing data. It provides a user interface for users to perform these activities.

GIS integrates various data capturing technologies such as satellite-based remote sensing, UAV-based remote sensing, GPS devices, and scanning of paper-based maps.

GIS uses multiple kinds of data models such as raster data model, vector data model to represent earth surfaces as a base map. On top of a base map layers are used to prepare thematic maps showing roads, land cover and population distributions. Few examples of GIS are ArcGIS [20], Google Earth [21], QGIS [22].

3.4 Deep Learning for Remotely Sensed Data Analytics

Images constitute the significant chunk of data acquired through the method of remote sensing. These images vary in terms of information representation and graphical resolution. Typically the images obtained through remote sensing represent information either in vector or in raster form. Depending on the resolution of the images, they can be of Low Resolution (LR), High Resolution (HR) and Very High Resolution (VHR) type.

Deep learning techniques have been found as a promising method to process images of all kinds. Especially for the purpose of image segmentation [23], image enhancement [24], and image classification [25], and to identify objects in images [26].

Applications of deep learning techniques for remotely sensed data analytics leverage these advancements in the field of image processing to develop novel applications for geospatial data analysis. This section briefly reviews some of these recent applications. A detailed survey appears in [27, 28].

3.4.1 Data Pre-processing

The images obtained through remote sensing techniques often are of poor quality due to atmospheric conditions. The quality of these images needs to be enhanced to extract useful information from these images.

A set of activities which include denoising, deblurring, super-resolution, pansharpening and image fusion are typically performed on these images. Recently, geoscientists have started applying deep learning techniques for this purpose. Table 3.1 shows application of deep learning techniques for image pre-processing activities.

From the Table 3.1, it can be observed that both supervised(e.g., CNN) and unsupervised (e.g., Auto-encoders) are used for image pre-processing. Automatic feature extraction and comparable performance with conventional methods are some of the motivating factors behind adopting DL techniques for image pre-processing.

Table 3.1 Application of deep learning image pre-processing

Pre-processing activity	Example	Deep learning Technique used
Denoising	To restore original clean image from the low quality image or image with irrelevant details	A combination of sparse coding with denoising Auto-encoders [29] denoising CNN [30]
Deblurring	Restoring original or sharp image from the blurred image. The blurring of images occurs due to atmospheric turbulence in remote sensing	CNN [31], a combination of Auto-encoders and Generative Artificial Neural Network (GAN) [32]
Pan sharpening, image fusion, super resolution	In many geospatial applications images with high spatial and high spectral resolutions are required. The *pan sharpening* is a method to combine a LR multi-spectral image with a HR panchromatic image. The *super-resolution* combines a LR hyper-spectral (HS) image and a HR MS image	Stacked Auto-encoders [33], CNN [34]

3.4.2   Feature Engineering

In the context of deep learning, feature engineering is simple as compared with conventional machine learning techniques because many deep learning techniques such as CNN automatically extract features. This is one reason to prefer deep learning techniques for the analysis of remote sensing imagery. However, there exists other feature engineering steps described below which need to be carried out for better performance and reuse of knowledge learnt on similar applications.

1. *Feature Selection*: Feature selection is one of the crucial steps in the feature engineering aimed to identify the most relevant features that contribute to establishing the relationship between input and output variables. Earlier studies observe that three key factors affect the performance of machine learning techniques [35]. These are (i) choice of data set, (ii) machine learning algorithm and (iii) features used for classification or prediction tasks. In conventional non-hierarchical machine learning methods, regression analysis is usually performed to select the most relevant features.

    Recently, a two stage DL-based technique to select features is proposed in [36]. It is based on a combination of supervised deep networks and autoencoders. Deep networks in the first stage learn complicated low-level representations. In the second stage, an unsupervised autoencoders learn simplified representations. Few special DL-techniques specific to geospatial application have also been developed in order to extract spatial and spectral features [37, 38] from remotely sensed imagery

2. *Feature Fusion*: Feature fusion is the process of identifying *discriminating features* so that an optimal set of features can be used to either classify or predict an output variable. For example, feature fusion is performed to recover a high-resolution image from a low-resolution image [39]. A remote sensing specific DL-technique is proposed in [40]. The method is designed to classify hyper-spectral images. In the first stage, an optimal set of multi-scale features for CNN are extracted from the input data. In the second stage, a collection of discriminative features are identified by fusing the multi-scale features.

3. *Transfer learning or domain Adaptation*: The problem of domain adaptation is formulated with respect to the classification of remotely sensed images in [41]. It is defined as adapting a classification model to spectral properties or spatial regions that are different from the images used for training purposes. Usually, in such circumstances, a trained model fails to accurately classify images because of different acquisition method or atmospheric conditions, and they are called to be sensitive to data shifts. Domain adaptation is a specialized technique to address the transfer learning problem. Transfer learning deals a general situation where the features of a trained classifier model are adapted to classify or predict data that is not stationary over a period of time or space. For example, images captured through a remote sensing device vary at day and night time.

A strategy called *transduction strategy* that adopts DL techniques (e.g., autoencoders) has been proposed in [42] to address the problem of transfer learning. The strategy suggests to pre-train a DL architecture on a test data set using unsupervised algorithms (e.g., autoencoders) to identify discrimating features. Here, the test data set is used to pre-train a DL architecture and the test data set is entirely different from the training data set used to train a DL architecture. The strategy works to learn abstract representations because it identifies a set of discriminating features responsible to generalize the inferences from the observed the data.

3.4.3 Geospatial Object Detection

The task of detecting or identifying an object of interest such as a road, or an airplane, or a building from aerial images acquired either through remote sensing or UAV is referred to as geospatial object detection.

It is a special case of general problem of object recognition from images. However with numerous challenges such as small size of the object to be detected, large number of objects in imagery, and complex environment [6, 43] make the task of object recognition more difficult. Hence, DL-methods have been found useful in such situations specially to extract low-level features and learn abstract representations.

When DL-based methods are applied, the models such as CNN are first trained with geospatial images labeled to separate out the objects of our interest. The images used for training purposes are optimized to detect a specific object. These steps can be repeated until the image analysis model accurately detect multiple types of objects from the given images [44].

3.4.4 Classification Tasks in Geospatial Analysis

Like in other fields such as Computer Vision and Image processing, the field of Geospatial Analysis primarily adopts DL-techniques to classify remotely sensed imagery in various ways to extract useful information and patterns. Some of these usecases are discussed below.

These classification tasks are either *pixel-based* or *object-based* [45]. In pixel-based classification, pixels are grouped based on spectral properties (e.g., resolution). In object based classification, a group pixels are classified using both spectral and spatial properties into various geometric shapes and patterns. Either a supervised (e.g., CNN) or unsupervised algorithms (e.g. Autoencoders) are applied for the purpose of classification. In comparison with conventional machine learning techniques, application of DL-based techniques has outperformed in both scenarios in terms of classification accuracy [6].

Some of the classification activities include.

1. *Land Cover Classification*:
   Observed bio-physical cover over the surface of earth is referred to as *land cover*. A set of 17 different categories have been identified by International Geosphere-Biosphere Programme (IGBP) to classify earth surface. Some of these broad categories include Water, Forest, Shrubland, Savannas, Grassland, Wetland, Cropland, Urban, Snow, and Barren [46].
   There are two primary methods of getting information about land cover: field survey and remotely sensed imagery. DL-techniques are applied when information is acquired through remote sensing. In [47–49] the pre-trained CNN model is used to classify different land cover types. CNN is the most preferred technique to classify earth surface according to land cover. However other techniques such as RNN are also applied in few cases [50].
2. *Land Use Classification*:
   In this mode of classification, earth surface can be classified according to their social, economical and political usages. For example, a piece of green-land may be used as a tennis court or a cricket pitch. A piece of urban can be used for industrial or residential purposes. So there are no fixed set of categories or a classification system to differentiate land uses.
   Like in case of land cover classification, CNN [51] is typically used to classify land uses and to detect change in land cover according their uses [52].
3. *Scene Classification*:
   Scenes in remotely sensed images are high-level semantic entities such as a densely populated area, a river, a freeway, a golf-course, an airport etc. Super-

vised and unsupervised deep learning techniques have been found effective for scene understanding and scene classification in remotely sensed imagery [53].

Further, DL-techniques have been applied in specific cases of land use and land cover classification such as fish species classification [54], crop type classification [47] and mangrove classification [55].

3.5 Deep Learning for GPS Data Analytics

GPS enabled devices, particularly vehicles, are generating data in the range of GB/Sec. Such data includes information about exact coordinates, i.e. longitude and latitude of the device, direction and speed with which the device is moving. Such information throws light on many behavioural patterns useful for targetted marketing.

One way to utilize this data is to identify the driving style of a human driver. Various methods of accelerating, braking, turning under different road and environmental conditions determine the driving style of a driver. This information is useful for multiple business purposes like insurance companies may use to mitigate driving risks, to design better vehicles, to train autonomous cars, and to identify an anonymous driver.

However, identifying features that precisely determine a driving style is a challenging task. Hence the techniques based on deep learning are useful to extract such features. Dong et al. [16] have applied DL-techniques to GPS data to characterize the driving styles. The technique is motivated by the applications of DL-methods in speech recognition. They interpreted GPS data as a time series and applied CNN and RNN to identify driving styles. The raw data collected from the GPS devices are transformed to extract statistical features (e.g., mean and standard deviation) and then it is fed for analysis to CNN and RNN. The method was validated through identifying the driver's identity.

In another interesting application, a combination of RNN and Restricted Boltzman machine [56] is adopted to analyze GPS data for predicting the congestion evolution in a transportation network. The technique effectively predicts how congestion at a place has a ripple effect at other sites.

GPS data has also been analyzed using DL techniques to manage resources efficiently. For example, in the construction industry, the usages of GPS enabled equipment are monitored and analyzed [57].

3.6 Deep Learning for RFID Data Analytics

RFID and GPS are emerging as technologies to ascertain the locations of devices or assets. However, RFID in comparison with GPS allows to build location-sensitive applications for shorter geographical coverage. Unlike GPS mechanism, RFID is a low cost alternative to measure locations of assets with respect to a RFID tracker. GPS enabled devices provide absolute location information in terms of longitude and latitude.

The RFID's potential to accurately measure the location of devices with respect to a RFID tracker is increasingly being used in automated manufacturing plants. By using the location data transmitted by a RFID tagged device, various intelligent applications have been designed by applying DL techniques. These applications are intended to identify patterns and recognise activities in a manufacturing or a business process.

DL techniques have been found specially effective when activities happens in a spatio-temporal dimension. For example, in the field of medicine, detecting activities during trauma resuscitation [18].

In such situations, activity recognition is represented as a multi-class classification problem. For example, in case of trauma resuscitation, activities are oxygen preparation, blood pressure measurement, temperature measurement, and cardiac lead placement. For detecting these activities, a hardware set up of RFID tagged device along-with RFID tracker is used to collect the data. The collected data is analysed using CNN to extract relevant features and recognise the activity during trauma resuscitation.

In another application from manufacturing processes, the deep learning techniques have been used to accurately predict the job completion time. The conventional methods of job completion time rely on use of historical data. Such predictions greatly vary from the actual job completion time. The method proposed in [19] adopts RFID tags and trackers to collect real-time data. The collected data then mapped to historical data using CNN for predicting job completion time.

3.7 Conclusion

The scope of the geospatial data analysis is very vast. The data collection methods vary from manual one to satellite-based remote sensing. Numerous data models have been developed to represent various aspects of earth surfaces and objects on earth. These models capture absolute and relative location-specific information. These models also reveal static and dynamic aspects, spectral and spatial resolutions. Various GIS tools are being used to integrate and manage geospatial information.

So far, harnessing this collected information for useful purposes such as to extract hidden information in the form patterns, behaviours and predictions were limited by the absence of powerful analysis methods. But the emergence of Deep learning-based data analysis techniques has opened up new areas for geospatial applications.

This chapter presents some of the emerging applications designed around DL-techniques in the field of geospatial data analysis. These applications are categorized based on the mode of data collection adopted.

The deep-learning architectures such as CNN and Autoencoders are increasingly used when the method of data collection is remote sensing and UAV. Applications of DL techniques realize the high-level classification tasks such as land uses and land covers.

When GPS is the primary method of data collection, the collected data needs to be interpreted as a sequence or as a series of data. In such situations, RNN, along with CNN, is increasingly used to identify hidden patterns and behaviours from the traffic and mobility data collected from GPS enabled devices.

The CNNs are primarily used to process the location-specific information gathered through RFID device over shorter geographic areas. These analyses lead to recognize a Spatio-temporal activity in a manufacturing or a business process and to predict the time required to complete these activities using real-time data, unlike using historical data for predictive analytics.

These novel applications of DL-techniques in the field of Geospatial Analysis are improving our knowledge of planet earth, creating new insights about our behaviour during mobility, assisting us to make decisions (e.g., route selection), and making us more environmentally conscious citizens (e.g., predictions about depleting levels of forest land and protections of mangroves).

Acknowledgements The authors acknowledge the funding provided by Ministry of Human Resource Development (MHRD), Government of India, under the Pandit Madan Mohan National Mission on Teachers Training (PMMMNMTT). The work presented in this chapter is based on the course material developed to train engineering teachers on the topics of Geospatial Analysis and Product Design Engineering.

References

1. A. Krizhevsky, I. Sutskever, G.E. Hinton, Imagenet classification with deep convolutional neural networks, in *Advances in Neural Information Processing Systems* (2012), pp. 1097–1105
2. I. Goodfellow, Y. Bengio, A. Courville, in *Deep Learning* (MIT Press, 2016)
3. L. Deng, A tutorial survey of architectures, algorithms, and applications for deep learning. APSIPA Trans. Signal Inform. Process. **3** (2014)
4. C. Cao, F. Liu, H. Tan, D. Song, W. Shu, W. Li, Y. Zhou, X. Bo, Z. Xie, Deep learning and its applications in biomedicine. Genom. Proteom. Bioinform. **16**(1), 17–32 (2018)
5. H. Chen, O. Engkvist, Y. Wang, M. Olivecrona, T. Blaschke, The rise of deep learning in drug discovery. Drug Discov Today **23**(6), 1241–1250 (2018)
6. L. Zhang, L. Zhang, D. Bo, Deep learning for remote sensing data: a technical tutorial on the state of the art. IEEE Geosci. Remote Sens. Mag. **4**(2), 22–40 (2016)
7. P. Domingos, A few useful things to know about machine learning. Commun. ACM **55**(10), 78–87 (2012)
8. Y. LeCun, Y. Bengio, G. Hinton, Deep learning. Nature **521**(7553), 436 (2015)
9. Q.V. Le et al., A tutorial on deep learning part 2: autoencoders, convolutional neural networks and recurrent neural networks. *Google Brain* (2015), pp. 1–20

10. S.K. Seelan, S. Laguette, G.M. Casady, G.A. Seielstad, Remote sensing applications for precision agriculture: a learning community approach. Remote Sens. Environ. **88**(1–2), 157–169 (2003)
11. T.J. Jackson, J. Schmugge, E.T. Engman, Remote sensing applications to hydrology: soil moisture. Hydrol. Sci. J. **41**(4), 517–530 (1996)
12. L. Wang, J.J. Qu, Satellite remote sensing applications for surface soil moisture monitoring: a review. Front. Earth Sci. China **3**(2), 237–247 (2009)
13. K. Themistocleous,The use of USV platforms for remote sensing applications: case studies in cyprus, in *Second International Conference on Remote Sensing and Geoinformation of the Environment (RSCy2014)*, vol. 9229 (International Society for Optics and Photonics, 2014), p. 92290S
14. C.A. Rokhmana, The potential of UAV-based remote sensing for supporting precision agriculture in Indonesia. Procedia Environ. Sci. **24**, 245–253 (2015)
15. K. Daniel, C. Wietfeld, Using public network infrastructures for USAV remote sensing in civilian security operations. Technical report, Dortmund University, Germany FR (2011)
16. W. Dong, J. Li, R. Yao, C. Li, T. Yuan, L. Wang, Characterizing driving styles with deep learning. arXiv preprint arXiv:1607.03611 (2016)
17. X. Niu, Y. Zhu, X. Zhang, Deepsense: a novel learning mechanism for traffic prediction with taxi GPS traces, in *2014 IEEE Global Communications Conference* (IEEE, 2014), pp. 2745–2750
18. X. Li, Y. Zhang, M. Li, I. Marsic, J.W. Yang, R.S. Burd, Deep neural network for RFID-based activity recognition, in *Proceedings of the Eighth Wireless of the Students, by the Students, and for the Students Workshop* (ACM, 2016), pp. 24–26
19. C. Wang, P. Jiang, Deep neural networks based order completion time prediction by using real-time job shop RFID data. J. Intell. Manuf. **30**(3), 1303–1318 (2019)
20. K. Johnston, J.M. Ver Hoef, K. Krivoruchko, N. Lucas, *Using ArcGIS Geostatistical Analyst*, vol. 380 (Esri Redlands, 2001)
21. N. Gorelick, M. Hancher, M. Dixon, S. Ilyushchenko, D. Thau, R. Moore, Google earth engine: planetary-scale geospatial analysis for everyone. Remote Sens. Environ. **202**, 18–27 (2017)
22. QGIS Developer Team et al., *QGIS geographic information system* (Open Source Geospatial Foundation, 2009)
23. V. Badrinarayanan, A. Kendall, R. Cipolla, Segnet: a deep convolutional encoder-decoder architecture for image segmentation. IEEE Trans. Pattern Anal. Mach. Intell. **39**(12), 2481–2495 (2017)
24. K.G. Lore, A. Akintayo, S. Sarkar, LLNet: a deep autoencoder approach to natural low-light image enhancement. Pattern Recogn. **61**, 650–662 (2017)
25. T.-H. Chan, K. Jia, S. Gao, L. Jiwen, Z. Zeng, Y. Ma, Pcanet: a simple deep learning baseline for image classification? IEEE Trans. Image Process. **24**(12), 5017–5032 (2015)
26. J. Wu, Y. Yu, C. Huang, K. Yu., Deep multiple instance learning for image classification and auto-annotation, in *Proceedings of the IEEE Conference on Computer Vision and Pattern Recognition* (2015), pp. 3460–3469
27. L. Ma, Y. Liu, X. Zhang, Y. Ye, G. Yin, B.A. Johnson, Deep learning in remote sensing applications: a meta-analysis and review. ISPRS J. Photogramm. Remote Sens. **152**, 166–177 (2019)
28. X.X.Zhu, D. Tuia, L. Mou, G.-S. Xia, L. Zhang, F. Xu, F. Fraundorfer, Deep learning in remote sensing: a comprehensive review and list of resources. IEEE Geosci. Remote Sens. Mag. **5**(4), 8–36 (2017)
29. J. Xie, L. Xu, E. Chen, Image denoising and inpainting with deep neural networks, in *Advances in Neural Information Processing Systems* (2012), pp. 341–349
30. K. Zhang, W. Zuo, Y. Chen, D. Meng, L. Zhang, Beyond a gaussian denoiser: residual learning of deep CNN for image denoising. IEEE Trans. Image Process. **26**(7), 3142–3155 (2017)
31. S. Nah, T.H. Kim, K.M. Lee, Deep multi-scale convolutional neural network for dynamic scene deblurring. in *Proceedings of the IEEE Conference on Computer Vision and Pattern Recognition* (2017), pp. 3883–3891

32. T.M. Nimisha, A.K. Singh, A.N. Rajagopalan, Blur-invariant deep learning for blind-deblurring, in *Proceedings of the IEEE International Conference on Computer Vision* (2017), pp. 4752–4760
33. Y. Liu, X. Chen, Z. Wang, Z.J. Wang, R.K. Ward, X. Wang, Deep learning for pixel-level image fusion: recent advances and future prospects. *Inform. Fusion* **42**, 158–173 (2018)
34. G. Masi, D. Cozzolino, L. Verdoliva, G. Scarpa, Pansharpening by convolutional neural networks. Remote Sens. **8**(7), 594 (2016)
35. D. Yan, C. Li, N. Cong, L. Yu, P. Gong, A structured approach to the analysis of remote sensing images. Int. J. Remote Sens. 1–24 (2019)
36. N.T. Dong, L. Winkler, M. Khosla, Revisiting feature selection with data complexity for biomedicine. bioRxiv, p. 754630 (2019)
37. W. Zhao, D. Shihong, Spectral-spatial feature extraction for hyperspectral image classification: dimension reduction and deep learning approach. IEEE Trans. Geosci. Remote Sens. **54**(8), 4544–4554 (2016)
38. Q. Zou, L. Ni, T. Zhang, Q. Wang, Deep learning based feature selection for remote sensing scene classification. IEEE Geosci. Remote Sens. Lett. **12**(11), 2321–2325 (2015)
39. W. Wang, R. Guo, Y. Tian, W. Yang, CFSNet: toward a controllable feature space for image restoration. arXiv preprint arXiv:1904.00634 (2019)
40. Z. Li, L. Huang, J. He, A multiscale deep middle-level feature fusion network for hyperspectral classification. Remote Sens. **11**(6), 695 (2019)
41. D. Tuia, C. Persello, L. Bruzzone, Domain adaptation for the classification of remote sensing data: an overview of recent advances. IEEE Geosci. Remote Sens. Mag. **4**(2), 41–57 (2016)
42. Y. Bengio, Deep learning of representations for unsupervised and transfer learning, in *Proceedings of ICML workshop on unsupervised and transfer learning* (2012), pp. 17–36
43. C.-I. Cira, R. Alcarria, M.-Á. Manso-Callejo, F. Serradilla, A deep convolutional neural network to detect the existence of geospatial elements in high-resolution aerial imagery, in *Multidisciplinary Digital Publishing Institute Proceedings*, vol. 19 (2019), p. 17
44. A. Estrada, A. Jenkins, B. Brock, C. Mangold, Broad area geospatial object detection using autogenerated deep learning models, US Patent App. 10/013,774, 3 July 2018
45. R.C. Weih Jr., N.D. Riggan Jr., Object-based classification vs. pixel-based classification: comparitive importance of multi-resolution imagery. in *Proceedings of GEOBIA 2010: Geographic Object-Based Image Analysis*, Ghent, Belgium, 29 June–2 July 2010; **38**, Part 4/C7, p. 6
46. A.D. Gregorio, *Land Cover Classification System: Classification Concepts and User Manual: LCCS*, vol. 2 (Food & Agriculture Org., 2005)
47. N. Kussul, M. Lavreniuk, S. Skakun, A. Shelestov, Deep learning classification of land cover and crop types using remote sensing data. IEEE Geosci. Remote Sens. Lett. **14**(5), 778–782 (2017)
48. G.J. Scott, M.R. England, W.A. Starms, R.A. Marcum, C.H. Davis, Training deep convolutional neural networks for land–cover classification of high-resolution imagery. IEEE Geosci. Remote Sens. Lett. **14**(4), 549–553 (2017)
49. X. Guang, X. Zhu, F. Dongjie, J. Dong, X. Xiao, Automatic land cover classification of geo-tagged field photos by deep learning. Environ. Model. Softw. **91**, 127–134 (2017)
50. Z. Sun, L. Di, H. Fang, Using long short-term memory recurrent neural network in land cover classification on landsat and cropland data layer time series. Int. J. Remote Sens. **40**(2), 593–614 (2019)
51. P. Helber, B. Bischke, A. Dengel, D. Borth, Eurosat: a novel dataset and deep learning benchmark for land use and land cover classification. IEEE J. Sel. Top. Appl. Earth Observ. Remote Sens. **12**(7), 2217–2226 (2019)
52. C. Cao, S. Dragićević, S. Li, Land-use change detection with convolutional neural network methods. Environments **6**(2), 25 (2019)
53. G. Cheng, J. Han, L. Xiaoqiang, Remote sensing image scene classification: benchmark and state of the art. Proc. IEEE **105**(10), 1865–1883 (2017)
54. A. Salman, A. Jalal, F. Shafait, A. Mian, M. Shortis, J. Seager, E. Harvey, Fish species classification in unconstrained underwater environments based on deep learning. Limnol. Oceanogr. Methods **14**(9), 570–585 (2016)

55. S. Faza, E.B. Nababan, S. Efendi, M. Basyuni, R.F. Rahmat, An initial study of deep learning for mangrove classification, in *IOP Conference Series: Materials Science and Engineering*, vol. 420 (IOP Publishing, 2018), p. 012093
56. X. Ma, Y. Haiyang, Y. Wang, Y. Wang, Large-scale transportation network congestion evolution prediction using deep learning theory. PLoS One **10**(3), e0119044 (2015)
57. N. Pradhananga, J. Teizer, Automatic spatio-temporal analysis of construction site equipment operations using GPS data. Autom. Constr. **29**, 107–122 (2013)

Chapter 4
Deep Learning Approaches in Food Recognition

Chairi Kiourt, George Pavlidis, and Stella Markantonatou

Abstract Automatic image-based food recognition is a particularly challenging task. Traditional image analysis approaches have achieved low classification accuracy in the past, whereas deep learning approaches enabled the identification of food types and their ingredients. The contents of food dishes are typically deformable objects, usually including complex semantics, which makes the task of defining their structure very difficult. Deep learning methods have already shown very promising results in such challenges, so this chapter focuses on the presentation of some popular approaches and techniques applied in image-based food recognition. The three main lines of solutions, namely the design from scratch, the transfer learning and the platform-based approaches, are outlined, particularly for the task at hand, and are tested and compared to reveal the inherent strengths and weaknesses. The chapter is complemented with basic background material, a section devoted to the relevant datasets that are crucial in light of the empirical approaches adopted, and some concluding remarks that underline the future directions.

Keywords Deep learning · Food recognition · Food image datasets · Convolutional neural networks

C. Kiourt (✉) · G. Pavlidis · S. Markantonatou
Athena Research Centre, University Campus at Kimmeria, Xanthi 67100, Greece
e-mail: chairiq@athenarc.gr

G. Pavlidis
e-mail: gpavlid@athenarc.gr

S. Markantonatou
e-mail: marks@athenarc.gr

G. A. Tsihrintzis and L. C. Jain (eds.), *Machine Learning Paradigms*, Learning and Analytics in Intelligent Systems 18, https://doi.org/10.1007/978-3-030-49724-8_4

4.1 Introduction

The advent of deep learning technology has boosted many scientific domains by providing advanced methods and techniques for better object prediction and recognition, based on images or video. In computer science, especially in computer vision and artificial intelligence, image classification is an essential task, with many recent advances coming from object recognition with deep learning approaches [18, 28, 36, 40–42].

Food, which is an important part of everyday life in all cultures, constitutes a particular challenge in the domain of image classification, due to its visually intricate complexity, as well as the semantic complexity stemming from the variation in the mixing of various ingredients practiced by regional communities. This challenge has been proved to have many complicated aspects [33, 46]. The abundance of food images provided by the social networks, dedicated photo sharing sites, mobile applications and powerful search engines [33] is considered to be an easy way for the development of new datasets for scientific purposes. The scientific community suggests that automatic recognition (classification) of dishes would not only help people effortlessly organize their enormous photo collections and help online photo repositories make their content more accessible, but it would also help to estimate and track daily food habits and calorie intake and plan nutritional strategies even outside a constrained clinical environment [33].

Despite the dozens of applications, algorithms and systems available, the problem of recognizing dishes (food) and their ingredients has not been fully addressed by the machine learning and computer vision communities [46]. This is due to the lack of distinctive and spatial layout of food images, typically found in images depicting scenes or objects. For example, an image of an outdoor scene can be typically decomposed into (a) a ground place, (b) a horizon, (c) a forest, and (d) the sky. Patterns like these, cannot be found in food images. Food ingredients, like those found in a salad, are mixtures that frequently come in different shapes and sizes, depending much on regional practices and cultural habits. It should be highlighted that the nature of the dishes is often defined by the different colors, shapes and textures of the various ingredients [12]. Nevertheless, the kind of features that describe food in most cases are easily recognizable by humans in a single image, regardless of the geometrical variations of the ingredients. Hence, it could be considered that food recognition is a specific and a very complex classification problem demanding the development of models that are able to exploit local information (features) within images, along with complex, higher level semantics. Food recognition is still a challenging problem that attracts the interest of the scientific community [12, 18].

Food recognition flourished with the improvement of computational performance and the computer vision and machine learning advances over the last decade. Matsuda et al. [32], achieved accuracy scores of 55.8% for multiple-item food images and 68.9% for single-item food images by utilizing two different methods [32]. The first method was based on the Felzenszwalb's deformable part model and the second was based on feature fusion. In 2014 one of the first works that employed deep learning was developed [27]. The researchers achieved a classification accuracy of 72.26% with a pre-trained (transfer learning) model similar to AlexNet [28]. At the same period, Bossard et al. achieved 50.76% classification accuracy with the use of random forest on the Food101 dataset [4]. The researchers noted that the random forest method cannot outperform deep learning approaches. Similar conclusions were drawn by Kagaya et al., who reported a classification accuracy of 73.70% using deep Convolutional Neural Networks (CNNs) [24]. The next year, deep CNNs, in combination with transfer learning, achieved 78.77% [47]. Christodoulidis et al. introduced a new method on deep CNNs achieving an accuracy of 84.90% on a custom dataset [9]. In 2016, Singla et al. adopted the GoogLeNet architecture [41] and with a pre-trained model they achieved an accuracy of 83.60% [37]. Lui et al. [30], developed DeepFood and achieved similar results by using optimized convolution techniques in a modified version of the Inception [30]. The researchers reported an accuracy of 76.30% on the UEC-Food100 dataset, an accuracy of 54.70% on the UEC-Food256 dataset and an accuracy of 77.40% on the Food-101 dataset. Hassannejad et al. scored an accuracy of 81.45% on the UEC-Food100 dataset, 76.17% on the UEC-Food256 dataset and 88.28% on the Food-101 dataset [17], using Google's image recognition architecture named Inception V3 [42].

In 2017, Ciocca et al. introduced a novel method that combined segmentation and recognition techniques based on deep CNNs on a new dataset achieving 78.30% accuracy [10]. Mezgec and Koroušić [33] adopted a modification of the popular AlexNet and introduced the NutriNet, which was trained with images acquired using web search engines. They reached a classification performance of 86.72% over 520 food and drinks classes. The NutriNet uses fewer parameters compared to the original structure of the AlexNet. In 2018, Ciocca et al. introduced a new dataset, which was acquired from the merging of other datasets [12]. They tested several popular architectures on the dataset, including a Residual Network [18] with 50 layers, setting the latter as the reference architecture. Apparently, many different research groups have worked vigorously on the topic of food recognition, using a variety of methods and techniques. Overall, the most successful methods are the variations of deep learning approaches.

This chapter serves the purpose of presenting the possible paths to follow in order to develop a machine learning technique for the complex task of food recognition. It starts with an introduction of general deep learning concepts to establish a basic level of understanding and an outline of the most popular deep learning frameworks available for the design and development of new models. Next, the three possible development approaches are being presented, including a totally new architecture, several transfer learning adaptations and the most relevant content prediction platforms. As the empirical learning employed in deep learning is data demanding, popular image datasets for food recognition are being presented in a separate subsection. A comparative study of the presented solutions follows, and the chapter concludes by providing information regarding the development of a deep learning model tasked to tackle the food recognition problem, as well as a list of future challenges in the relevant scientific area.

4.2 Background

The most recent food recognition systems are developed based on deep convolutional neural network architectures [11, 12, 19, 46, 48], thus, this section presents some "must-know" techniques and methods for the development and utilization of an efficient food recognition CNN model.

A CNN is a multi-layered neural network with a unique architecture designed to extract increasingly complex features of the data at each layer, in order to correctly determine the output. CNN's are well suited for perceptual tasks. Figure 4.1 depicts a simple CNN architecture with two convolution layers and two pooling layers at the feature extraction level of the network. The classification level of the network (top layers) is composed of three layers, (a) a flattening layer, which transforms the multiple feature maps, namely a multi-dimensional array (tensor) of the last convolution layer to a one-dimension array, (b) a dense layer and (c) a prediction layer. The output of the prediction layer is, in most cases, *one-hot encoded*, which means that the output is a binary vector representation of the categorical classes. The cell in this vector which is triggered (is true or "1") shows the prediction of the CNN, thus, each class is represented by a vector of "0"s and a single "1", the position of which determines the class.

A CNN is mostly used when there is an unstructured dataset (e.g., images) and the model needs to extract information from it. For instance, if the task is to predict an image caption:

- the CNN receives a color image (i.e. a salad) as a 3D matrix of pixels (three matrices of 2D intensity images);
- during the training, the hidden layers enable the CNN to identify unique features by shaping suitable filters that extract information from the images;
- when the network learning process converges, then the CNN it is able to provide a prediction regarding the class an image belongs to.

Each convolution in a network consist of, at least, an *input*, a *kernel* (filter) and a *feature map* (output).

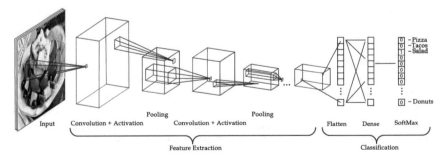

Fig. 4.1 A simple CNN architecture

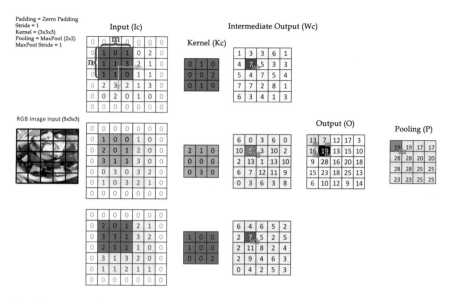

Fig. 4.2 An analysis of the convolution process

- The *input* (or the input feature map) for a CNN is typically a tensor.[1] A tensor is a multidimensional array containing data for the validation of training of the CNN. An input, as a tensor, can be represented by (i, h, w, c), where: i is the number of images, h is the height of the image, w is the width of the image and c is the number of channels in the image. For example, $(64, 229, 229, 3)$ represents a batch of 64 images of 229×299 pixels and 3 color channels (RGB).
- A *kernel* (also called filter and feature detector) is a small array with fixed values. Convolution of this array with the input tensor reveals visual features (such as edges). The kernel moves with specific steps and direction in order to cover the entire surface of each image. These step patterns are called the *strides*. A stride denotes the number of pixels by which the kernel moves after each operation.
- The *output feature map* or simply the output of a convolution is the group of features extracted by the filter (kernel) from the input, and is practically a transformed form of the input feature map. The size of the output feature map's tensor depends mainly on the input, the kernel size and the stride.

Figure 4.2 depicts the entire process (a single step) for an RGB image, over $5 \times 5 \times 3$ image blocks. For each color channel c of the input tensor an intermediate output ($\mathbf{W}_c$) is calculated as

$$\mathbf{W}_c = \mathbf{K}_c \otimes \mathbf{I}_{c_{(m,n)}} \tag{4.1}$$

---

[1] It should be noted that there is an inconsistency across the scientific domains in the usage of the word "tensor", which has a totally different meaning in physics than in computer science and in particular in machine learning.

where $\otimes$ denotes the convolution, $\mathbf{K}_c$ is the convolution kernel and $\mathbf{I}_{c(m,n)}$ is the input image block.

The final feature map $\mathbf{O}$ is calculated by adding the intermediate outputs ($\mathbf{W}_c$) of each channel:

$$\mathbf{O} = \sum_c \mathbf{W}_c \qquad (4.2)$$

In order to reduce the size of the feature maps, a pooling method is commonly used. Pooling reduces the size of a feature map by using some functions to summarize subregions. Typical functions used for the task are the maximum (max pooling) or the average (average pooling) of a subregion (or submap) x of the feature map. In the example of Fig. 4.2 max pooling is used, which is expressed as:

$$\mathbf{O}_x = \begin{bmatrix} o_{1,1} & o_{1,2} \\ o_{2,1} & o_{2,2} \end{bmatrix}, \mathbf{P}_x = \max_{i,j=1,2} o_x^{i,j} \qquad (4.3)$$

where $o_x^{i,j}$ denotes the $i-$th, $j-$th element of the submap $\mathbf{O}_x$. Generally, the pooling window (subregion of the feature map) moves with a specific stride.

Most of the times, the size of a kernel is not consistent with the size of the input tensor, so to maintain the dimensions of the output (feature map) and in order for the kernel to be able to pass correctly over the entire input tensor, a data padding strategy is required. This is typical in any image convolution application. Padding is a process of adding extra values around the input tensor. One of the most frequent padding methods is *zero-padding*, which adds z zeroes to each side of the boundaries of the input tensor. There are many different ways to calculate the number z. A simple way, if a stride of 1 is used, is to compute the size of zero padding as $z = \left\lceil \frac{(k-1)}{2} \right\rceil$, where k is the filter size and $\lceil * \rceil$ the ceiling operation. For example, a kernel of size 3×3 imposes a zero-padding in each dimension of $z = \frac{(3-1)}{2} = 1$ zero, which adds a zero at the beginning and a zero at the end of each image dimension. Based on these values the size of the output of a convolution can be calculated easily. Supposing a one-dimensional case or a square input image and filter kernels, the formula to calculate the output size of any convolutional layer is given as

$$o = \left\lfloor \frac{i - k + 2z}{s} \right\rfloor + 1 \qquad (4.4)$$

where o is the dimension of the output, i is the input dimension, k is the filter size, z is the padding, s is the stride and $\lfloor * \rfloor$ the floor operation. For example, in a typical case in which $i = 5, k = 3, z = \left\lceil \frac{(k-1)}{2} \right\rceil = 1$ and $s = 2$, then the expected output should be of size $o = \left\lfloor \frac{5-3+2}{2} \right\rfloor + 1 = 2 + 1 = 3$. This example is graphically depicted in Fig. 4.3, which shows the first two steps in the process of a square image convolution with a square filter kernel with the appropriate zero-padding.

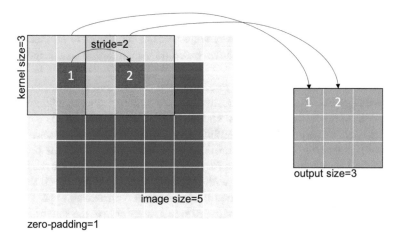

Fig. 4.3 Example of a square image convolution with zero-padding

While training a CNN there are many parameters that should be taken into consideration for an effective training process, which will eventually produce a powerful and efficient deep learning model. Especially when a model has to handle a couple of hundreds to a million of trainable parameters, there are many challenges to be met. Among the most prominent issues are *overfitting* and *underfitting* [16]. There are a few ways to mitigate these issues. Underfitting occurs when the model fails to sufficiently learn the problem and performs poorly both on the training and the validation samples (lacks adaptation). Overfitting occurs when the model adapts too well to the training data but does not perform well on the validation samples (lacks generalization). Apparently, the ideal situation is when the model performs well on both training and validation data. So, it can be said that when a model learns suitably the training samples of the dataset and generalizes well, and if the model performs well in the validation samples, then there is high probability the process resulted in a *good-fitting model*. Simplistic examples of these issues are depicted in the graphs of Fig. 4.4. The leftmost graph represents the dataset, including training and validation samples. Second from the left is an example of underfitting, whereas third from the

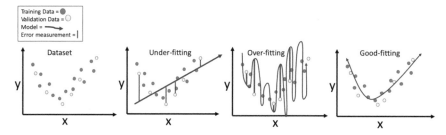

Fig. 4.4 Underfitting, overfitting and good fitting examples

left is an example of overfitting. The rightmost graph represents a good-fitting model. It should be noted that there are many ways to calculate the *error* in these issues, which in the graphs of Fig. 4.4 are shown based on the validation data as red vertical lines.

Some very common approaches to address the underfitting issue are to increase the dataset and train for a longer time, to check for a more proper regularization and to check if the model is not powerful enough (the architecture needs to change). On the other hand, using popular CNN architectures, which contain many convolutional layers with potentially hundreds to millions of trainable parameters, the probability of overfitting is very high. A solution for this problem is to increase the flexibility of the model, which may be achieved with regularization techniques, such the L1/L2 regularization [3], the Dropout method [38] and Batch Normalization [21].

The *L1 regularization*, uses a penalty technique to minimize the sum of the absolute values of the parameters, while the *L2 regularization* minimizes the sum of the squares of the parameters. Both L1 and L2 regularizations are known as "weight decay" methods because they try to keep the weights of the network as small as possible. The *Dropout* [38], randomly removes neurons during training (units are randomly dropped, along with their connections), so the network does not over-rely on any particular neuron. *Batch Normalization* [21] is mainly used to increase the training speed and the performance of the network, but it can also be used to perform a regularization similar to dropout. Batch normalization is used to normalize the inputs of each layer (intermediate layers), in order to tackle the *internal covariate shift*[2] problem. It also allows the use of higher learning rates and, many times, acts as a powerful regularizer, in some cases eliminating the need for Dropout.

Apart from the methods presented in the previous paragraphs, data augmentation is another popular technique that effectively addresses both underfitting or overfitting issues. Deep learning models usually need a vast amount of data to be properly trained. So, the data augmentation technique is a very important part of the training process. Some of the most popular operations that can be applied in images to perform data augmentation include (a) flipping, (b) random cropping, (c) tilting, (d) color shifting, (e) rotation, (f) corruption with noise, and (g) changing of contrast. It should be noted that when the number of classes is large and the network is too complex, a combination of these methods may be the appropriate approach to obtain a good-fitting model.

For an in-depth treatment of the convolutional neural networks and the apprehension of the mathematical background the interested reader is advised to consider [8, 13, 16, 50], or the available vast relevant literature and online sources.

---

[2]Covariate shift relates to a change in the distribution of the input variables of the training and the test data.

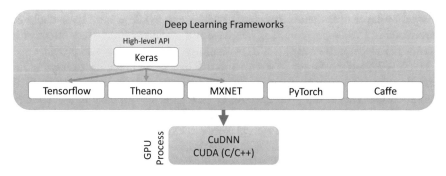

Fig. 4.5 Most popular deep learning frameworks

4.2.1   Popular Deep Learning Frameworks

A deep learning framework is a platform containing interfaces, libraries, tools and applications, which allows developers and scientists to easily build and deploy complex deep learning models, without directly being involved in designing their architectures and handling the math. Also, deep learning frameworks offer building blocks for designing, training and validating deep neural network models, through high level programming interfaces. Most of these frameworks rely on GPU-accelerated solutions such as CuDNN to deliver high-performance multi-GPU accelerated training. Some key features of a good deep learning framework are:

- optimization for performance
- GPU-acceleration
- good documentation and good community support
- easy to code
- parallelized processes
- ready to use deep models/architectures (applications)
- pre-trained models
- multiple, ready to use functions (e.g. optimizers, activation functions, layers etc.)
- popularity
- continues update.

Based on these key features six deep learning frameworks stand out and are presented in this section. Figure 4.5 presents a simple hierarchical view of these frameworks, highlighting how all of them are GPU accelerated. It should be stressed that only the Keras framework can be considered as high-level API developed over other libraries, mainly focusing on the provision of a user-friendly and easy to extend environment (framework).

Caffe[3] [22] is the oldest and the most popular deep learning framework developed by Yangqing Jia at BAIR (Berkeley Artificial Intelligence Research). Caffe supports

---

[3]http://caffe.berkeleyvision.org/.

a layer-by-layer deep neural network development. While a new layer could be developed in Python programming language it is considered to be less effective in contrast to a layer written in C++ CUDA. Caffe besides the Python API provides also a MATLAB interface. Several years after its first release, its godfather (Yangqing Jia) and his team at Facebook developed the next generation Caffe framework in 2018, called Caffe2. This version was improved towards the development of a more user-friendly framework. In addition, it provides an API that supports mobile applications. Also, Caffe2 supports the Open Neural Network Exchange (ONNX) format, which allows easy integration of models with other frameworks, such as MXNet, PyTorch etc. In fact, since May 2, 2018, Caffe2 was integrated in PyTorch.

TensorFlow[4] [1] is an open source framework developed and backed by a machine learning team in Google. Its development and design are based on numerical computations using data flow graphs focusing on large-scale distributed training models. It consists of more than two hundred standard operations, including a variety of mathematical operations, multi-dimensional array manipulation, flow control and many more, written in C++ programming language. This framework is GPU-accelerated and may also run in CPUs and mobile devices. For this reason, it could be considered as a framework for use both in scientific research and in product development. Its distinctive drawback is that it requires a low-level API and its use seems quite difficult to inexperienced users. However, it has been integrated into several high-level APIs, such as Keras, Estimator and Eager.

Keras[5] [7] is an open source neural network library running on top of some popular frameworks such as TensorFlow and Theano. It is considered to be a high-level API (interface), written in Python, rather than a standalone machine-learning framework, focusing on its extensibility, its modularity and its user-friendly environment. It requires minimum knowledge and technical skills. Since 2017, the TensorFlow team includes Keras in TensorFlow's core library.

PyTorch[6] [2] is a GPU-accelerated deep learning framework written in a combination of programming languages, Python, C and CUDA. It was developed by the Facebook AI research team. PyTorch is younger than TensorFlow but it has grown rapidly in popularity, because of its flexibility and the ease of development when it comes to complex architectures. Additionally, it allows customizations that Tensor-Flow does not. The framework is used by both scientific and industrial communities, for example Uber has been using PyTorch for its applications. PyTorch is backed by some of the most high-tech companies such as Facebook, Twitter and NVIDIA. It should be highlighted that PyTorch supports ONNX models, allowing an easy exchange of models among a variety of frameworks.

Theano[7] [43] was first released in 2007 and it has been developed at the University of Montreal. It is the second oldest significant Python deep learning framework. Theano compiles mathematical expressions by using NumPy in Python and some

---

[4]https://www.tensorflow.org/.

[5]https://keras.io/.

[6]https://pytorch.org/.

[7]http://www.deeplearning.net/software/theano/.

Table 4.1 Most popular deep learning frameworks

Name	Released	Programming Interfaces	Citations[a]
Caffe	2014	Python, MATLAB, C++	11,990
TensorFlow	2016	Python, C++[b], Java[b], Go[b]	9250
Keras	2015	Python	5399
PyTorch	2017	Python, ONNX, C++	2718
MXNET	2015	Python, C++, R, Java,...[c]	1077
Theano	2007	Python, C++	572

[a]Google Scholar: Last accessed 7 November 2019
[b]Not fully covered
[c]Many more languages

efficient native libraries such as BLAS to run the experiments in multiple GPUs and CPUs. In November 2017, Theano stopped to be actively upgrading, but it is maintained continuously. Note, that Keras runs also on top of Theano.

MXNet[8] [6] is incubated by Apache and used by Amazon. It is the fifth most popular deep learning library. One important benefit of MXNet is its portability, its lightweight development, and its scalability, which allows for an effective adaptation in a variety of computer systems with multiple GPUs and CPUs. This key point makes MXNet very powerful and useful for enterprises. Also, its portability supports a wide range of smart devices, such as mobile devices (using Amalgamation), IoT devices (supporting AWS Greengrass) and Serverless (using AWS Lambda). Furthermore, it is cloud-friendly and directly compatible with a variety of cloud services. MXNet supports Keras as a top layer interface, which allows easy and fast prototyping. MXNet supports ONNX format models.

There are several comprehensive studies [23, 34] presenting the benefits and drawbacks of various frameworks/libraries focused on machine learning. It should be mentioned that most frameworks have a scientifically documented presence. A brief presentation of the six aforementioned frameworks is provided in the following paragraphs. All of them mainly focus on deep learning neural networks and are shorted based on their popularity according to the literature in Table 4.1. Observe that, while Theano is the oldest framework, it is not the most popular. The most popular framework in the scientific community is Caffe. Moreover, Python programming interface is common for all the frameworks. An interested reader is advised to consider the extensive survey published in 2019 by Nguyen et al. that considers almost all machine learning frameworks and libraries [34].

---

[8]https://mxnet.apache.org/.

4.3 Deep Learning Methods for Food Recognition

Nowadays, practitioners of deep learning debate on choosing to develop new models or to adopt existing ones and sometimes, opinions are sharply divided. In summary, the development of a new deep learning model (architecture) applied to a new domain (problem—task) is a complex and time-consuming process that requires programming and mathematical skills beyond the average, along an understanding of data-driven (empirical) model learning. On the other hand, transfer learning with fine-tuning should be adopted when an immediate solution is required. This approach is less demanding in technical skills and theoretical knowledge that could be basically limited to the development of a simple classification method. In addition, there are today a number of online platforms that enable the rapid development of new models based on proprietary technologies, which basically act as black boxes to the developers. These platforms provide a third alternative and are typically machine learning systems pre-trained with large datasets spanning a wide spectrum of different classes ready to provide solutions to a wide variety of problems, through the use of Application Programming Interfaces (APIs). This section reviews these three approaches in the context of food recognition in mobile tourism applications. The study begins with the presentation of the relevant image datasets.

4.3.1 Food Image Datasets

Deep learning approaches, especially CNN methods, require large datasets to build an efficient classification model. Nowadays, even if there are hundreds of datasets available, with a variety of content, datasets focusing on nutritional content (foods, drinks, ingredients etc.) are very limited and quite small [11]. In this section we present some of the most popular and large datasets focusing in food recognition used in the literature. Table 4.2 summarizes these datasets and ranks them by the number of the images they contain and the year they have been introduced. It should be noted, that the last five years numerus teams have worked on the development

Table 4.2 Most popular large datasets focusing on food categories

Name	#Classes	#Images	Year	Description
UECFood100	100	14,461	2012	Most categories are popular foods in Japan
UECFood256	256	31,651	2014	
Food-101	101	101,000	2015	More general food categories
UMPCFood-101	101	100,000	2015	
VireoFood-172	172	110,241	2016	
Food524DB	524	247,636	2017	
Food-101N	101	310,000	2018	

of larger food datasets, since the recognition of foods and their ingredients is a very interesting and challenging problem with many real-life applications [10, 11, 33, 29]

Deep Learning approaches mainly provide a prediction for each sample (image) based on mathematical probabilities. In most cases, the evaluation of a prediction is based on the comparison of the network result (prediction) and the true class label. The *top-1 accuracy* is a sorting of the results of a prediction that reveals if the target output (class) is the same with the top class (the one with the highest calculated probability) of the prediction made for an image by the model. *Top-k accuracy* [51] can be defined as

$$\text{top-k} = \frac{1}{N} \sum_{i=1}^{N} 1\big[a_i \in C_i^k\big] \tag{4.5}$$

where $1[\cdot] \rightarrow \{0, 1\}$ denotes the indicator function. If the condition in the square brackets is satisfied, which stands for a_i as the ground truth answer to a question q_i, and C_i^k as the candidate set with top-k of the highest similar answers, then the function returns 1, or 0 otherwise.

The *UECFood100*[9] dataset [32] was first introduced in 2012 and contains 100 classes of food images. In each image the food item is located within a bounding box. Most of the food categories (classes) are popular Japanese foods. For this reason, most of the food classes might not be familiar to everyone. The main development reason of this dataset was to implement a practical food recognition system to use in Japan. In 2014 Kawano and Yanai used this dataset to score a classification rate of 59.6% for the top-1 accuracy and 82.9% for the top-5 accuracy, respectively [25]. The same team improved their classification results, achieving 72.26% in the top-1 accuracy and 92.00% in the top-5 accuracy when using a Deep CNN pre-trained with the ILSVRC2010[10] dataset [26].

The *UECFood256*[11] dataset [25] consists of 256 classes of food images. The structure of the classes and the images arew similar to the *UECFood100*. Again, most food categories are focused on Japanese foods. This dataset was used in order to develop a mobile application for food recognition. In 2018 Martinel, Foresti and Micheloni achieved a classification performance of 83.15% for the top-1accuracy and 95.45% for the top-5 accuracy on this dataset [31].

The *Food-101*[12] dataset [4] was introduced in 2014 from the Computer Vision Lab, ETH Zurich, Switzerland. In total, the dataset has a thousand images for each of the 101 food classes (categories), comprising a dataset of 101,100 images. For each class, 250 test images and 750 training images are provided. All images were manually reviewed and training images contain some noise on purpose, such as intense image color shift and wrong labels. All images were rescaled to have a maximum side

---

[9]http://foodcam.mobi/dataset100.html.

[10]http://www.image-net.org/challenges/LSVRC/2010/.

[11]http://foodcam.mobi/dataset256.html.

[12]https://www.vision.ee.ethz.ch/datasets_extra/food-101/.

length of 512 pixels. In 2018 Martinel, Foresti and Micheloni reported 90.27% for the top-1 and 98.71% for the top-5 on the food-101 dataset [31].

The *UPMC Food-101* [45] was developed in 2015 and is considered as a large multimodal dataset consisted of 100,000 food items classified in 101 categories. This dataset was collected from the web and can be considered as a "twin dataset" to the Food-101, since they share the same 101 classes and they have approximately the same size.

VireoFood-172[13] [5] is a dataset with images depicting food types and ingredients. This data set consist of 110,241 food images from 172 categories. All images are manually annotated according to 353 ingredients. The creators of this dataset used a MultiTaskDCNN and achieved classification rates of 82.05% for the top-1 accuracy and 95.88% for the top-5 accuracy [5].

The *Food524DB*[14] [11] was introduced 2017 and is considered to be one of the largest publicly available food datasets, with 524 food classes and 247,636 images. This dataset was developed by merging food classes from existing datasets. The dataset has been constructed by merging four benchmark datasets: VireoFood-172, Food-101, Food50, and a modified version of UECFood256. The creators of this dataset used the popular RestNet-50 architecture and scored an 81.34% for the top-1 accuracy and an 95.45% for the top-5 accuracy [11]. A modification of this dataset was released in 2018 as Food-475, introducing small differences [12].

Food-101N[15] [29] was introduced in 2018 in CVPR from Microsoft AI & Research. The dataset was designed to address label noise with minimum human supervision. This dataset contains 310,000 images of food recipes classified in 101 classes (categories). Food-101and Food-101N and UPMC Food-101 datasets, share the same 101 classes. Additionally, the images of Food-101N are much noisier. Kuang-Huei, Xiaodong and Linjun provided an image classification of 83.95% for the top-1 accuracy by using the CleanNet model [29].

Note that the development of a new large datasets with world-wide known food categories, is a very difficult and time-consuming process. Nowadays, multiple research teams focus on automated processes for the development of larger datasets. Automated processes employ powerful search image engines, crowdsourcing and social media content to perform the task of data collection. For example, Mezgec and Koroušic [33] used an image search engine to develop a food dataset and the result was quite interesting.

---

[13]http://vireo.cs.cityu.edu.hk/VireoFood172/.

[14]http://www.ivl.disco.unimib.it/activities/food524db/.

[15]https://kuanghuei.github.io/Food-101N.

4.3.2 Approach #1: New Architecture Development

This section presents a CNN model (named *PureFoodNet*) that was developed for the purposes of an R&D project in Greece that targets regional Greek food recognition for mobile tourism applications. This model is based on the building blocks and principles (simple homogenous topology) of the VGG architecture [36]. A common practice for developing new CNN architectures (models) is to add convolutional layers, until the model starts performing well and overfits. After that, the tuning process is applied on the hyperparameters of the model until a good-fitting, high accuracy model emerges. This process is time-consuming and needs vast computational resources. For example, the new model presented, the PureFoodNet, required a full month's time to be developed and properly configured on a high-end computing system.

The architecture of the PureFoodNet is depicted in Fig. 4.6. It consists of three convolutional blocks and a classification block. The first block contains two convolutional layers each having 128 filters. The second block contains three convolutional layers with 256 filters each and the third block contains three convolutional layers with 512 filters each. The last block which is usually called the "Top Layers" or the "Classification layers" contains a dense layer with 512 neurons and a predictor layer (Dense) with 101 outputs. For the initial development of this model, note that, the output of the prediction layer was designed to match the 101 classes (food categories) of the Food101 dataset. In order to avoid overfitting in the PureFoodNet model, the

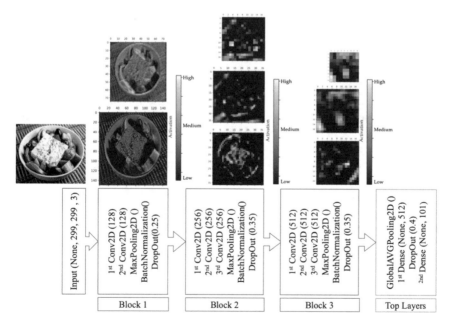

Fig. 4.6 PureFoodNet: a simple pure CNN architecture for food recognition

architecture includes classical regularization mechanisms such as Dropout, L2 regularization and Batch Normalization layers. Also, for the purpose of the PureFoodNet's tuning, a weight initialization method was used [14, 39]. To opt for a fast and accurate training, an optimizer with Nestorov momentum was employed [39].

In food recognition and in many other cases, it is important to know what each layer (or even each neuron) learns. One method to do this is by visualizing the layers' values (local "knowledge"). Hence, the issue of the unused (dead) filters can be addressed easily. This issue can be noticed by checking if some activation maps are all zero for many different inputs, which is a symptom caused by a high learning rate. The most commonly used method is to visualize the activations of the layers during the forward pass. Networks that use ReLU activation functions [15] usually have activations that look too noisy (relatively blobby and dense) in the beginning of the training. After several training epochs and after the network starts to perform well, the visualization of the activations becomes more intuitive, sparse and localized. In Fig. 4.6, above each block, a few activations of some indicative layers are shown in order to highlight the information learnt in each layer. It is clear that the lower one looks (Block 1) at the network the more general information the layer keeps, in contrast to the high-level layers (Block 3) where more specific/targeted information is learnt by the network. This is because in many images of food dishes a self-similarity exists in the form of repeating patterns of the same shapes throughout the surface of the image. This is why it is expected that the tiling of an image caused by the convolutional layers improves the localization of the ingredients, which, in turn, results a better learning of the dishes and improved predictions.

4.3.3 Approach #2: Transfer Learning and Fine-Tuning

The second alternative path in deep learning-based food recognition approaches, relies on using popular pre-trained deep learning models, implementing the transfer learning technique. Transfer learning in deep learning theory is a popular machine learning technique, in which a model developed and trained for a specific task is reused for another similar task under some parametrization (fine tuning). An intuitive definition of transfer learning was provided by Torrey and Shavlink [44]: *"Transfer learning is the improvement of learning in a new task through the transfer of knowledge from a related task that has already been learned"*. Fine tuning is the process in which a pre-trained model is used without its top layers, which are replaced by new layers, more appropriate to the new task (dataset). For example, if the pre-trained PureFoodNet model were to be used, the Top Layer should be removed (i.e. the layers inside the green rectangular in Fig. 4.6) and only the convolutional layers should be kept with their weights. New top layers should be added according to the needs of the new task (new dataset) and the number of the outputs in the prediction layer should be associated with the number of the new classes. It is a common practice, to make an appropriate recombination of the top layers, a reconfiguration of the parameters, and many repetitions in order to achieve a desirable performance.

Because of the ease of use of transfer learning, nowadays, very few people develop and train new CNN models (architectures) from scratch. This is due to the fact that it is difficult to have a dataset of sufficient size for the specific task. Also, as mentioned before, the development of a new model is complex and time-consuming. Instead, it is more convenient to reuse a model that was trained on a very large dataset, like the ImageNet [49] that contains more than 14 million images spanning more than 20,000 categories altogether. Also note that, the ImageNet contains a lot of categories with images depicting foods or edible items in general, thus a pre-trained model on the ImageNet dataset is expected to be quite effective and easily tuned to improve the performance in food recognition applications.

4.3.4 Approach #3: Deep Learning Platforms

The need for the development of a new model or the fine-tuning of an existing pre-trained model for a specific task is sometimes questionable, due to the existence of a third alternative in deep learning practice: the powerful online platforms, which are pre-trained with large datasets on a wide spectrum of classes. In this section some popular platforms for image-based food recognition are presented, along with an assessment on their efficacy and a summary of their benefits and drawbacks. These platforms keep on training on a daily basis and thus are improving over time. Some of these platforms belong to well-established technology-providing enterprises, who benefit from the deep learning services and applications. These enterprises offer the option for commercial use of their infrastructures to the scientific and industrial communities either in the form of a paid subscription or under limited free trial usage, which is, in most cases, quite enough for experimentation. It is emphasized, though, that the pre-trained models offered by these platforms work as "black boxes", since they do not provide enough details about their prediction mechanism, architectures etc. and their parameterization is restricted. All these platforms provide several tools and programming interfaces through APIs.

The *Vision AI*[16] (VAI) was developed by Google as a cloud service in 2016. It includes multiple different features detection, such as Optical Character Recognition (OCR) boosted with a mechanism focusing on handwritten text recognition in multiple languages, face detection, moderate content (explicit content such as adult, violent, etc.), landmarks and logo detections.

The *Clarifai*[17] (CAI) was developed in 2013 for the ImageNet Large Scale Visual Recognition Competition (ILSVRC) and won the top 5 places that year [49]. Clarifai is the only platform that provides options regarding the model selection for the prediction. It offers quite an interesting range of models applied for different purposes, such as generic image recognition, face identification, Not Safe for Watching (NSFW) recognition, color detection, celebrity recognition, food recognition, etc.

---

[16]https://cloud.google.com/vision/.

[17]https://clarifai.com.

Amazon Rekognition[18] (AR) was first introduced in 2016 by Amazon as a service after the acquisition of Orbeus, a deep learning startup. Amazon Rekognition provides identification of objects, people, text, scenes, and activities, and it is able to detect any inappropriate content, with considerable capabilities in facial detection and analysis (emotion, age range, eyes open, glasses, facial hair, etc.).

Computer Vision[19] (CV) was developed by Microsoft as part of the Cognitive Services. This platform offers detection of color, face, emotions, celebrities, text and NSFW content in images. An important feature of Computer Vision is the OCR system, which identifies text in 25 different languages, along with a powerful content interpretation system.

4.4 Comparative Study

This section presents a comparative study for the three aforementioned alternatives in developing deep learning models for image-based food recognition. The introduced architecture of the PureFoodNet is tested against the most popular pre-trained models on the Food101 dataset, whereas the deep learning platforms are tested against each other in a toy experiment that reveals their strengths and weaknesses.

4.4.1 New Architecture Against Pre-trained Models

Before anything, the PureFoodNet has a much shallower architecture than the rest of the pre-trained models against which it is compared. In the development of the PureFoodNet random weight initialization was used, in order to speed up the training process, and a custom adaptation of the learning rate. For the experiments presented in this section VGG16 [36], InceptionV3 [42], ResNet50 [18], InceptionResNetV2 [40], MobileNetV2 [35], DenseNet121 [20], and NASNetLarge [52] deep CNN architectures are being considered. The python code for the experiments can be accessed through a source code management repository.[20]

In these experiments, each model used exactly the same top layers (the layers in the green rectangle in Fig. 4.6), in order to have a fair comparison. Figure 4.7 presents a bar graph of the accuracy of each model, as well as a bar graph of the number of epochs needed for the models to achieve these results. The accuracy bar graph depicts the categorical top-1 and top-5 performances, both for the training (top-1, top-5) and for the validation (top-1 val, top-5 val) of each model. Training of the models, in these experiments, halted the moment the validation accuracy stopped improving, thus the difference in the required training epochs reported for the models. It is evident

---

[18]https://aws.amazon.com/rekognition/.

[19]https://www.microsoft.com/cognitive-services.

[20]https://github.com/chairiq/FoodCNNs.

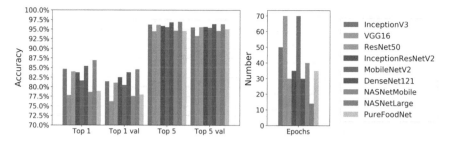

Fig. 4.7 Overview of the comparison of popular deep architectures against PureFoodNet in food recognition

that the majority of the models performed quite well. The worst validation accuracy observed was 76% for the VGG16 model, which was also among the slowest to learn. An interesting observation, which is not apparent in the graphs, is related to the performance of the NASNetLarge that achieved the best score with a very few training epochs, however, requiring approximately 83.33% more time than any other pre-trained network used in this experiment and around 92% more time than the PureFoodNet. Clearly PureFoodNet performed well even though it was not the best. There are cases in which the difference in performance between PureFoodNet and NASNetLarge was insignificant, especially from the perspective of top-5 accuracy.

Depending on the application, one should always consider the complexity of each model, thus there is a trade-off between performance and requirements/specifications, and a balance should be reached. The model size produced by NASNetLarge is about 86% larger than the size of PureFoodNet. In addition, in a study that examines the task of food recognition, the analysis that each layer provides should be in line with the semantic meaning in the nature of food's structure, namely the ingredients. Due to this, the efficiency of a model in this task is not precisely quantified by typical measures and metrics. Consequently, the choice of a model (architecture) cannot solely rely on typical performance scores (accuracy achieved during the training process). An elegant model selection is a complicated procedure, which should take into account more criteria, besides class prediction.

4.4.2 Deep Learning Platforms Against Each Other

In the following paragraphs a toy experiment is unfolded to provide intuition on the evaluation of popular deep learning platforms in image-based food recognition. Two different test images (Fig. 4.8) with complicated content were selected for the experiment. Both images contain multiple ingredients. Image A (Fig. 4.8 left) presents a Greek salad. This image contains some noticeable ingredients, like the sliced tomatoes, the cucumber, the onion, the cheese (feta), the olives, the dill and the parsley. Image B (Fig. 4.8 right) presents a fruit dessert, in which there are

Fig. 4.8 Test images: **a** Greek salad and **b** Fruit dessert

strawberries, blueberries, kiwis, grapefruits, oranges and a cream at the bottom of the cup.

Figures 4.9 and 4.10 present the prediction results for the two test images respectively by all reviewed platforms, with (a) green bars corresponding to true positive (correctly predicted) classes (in terms of ingredients, food type, etc.), (b) red bars corresponding to false positive (wrongly predicted) classes and (c) blue bars corresponding to neutral, partially correct or unimportant (to this case study) results

Even a superficial comparison of the results presented in Fig. 4.9, clearly shows that both CV and AR perform poor, both in term of the ingredients and in terms of the overall results. VAI was somewhat better but still identified some non-existing (or irrelevant to the context) classes. CAI performed significantly better but also reported non-existing food ingredients within the image. Apparently, the number of the correctly identified classes in combination with their accuracy, clearly rank CAI as the best platform for image A. It should be noted that CAI was the only platform that recognized the existence of parsley, which was out of the plate (right bottom in image A). VAI predicted with quite high accuracy that the food type might be a "Caprese Salad" (64% accuracy) or Israeli Salad (63% accuracy), and this is clearly justifiable because of the high similarity of these three salads.

In Fig. 4.10, in which the results on image B are shown, it is once again clear that CV identified the smallest number of classes, including two very general classes (Fresh, Food). However, it is also important that CV is the only platform that spotted the class "Sliced" with very little confidence, though. AR did not perform well also, recognizing only a few correct classes and with a very low confidence. It is clear that CAI and VAI performed well in this case. Although CAI presented some false positive (but still somehow relevant) results the high confidence in true positives and the more classes reported render it a valuable tool for the task. On the other hand, VAI seems to have managed avoid false positives, but the number of classes and the confidence are considerably lower.

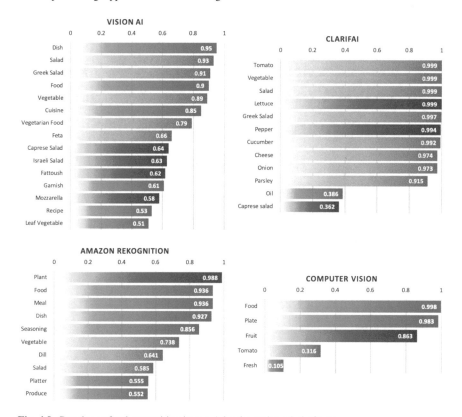

Fig. 4.9 Results on food recognition image A by the reviewed platforms

Overall, from just a toy experiment, CAI and VAI seem to perform better among the tested content identification platforms in the domain of food recognition, with slightly better results provided by CAI, apparently due to its specialized model on food. What is important from the developer perspective is that none of these platforms requires advanced programming skills to develop an image content analysis system.

4.5   Conclusions

Food recognition based on images is a challenging task that ignited the interest of the scientific communities of computer vision and artificial intelligence in the past decade. It is certain that a dish of food may contain a variety of ingredients, which may differ in shapes and sizes depending on the regional traditions and habits of the local communities. For example, it is worldwide known today that a salad with tomatoes, olives and onions relates to a Mediterranean diet. Thus, the distinction among a Greek, a Caprese or an Israeli salad might be a difficult task, even for

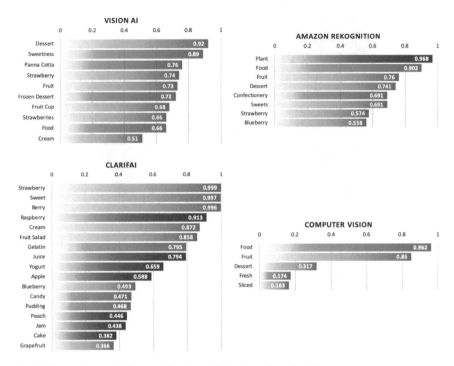

Fig. 4.10   Results on food recognition image B by the reviewed platforms

humans. The way ingredients are sliced and placed on a plate may be a lead for a distinction, but this can also be debated.

In any modern machine learning task, it is often debatable whether to design a new deep learning model (architecture), or to utilize an existing model through transfer learning, or to use an available image-based content prediction platform (typically through APIs). Designing a new architecture requires extensive knowledge and intuition, whereas transfer learning, basically, requires specific knowledge and platform-based application development is just a technical solution. On the other hand, from the productivity and ease of deployment perspective (with a compromise in efficiency), the approaches are ranked in the opposite direction, with platforms easily ranking first.

This chapter focused on the particularly interesting and complex problem of image-based food recognition and tried to build on those three lines of machine learning solutions to reveal the strength and weaknesses of each approach and highlight the opportunities being offered and the inherent threats both in the approaches and the prerequisites. Furthermore, as machine learning approaches are data hungry, popular large datasets for food recognition were presented and the need for even better datasets with more specific and world-wide known food categories was identified. The study highlighted a need for the labeling of food ingredients, which would enable deep learning approaches to attain better performances and empower more useful real-world applications (like in nutrition and tourism). Overall, this chapter suggests a bright future in the domain of food recognition with deep learning approaches, and proposes a number of future directions such as:

- creation of large-scale food datasets, with specific content, world-wide known food data categories
- labeling of datasets not only with food types (classes), but also with food ingredients (multi labelled classes)
- development of hybrid methods for better results, such as combinations of new models with prediction platforms
- development and integration of food and ingredient ontologies

  - to assist more accurate predictions in general
  - to develop rules based on contextual factors that (a) may lead to the development of similar food separation rules based the origin, habits, culture, etc., (b) may assist in targeted predictions based on geographical locations, personal preferences, etc.

Acknowledgements This research has been co-financed by the European Regional Development Fund of the European Union and Greek national funds through the Operational Program Competitiveness, Entrepreneurship and Innovation, under the call RESEARCH—CREATE—INNOVATE (project code: T1EDK-02015). We also gratefully acknowledge the support of NVIDIA Corporation with the donation of the Titan Xp GPU used for the experiments of this research.

References

1. M. Abadi, P. Barham, J. Chen, Z. Chen, A. Davis, J. Dean, … X. Zheng, Tensorflow: a system for large-scale machine learning, in *12th USENIX Symposium on Operating Systems Design and Implementation,* Savannah, GA, USA (2016), pp. 265–283
2. S.G. Adam Paszke, Automatic differentiation in PyTorch, in *31st Conference on Neural Information Processing Systems,* Long Beach, CA, USA (2017)
3. Y. Andrew, Feature selection, L1 vs. L2 regularization, and rotational invariance, in *Proceedings of the Twenty-First International Conference on Machine Learning* (ACM, 2004)
4. L. Bossard, M. Guillaumin, & L. Van Gool, Food-101—mining discriminative components with random forests, in *European Conference on Computer Vision* (Springer, Cham, 2014), pp. 446–461
5. J. Chen, & W.C. Ngo, Deep-based ingredient recognition for cooking recipe retrival, in *ACM Multimedia* (2016), pp. 32–41
6. T. Chen, M. Li, Y. Li, M. Lin, M. Wang, M. Wang, … Z. Zhang, MXNet: a flexible and efficient machine learning library for heterogeneous distributed systems, in *NIPS Workshop on Machine Learning Systems (LearningSys)* (2015)
7. F. Chollet, *François Chollet.* Keras.io. (2015). Accessed Keras: https://keras.io
8. F. Chollet, *Deep Learning with Python* (Manning Publications, 2018)
9. S. Christodoulidis, M. Anthimopoulos, S. Mougiakakou, Food recognition for dietary assessment using deep convolutional neural networks, in *International Conference on Image Analysis and Processing* (Springer, Cham, 2015), pp. 458–465

10. G. Ciocca, P. Napoletano, R. Schettini, Food recognition: a new dataset, experiments, and results. IEEE J. Biomed. Health Inform. **21**(3), 588–598 (2017)
11. G. Ciocca, P. Napoletano, R. Schettini, Learning CNN-based features for retrieval of food images, *New Trends in Image Analysis and Processing—ICIAP 2017: ICIAP International Workshops, WBICV, SSPandBE, 3AS, RGBD, NIVAR, IWBAAS, and MADiMa 2017* (Springer International Publishing, Catania, Italy, 2017), pp. 426–434
12. G. Ciocca, P. Napoletano, R. Schettini, CNN-based features for retrieval and classification of food images. Comput. Vis. Image Underst. **176–177**, 70–77 (2018)
13. V. Dumoulin, F. Visin, A guide to convolution arithmetic for deep learning (2016). arXiv arXiv: 1603.07285
14. X. Glorot, Y. Bengio, Understanding the difficulty of training deep feedforward neural networks, in *Proceedings of the Thirteenth International Conference on Artificial Intelligence and Statistic* (2010), pp. 249–256
15. X. Glorot, A. Bordes, Y. Bengio, Deep sparse rectifier neural networks, in *International Conference on Artificial Intelligence and Statistics,* ed. by G. Gordon, D. Dunson, M. Dudk (2011), pp. 315–323
16. I. Goodfellow, Y. Bengio, & A. Courville, *Deep Learning* (MIT Press, 2016)
17. H. Hassannejad, G. Matrella, P. Ciampolini, I. DeMunari, M. Mordonini, S. Cagnoni, Food image recognition using very deep convolutional networks, in *2nd International Workshop on Multimedia Assisted Dietary Management*, Amsterdam, The Netherlands (2016), pp. 41–49
18. K. He, X. Zhang, S. Ren, J. Sun, Deep residual learning for image recognition, in *2016 IEEE Conference on Computer Vision and Pattern Recognition (CVPR),* Las Vegas, NV (2016), pp. 770–778
19. S. Horiguchi, S. Amano, M. Ogawa, K. Aizawa, Personalized classifier for food image recognition. IEEE Trans. Multimedia **20**(10), 2836–2848 (2018)
20. G. Huang, Z. Liu, L. van der Maaten, K. Weinberger, Densely connected convolutional networks, in *IEEE Conference on Pattern Recognition and Computer Vision* (2017), pp. 4700–4708
21. S. Ioffe, C. Szegedy, Batch normalization: accelerating deep network training by reducing internal covariate shift, in *Proceeding ICML'15 Proceedings of the 32nd International Conference on International Conference on Machine Learning* (2015), pp. 448–4456
22. Y. Jia, E. Shelhamer, J. Donahue, S. Karayev, J. Long, R. Girshick, … T. Darrell, Caffe: convolutional architecture for fast feature embedding, in *22nd ACM International Conference on Multimedia* (ACM, Orlando, Florida, USA, 2014), pp. 675–678
23. A. Jovic, K. Brkic, N. Bogunovic, An overview of free software tools for general data mining, in *37th International Convention on Information and Communication Technology, Electronics and Microelectronics* (2014), pp. 1112–1117
24. H. Kagaya, K. Aizawa, M. Ogawa, Food detection and recognition using convolutional neural network, in *22nd ACM international conference on Multimedia,* Orlando, FL, USA (2014), pp. 1055–1088
25. Y. Kawano, K. Yanai, FoodCam: a real-time food recognition system on a smartphone. Multimedia Tools Appl. **74**(14), 5263–5287 (2014)
26. Y. Kawano, K. Yanai, Automatic expansion of a food image dataset leveraging existing categories with domain adaptation, in *Proceedings of ECCV Workshop on Transferring and Adapting Source Knowledge in Computer Vision (TASK-CV)* (2014), pp. 3–17
27. Y. Kawano, K. Yanai, Food image recognition with deep convolutional features, in *Proceedings of ACM UbiComp Workshop on Cooking and Eating Activities (CEA)* (2014c), pp. 589–593
28. A. Krizhevsky, I. Sutskever, G.E. Hinton, ImageNet classification with deep convolutional neural networks, in *25th International Conference on Neural Information Processing Systems,* Lake Tahoe, Nevada (2012), pp. 1097–1105
29. L. Kuang-Huei, H. Xiaodong, Z. Lei, Y. Linjun, CleanNet: transfer learning for scalable image classifier training with label noise, in *Proceedings of the IEEE Conference on Computer Vision and Pattern Recognition* (2018)

30. C. Liu, Y. Cao, Y. Luo, G. Chen, V. Vokkarane, Y. Ma, DeepFood: deep learning-based food image recognition for computer-aided dietary assessment, in *14th International Conference on Inclusive Smart Cities and Digital Health,* Wuhan, China (2016), pp. 37–48
31. N. Martinel, G. Foresti, C. Micheloni, Wide-slice residual networks for food recognition, in *2018 IEEE Winter Conference on Applications of Computer Vision (WACV),* Lake Tahoe, NV (2018), pp. 567–576
32. Y. Matsuda, H. Hoashi, K. Yanai, Recognition of multiple-food images by detecting candidate regions, in *Proceedings of IEEE International Conference on Multimedia and Expo (ICME)* (2012)
33. S. Mezgec, S. Koroušić, NutriNet: a deep learning food and drink image recognition system for dietary assessment. Nutrients **9**(7), 657 (2017)
34. G. Nguyen, S. Dlugolinsky, M. Bobák, V. Tran, A.L. García, I. Heredia, L. Hluchý, Machine learning and deep learning frameworks and libraries for large-scale data mining: a survey. Artif. Intell. Rev. **52**(1), 77–124 (2019)
35. M. Sandler, A. Howard, M. Zhu, A. Zhmoginov, L.C. Chen, MobileNetV2: inverted residuals and linear bottlenecks, in *The IEEE Conference on Computer Vision and Pattern Recognition* (2018), pp. 4510–4520
36. K. Simonyan, A. Zisserman, Very deep convolutional networks for large-scale image recognition, in *3rd IAPR Asian Conference on Pattern Recognition (ACPR),* Kuala Lumpur (2015), pp. 730–734
37. A. Singla, L. Yuan, T. Ebrahimi, Food/non-food image classification and food categorization using pre-trained GoogLeNet model, in *2nd International Workshop on Multimedia Assisted Dietary Management,* Amsterdam, The Netherlands (2016), pp. 3–11
38. N. Srivastava, G. Hinton, A. Krizhevsky, I. Sutskever, R. Salakhutdinov, Dropout: a simple way to prevent neural networks from overfitting. J. Mach. Learn. Res. **15**(1), 1929–1958 (2014)
39. I. Sutskever, J. Martens, G. Dahl, G. Hinton, On the importance of initialization and momentum in deep learning, in *International Conference on Machine Learning* (2013), pp. 1139–1147
40. C. Szegedy, S. Ioffe, V. Vanhoucke, A. Alemi, Inception-v4, inception-resnet and the impact of residual connections on learning, in *Thirty-First AAAI Conference on Artificial Intelligence* (2017)
41. C. Szegedy, W. Liu, Y. Jia, P. Sermanet, S. Reed, D. Anguelov, … A. Rabinovich, Going deeper with convolutions, in *2015 IEEE Conference on Computer Vision and Pattern Recognition (CVPR),* Boston, MA (2015), pp. 1–9
42. C. Szegedy, V. Vanhoucke, S. Ioffe, J. Shlens, Z. Wojna, Rethinking the inception architecture for computer vision, in *Proceedings of the IEEE Conference on Computer Vision and Pattern Recognition* (2016), pp. 2818–2826
43. D. TheanoTeam, Theano: a Python framework for fast computation of mathematical expressions (2016). arXiv preprint arXiv:1605.02688
44. L. Torrey, J. Shavlik, Transfer learning, in *Handbook of Research on Machine Learning Applications,* ed. by E. Soria, J. Martin, R. Magdalena, M. Martinez, A. Serrano (IGI Global, 2009), pp. 242–264
45. X. Wang, D. Kumar, N. Thome, M. Cord, F. Precioso, Recipe recognition with large multi-modal food dataset, in *2015 IEEE International Conference on Multimedia & Expo Workshops (ICMEW),* Turin (2015), pp. 1–6
46. M. Weiqing, J. Shuqiang, L. Linhu, R. Yong, J. Ramesh, A Survey on Food Computing (ACM Computing Surveys, 2019)
47. K. Yanai, Y. Kawano, Food image recognition using deep convolutional network with pre-training and fine-tuning, in *IEEE International Conference on Multimedia & Expo Workshops,* Turin, Italy (2015), pp. 1–6
48. Q. Yu, M. Anzawa, S. Amano, M. Ogawa, K. Aizawa, Food image recognition by personalized classifier, in *25th IEEE International Conference on Image Processing,* Athens (2018), pp. 171–175
49. M. Zeiler (2013). Accessed http://www.image-net.org/challenges/LSVRC/2013/results.php

50. M. Zeiler, R. Fergus, Visualizing and understanding convolutional networks, in *IEEE European Conference on Computer Vision* (2014), pp. 818–833
51. S. Zhang, X. Zhang, H. Wang, J. Cheng, P. Li, Z. Ding, Chinese medical question answer matching using end-to-end character-level multi-scale CNNs. Appl. Sci. **7**(8), 767 (2017)
52. B. Zoph, V. Vasudevan, J. Shlens, V. Le, Learning transferable architectures for scalable image recognition, in *Proceedings of the IEEE Conference on Computer Vision and Pattern Recognition* (2018), pp. 8697–8710

Part II
Deep Learning in Social Media and IOT

Chapter 5
Deep Learning for Twitter Sentiment Analysis: The Effect of Pre-trained Word Embedding

Akrivi Krouska, Christos Troussas, and Maria Virvou

Abstract Twitter is the most popular microblogging platform, with millions of users exchanging daily a huge volume of text messages, called "tweets". This has resulted in an enormous source of unstructured data, Big Data. Such a Big Data can be analyzed by companies or organizations with the purpose of extracting customer perspective about their products or services and monitoring marketing trends. Understanding automatically the opinions behind user-generated content, called "Big Data Analytics", is of great concern. Deep learning can be used to make discriminative tasks of Big Data Analytics easier and with higher performance. Deep learning is an aspect of machine learning which refers to an artificial neural network with multiple layers and has been extensively used to address Big Data challenges, like semantic indexing, data tagging and immediate information retrieval. Deep learning requires its input to be represented as word embeddings, i.e. as a real-value vector in a high-dimensional space. However, word embedding models need large corpuses for training and presenting a reliable word vector. Thus, there are a number of pre-trained word embeddings freely available to leverage. In effect, these are words and their corresponding n-dimensional word vectors, made by different research teams. In this work, we have made data analysis with huge numbers of tweets taken as big data and thereby classifying their polarity using a deep learning approach with four notable pre-trained word vectors, namely Google's Word2Vec, Stanford's Crawl GloVe, Stanford's Twitter GloVe, and Facebook's FastText. One major conclusion is that tweet classification using deep learning outperforms the baseline machine

A. Krouska · C. Troussas · M. Virvou (✉)
Software Engineering Laboratory, Department of Informatics, University of Piraeus, Piraeus, Greece
e-mail: mvirvou@unipi.gr

A. Krouska
e-mail: akrouska@unipi.gr

C. Troussas
e-mail: ctrouss@unipi.gr

G. A. Tsihrintzis and L. C. Jain (eds.), *Machine Learning Paradigms*, Learning and Analytics in Intelligent Systems 18, https://doi.org/10.1007/978-3-030-49724-8_5

learning algorithms. At the same time and with regard to pre-trained word embeddings, FastText provides more consistent results across datasets, while Twitter GloVe obtains very good accuracy rates despite its lower dimensionality.

Keywords Deep learning · Big data · Twitter · Sentiment analysis · Word embeddings · CNN

5.1 Introduction

The proliferation of Web 2.0 has led to a digital landscape where people are able to socialize, by expressing and sharing their thoughts and opinions on various issues in public, through a variety of means and applications. Indicative of this innovation of social interaction is microblogging, an online broadcast medium that allows users to post and share short messages with an audience online. Twitter is the most popular of such services with million users exchanging daily a huge volume of text messages, called tweets. This has resulted in an enormous source of unstructured data, Big Data, which can be integrated into the decision-making process, Big Data Analytics. Big data analytics is a complex process of examining large and varied data sets (Big Data) to uncover information including hidden patterns, unknown correlations, market trends and customer preferences that can help organizations make better-informed business decisions [1]. Twitter is a valuable media for this process. This paper focuses on the analysis of opinions of tweets, called Twitter Sentiment Analysis.

Sentiment Analysis is the process aiming to detect sentiment content of a text unit in order to identify people's attitudes and opinions towards various topics [2]. Twitter, with nearly 600 million users and over 250 million messages per day, has become one of the largest and most dynamic datasets for sentiment analysis. Twitter Sentiment Analysis refers to the classification of tweets based on the emotion or polarity that the user intends to transmit. This information is extremely valuable to numerous circumstances where the decision making is crucial. For instance, the opinion and sentiment detection is useful in politics to forecast election outcomes or estimate the acceptance of politicians [3], in marketing for sales predictions, product recommendations and investors' choices [1], and in the educational context for incorporating learner's emotional state to the student model providing an affective learning environment [4].

Twitter Sentiment analysis is becoming increasingly important for social media mining, as it gives the access to valuable opinions of numerous participants on various business and social issues. Consequently, many researchers have focused on the study of applications and enhancements on sentiment analysis algorithms that provide more efficient and accurate results [5]. There are three main sentiment analysis approaches: a. machine learning-based, which uses classification technique to classify the text entity, b. lexicon-based, which uses sentiment dictionary with opinion words and weights determined the polarity, and c. hybrid, where different approaches are combined [4, 6, 7]. Experimental results have shown that machine learning methods have higher precision, while lexicon-based methods are competitive in case the training dataset lacks quantity and quality, as they require few efforts in human-labeled document [5, 7]. On the other hand, combing the proper techniques foster better results, as the hybrid approaches could collectively exhibit the accuracy of a machine learning algorithms and the speed of lexicons [6]. However, recently a new approach has arisen from machine learning field, namely deep learning, overshadowing the aforementioned methods.

Deep learning is an aspect of machine learning which refers to an artificial neural network with multiple layers: an input, an output and at least one hidden [8]. The "deep" in deep learning refers to having more than one hidden layer. Replicating the human brain, neural networks consist of a large number of information processing units, called neurons, organized in layers, which work in unison. Deep learning is a hierarchical feature learning, where each layer learns to transform its input into a slightly more abstract and composite representation through a nonlinear processing, using a weight-based model with an activation function to each neuron in order to dictate the importance of the input value. Iterations continue until an acceptable rate of accuracy reached at output data.

Deep learning can be used to extract incredible information that buried in a Big Data [1]. It has been extensively applied in artificial intelligence field, like computer vision, transfer learning, semantic parsing, and natural language processing. Thus, for more accurate sentiment analysis, many researchers have implemented deferent models of this approach, namely CNN (convolutional neural networks), DNN (deep neural networks), RNN (recurrent neural networks) and DBN (deep belief networks) [8]. The results show that deep learning is very beneficial in the sentiment classification. Except the higher performance, there is no need for carefully optimized hand-crafted features and feature engineering, one of the most time-consuming parts of machine learning practice. Big Data challenges, like semantic indexing, data tagging and immediate information retrieval, can be addressed better using Deep learning. However, it requires large data sets and is extremely computationally expensive to train.

In this study, a convolutional neural network was applied on three well-known Twitter datasets, namely Obama-McCain Debate (OMD), Health Care Reform (HCR) and Stanford Twitter Sentiment Gold Standard (STS-Gold, for sentiment classification, using different pre-trained word embeddings. Word Embeddings is one of the most useful deep learning methods used for constructing vector representations of words and documents. Their novelty is the ability to capture the syntactic and semantic relations among words. The most successful deep learning methods of word embeddings are Word2Vec [9, 10], GloVe [11] and FastText [12]. Many researchers have used these methods in their sentiment analysis experiments [13, 14]. Despite their effectiveness, one of their limitations is the need of large corpuses for training and presenting an acceptable word vector. Thus, researches have to use pre-trained word vectors, due to the small size of some datasets. In this research, we experiment on four notable pre-trained word embeddings, namely Google's Word2Vec, Stanford's Crawl GloVe, Stanford's Twitter GloVe, and Facebook's FastText, and examine their effect on the performance of sentiment classification.

The structure of this paper is as follows. First, we present the related work in Twitter sentiment analysis using deep learning techniques. Following, we present the evaluation procedure of this research, describing the datasets, the pre-trained word embeddings and the deep learning algorithm used. Section 5.4 deals with the comparative analysis and discussion on the experiment results. Finally, we present our conclusions and future work.

5.2 Related Work

The need for analyzing the big data originated from the proliferation of social media and classifying sentiment efficiently lead many researchers in the adoption of deep learning models. The models used vary in the dimensions of the datasets employed, the deep learning techniques applied regarding the word embeddings and algorithm, and the purpose served. This section briefly describes representative studies related to Twitter sentiment analysis using deep learning approaches, and tables them based on the aforementioned dimensions.

In [13], the authors experiment with deep learning models along with modern training strategies in order to achieve better sentiment classifier for tweets. The proposed model was an ensemble of 10 CNNs and 10 LSTMs together through soft voting. The models ensembled were initialized with different random weights and used different number of epochs, filter sizes and embedding pre-training algorithms, i.e. Word2Vec or FastText. The GloVe variation is excluded from ensembled model, as it gave a lower score than both the other two embeddings.

In [15], the authors propose a three-step process to train their deep learning model for predicting polarities at both message and phrase levels: i. word embeddings are initialized using Word2Vec neural language model, which is trained on a large unsupervised tweet dataset; ii. a CNN is used to further refine the embeddings on a large distant supervised dataset; iii. the word embeddings and other parameters of the network obtained at the previous stage are used to initialize the network with the same architecture, which is then trained on a supervised dataset from Semeval-2015.

In [16], the authors develop a Bidirectional Long Short-Term Memory (BLSTM) RNN, which combines 2 RNNs: a forward RNN where the sequence is processed from left to right, and a reverse RNN where the sequence is processed from right to left. The average of both RNNs outputs for each token is used to compute the model's final label for adverse drug reaction (ADR) identification. For pre-trained word embeddings, they utilized the Word2Vec 400 M Twitter model (by Frederic Godin),[1] which is a set of 400-dimensional word embeddings trained using the skip-gram algorithm on more than 400 million tweets.

In [17], the authors propose a new classification method based on deep learning algorithms to address issues on Twitter spam detection. They firstly collected a 10-day

---

[1]https://github.com/loretoparisi/word2vec-twitter.

real tweets corpus and applied Word2Vec to pre-process it, instead of feature extraction. Afterwards, they implemented a binary detection using MLP and compared their approach to other techniques, such as traditional text-based methods and feature-based methods supported by machine learning.

In [18], the authors apply convolution algorithm on Twitter sentiment analysis to train deep neural network, in order to improve the accuracy and analysis speed. Firstly, they used GloVe model to implement unsupervised learning of word-level embeddings on a 20 billion twitter corpus. Afterwards, they combined this word representation with the prior polarity score feature and sentiment feature set, and used it as input in a deep convolution neural network to train and predict the sentiment classification labels of five Twitter datasets. Their method was compared to baseline approaches regarding preprocessing techniques and classification algorithms used.

In [19], the authors present an effective deep neural architecture for performing language-agnostic sentiment analysis over tweets in four languages. The proposed model is a convolutional network with character-level embedding, which is designed for solving the inherent problem of word-embedding and n-gram based approaches. This model was compared to other CNN and LSTM models with different embedding, and a traditional SVM-based approach.

In [14], the authors, firstly, construct a tweet processor using semantic rules, and afterwards, train character embedding DeepCNN to produce feature maps capturing the morphological and shape information of a word. Moreover, they integrated global fixed-size character feature vectors and word-level embedding for Bi-LSTM. As pre-trained word vectors, they used the public available Word2Vec with 300 dimensionality and Twitter GloVe of Stanford with 200 dimensionality. The results showed that the use of character embedding through a DeepCNN to enhance information for word embedding built on top of Word2Vec or GloVe improves classification accuracy in Twitter sentiment analysis.

In [20], the authors apply deep learning techniques to classify sentiment of Thai Twitter data. They used Word2Vec to train initial word vectors for LSTM and Dynamic CNN methods. Afterwards, they studied the effect of deep neural network parameters in order to find the best settings, and compared these two methods to other classic techniques. Finally, they demonstrated that the sequence of words influences sentiment analysis.

Feeding the network with the proper word embeddings is a key factor for achieving an effective and punctual sentiment analysis. Therefore, this article is concentrated on the evaluation of the most popular pre-trained word embeddings, namely Word2Vec, Crawl GloVe, Twitter GloVe, FastText. The experiments were conducted using CNN method in the context of sentiment analysis on three Twitter datasets.

Table 5.1 illustrates the analysis of deep learning researches described in this section, analyzing them regarding the datasets used, the deep learning techniques applied, and the scope served. It is worth noting that the authors mainly used the well-known word vectors Word2Vec and GloVe for training their network and CNN as base algorithm. Moreover, in the vast majority of cases, the purpose is to present a comparative analysis of deep learning approaches reporting their performance metrics.

Table 5.1 Analysis of deep learning researches

Research	Tweets dataset	Num. of tweets	Deep learning techniques		Scope
			Word embeddings	Classification	
Cliche [13]	SemEval-2017	12.379 test set	Word2Vec, FastText, GloVe	Ensemble of CNN and LSTM	Algorithms evaluation
Severyn and Moschitti [15]	Twitter'15	5.482 test set	Word2Vec	CNN	Algorithm evaluation
Cocos [16]	Combination of Adverse Drug Reaction (ADR) dataset and Attention Deficit Hyperactivity Disorder (ADHD) dataset	844 (634 training set–210 test set)	Word2Vec 400 M Twitter model (by Frederic Godin)[a]	BLSTM	Adverse Drug Reaction (ADR) detection
Wu et al. [17]	Real-life 10-day ground-truth dataset	1.376.206 spam tweets + 673.836 non-spam	Word2Vec	MLP	Spam detection
Jianqiang et al. [18]	STSTd, SE2014, STSGd, SSTd, SED	359, 5.892, 2.034, 3.326, 2.648 correspondingly	GloVe	DeepCNN	Algorithm evaluation
Wehrmann et al. [19]	Tweets from four European languages	128.189 (89.732 training, 12.819 validation, 25.638 test)	Word and character-level embedding	CNN, LSTM	Language-agnostic translation
Nguyen and Nguyen [14]	STS Corpus, Sanders, HCR	80 K train + 16 K dev + 359 test, 991 train + 110 dev + 122 test, 621 train + 636 dev + 665 test correspondingly	Word2Vec, Twitter GloVe	DeepCNN, BLSTM	Algorithm evaluation
Vateekul and Koomsubha [20]	Thai Twitter Data	3.813.173	Word2Vec	Dynamic CNN, LSTM	Algorithm evaluation
Ours	OMD, HCR, STS-Gold	1.904, 1.922, 2.034 correspondingly	Word2Vec, Crawl GloVe, Twitter GloVe, FastText	CNN	Word embeddings evaluation

[a]https://github.com/loretoparisi/word2vec-twitter

5.3 Evaluation Procedure

The scope of the current research is to investigate the effect of the top pre-trained word embeddings obtained by unsupervised learning on large text corpora on training a deep convolutional neural network for Twitter sentiment analysis. Figure 5.1 illustrates the steps followed.

Firstly, word embeddings is used to transform tweets into its corresponding vectors to build up the sentences vectors. Then, the generated vectors are divided into training and test set. The training set is used as input to CNN in order to be trained and make predictions on the test set. Finally, the experiments' outcomes have been tabulated and a descriptive analysis has been conducted, compared CNN results with baseline machine learning approaches. All the preprocessing settings and the classification were employed in Weka data mining package.

5.3.1 Datasets

The experiments described in this paper were performed in three well-known and freely-available on the Web Twitter datasets. The reason they have been chosen is that they consist of a significant volume of tweets, created by reputable universities for academic scope, and they have been used in various researches. STS-Gold dataset [21] consists of random tweets with no particular topic focus, whereas OMD [22] and HCR [23] include tweets on specific domains. In particular, OMD dataset contains tweets related with Obama-McCain Debate and HCR dataset focuses on Health Care Reform. The statistics of the datasets are shown in Table 5.2 and examples of tweets are given in Table 5.3.

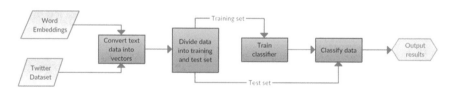

Fig. 5.1 Twitter classification using word embeddings to train CNN

Table 5.2 Statistics of the three Twitter datasets used

Dataset	Tweets	Positive	Negative
Obama-McCain Debate (OMD)	1904	709	1195
Health Care Reform (HCR)	1922	541	1381
Stanford Sentiment Gold Standard (STS-Gold)	2034	632	1402

Table 5.3 Examples of tweets

Dataset	Tweet	Polarity
OMD	the debate was a draw. I thought Mccain did come out slightly on top because he sounded more educated on the issues. good closing. #debate08	Positive
	Good debate. Job well done to both candidates! #current, #debate08, #tweetdebate	
	I am disappointed. #tweetdebate	Negative
	#current #tweetdebate It's sad that you can watch this and think that Mc has been the only one saying anything right tonight!	
HCR	Healthcare is not a privilege, it is a right! That is why I support healthcare reform. #hcr	Positive
	Maybe you healthy and happy with your healthcare plan and that's ok. But don't stand in the way of the people who are sick and in need #HCR	
	You know its a bad bill when you have to payoff your own party #tcot #tweetcongress #killthebill #HCR #handsoff	Negative
	Retweet this if you are against a government takeover of health care #hcr #tcot #gop #txgop #handsoff #codered	
STS-Gold	I love my new tiny cute little iPod! Thank you @Santino_gq! Xoxoxox!!!	Positive
	sooo excited....i just bought tickets to Taylor Swift. ahhhhhh	
	So disapointed Taylor Swift doesn't have a Twitter	Negative
	Trying to set up Xbox Live and failing tremendously...brain ache	

5.3.2 Data Preprocessing

Due to the short length of tweets and their informal aspect, tweets are usually consisted of a variety of noise and language irregularities. This noise and unstructured sentences will affect the performance of sentiment classification [24]. Thus, before feature selection, a stage of preprocessing is essential to be involved in the classification task. The preprocessing includes:

- Stemming (reducing inflected words to their word stem, base or root form).
- Remove all numbers, punctuation symbols and some special symbols.
- Force lower case for all tokens.
- Remove stopwords (referred to the most common words in a language).
- Replace the emoticons and emoji with their textual form using an emoticon dictionary.
- Tokenize the sentence.

In our experiments, we use Wikipedia emoticons list,[2] Weka Stemmer, Rainbow stopwords list, and Tweet-NLP.

---

[2]http://en.wikipedia.org/wiki/List_of_emoticons.

5.3.3 Pre-trained Word Embeddings

Word embeddings are a set of natural language processing techniques where individual words are mapped to a real-value vector in a high-dimensional space [9]. The vectors are learned in such a way that words that have similar meanings will have similar representation in the vector space. This is a more semantic representation for text than more classical methods like bag-of-words, where relationships between words or tokens are ignored, or forced in bigram and trigram approaches [10]. This numeric representation is essential in sentiment analysis, as many machine learning algorithms and deep learning approaches are incapable of processing strings in their raw form and require numeric values as input to perform their tasks. Thus, there are a variety of word embeddings methods which can be applied on any corpus in order to build a vocabulary with hundred of dimensions word vectors, capturing a great deal of semantic patters.

In 2013, Word2Vec [9, 10] has been proposed by Mikolov et al., which has now become the mainstream of word embedding. Word2Vec employs a two-level neural network, where Huffman techniques is used as hierarchical softmax to allocate codes to frequent words. The model is trained through stochastic gradient descent and the gradient is achieved by backpropagation. Moreover, optimal vectors are obtained for each word by CBOW or Skip-gram.

In 2014, Pennington et al. introduced GloVe [11], which become also very popular. GloVe is essentially a log-bilinear model with a weighted least-squares objective. The model is based on the fact that ratios of word-word co-occurrence probabilities have the potential for encoding some form of meaning.

In 2016, based on Word2Vec, Bojanowski et al. proposed FastText [12], which can handle subword units and has fast computing. FastText extends Word2Vec by introducing subword modeling. Instead of feeding individual words into the neural network, it represents each word as a bag of character n-gram. The embedding vector for a word is the sum of all its n-grams.

Word embedding is a demanding process, requiring large corpuses for training and presenting an acceptable word vector. Thus, rather than training the word vectors from scratch, it can be used pre-trained word embeddings, public available in Internet, for sentiment analysis through deep learning architectures. In this paper, we use four well-known pre-trained word vectors, namely Google's Word2Vec,[3] Stanford's Crawl GloVe,[4] Stanford's Twitter GloVe, and Facebook's FastText,[5] with the intention of evaluating their effect in sentiment analysis. A briefly description of these vectors is following.

Google's Word2Vec trained on part of Google News dataset, about 100 billion words. The model contains 300-dimensional vectors for 3 million words and phrases. Crawl GloVe was trained on a Common Crawl dataset of 42 billion tokens (words), providing a vocabulary of 2 million words with an embedding vector size of 300

---

[3] https://code.google.com/archive/p/word2vec/.
[4] https://nlp.stanford.edu/projects/glove/.
[5] https://github.com/facebookresearch/fastText/blob/master/pretrained-vectors.md.

Table 5.4 Statistics of pre-trained word embeddings

Pre-trained word vector	Dataset	Dimension	Corpora	Vocabulary
Word2Vec	Google news	300	100 billion	3 million
Crawl GloVe	Common crawl	300	42 billion	2 million
Twitter GloVe	Twitter	200	27 billion	2 million
FastText	Wikipedia 2017, UMBC webbase corpus and statmt.org news	300	16 billions	1 million

dimensions. On the other hand, Twitter GloVe was trained on a dataset of 2 billion tweets with 27 billion tokens. This representation has a vocabulary of 2 million words and embedding vector sizes, including 50, 100 and 200 dimensions. Finally, FastText consists of 1 million-word vectors of 300 dimensions, trained on Wikipedia 2017, UMBC webbase corpus and statmt.org news dataset (16 billions of tokens) (Table 5.4).

5.3.4 Deep Learning

In this paper, we use a deep Convolutional Neural Network (CNN) for tweets classification. CNN is a class of deep learning approaches which use a variation of multilayer perceptrons designed to require minimal preprocessing [25]. First, the tweet is tokenized and transformed into a list of word embeddings. Afterwards, the created sentence matrix is combined using multiple filters with variable window size. Thus, local sentiment feature vectors are generated for each possible word window size. Then, the feature maps activate a 1-max-pooling layer via a non-linear activation function. Finally, this pooling layer is densely connected to the output layer using softmax activation to generate probability value of sentiment classification, and optional dropout regularization to prevent over-fitting. The architecture of the convolutional neural network used for sentiment classification is shown on Fig. 5.2.

5.4   Comparative Analysis and Discussion

In this paper, we experimented on CNN for the sentiment analysis of three well-known dataset, using four different word embeddings. The scope of the paper is to evaluate the effect of pre-trained word embeddings, namely Word2Vec, Crawl GloVe, Twitter GloVe, and FastText, on the performance of Twitter classification. For all datasets, the same preprocessing steps were applied. Many experiments were made in order to define the best settings for CNN. Thus, we set a mini-batch size of 50. Learning rate technique was applied for stochastic gradient descent with a maximum

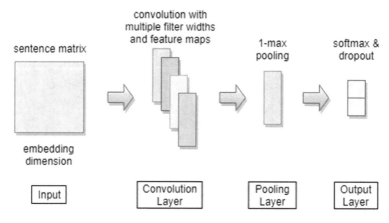

Fig. 5.2 Architecture of Convolutional Neural Network (CNN) used

of 100 training epochs. We tested with various filter windows and concluded to set the filter windows to 5. Moreover, we use the activation function ReLu for the convolution layer and SoftMax for the output one. The experimental evaluation metric is the accuracy in the classification of positive and negative tweets, and the best CNN results are compared to the performance of baseline classifiers emerged from other previous studies [5].

Table 5.5 illustrates the percentage of accuracy achieved by CNN regarding the different word embeddings used. We observe that using FastText gives better performance than other approaches in the majority of experiments. Moreover, Twitter Glove performs satisfactory, despite having lower dimensionality than other word vectors and training on a considerably smaller dataset. Its good accuracy rates may be due

Table 5.5 CNN accuracy using different word vectors on Twitter datasets

Dataset	Word embeddings	CNN accuracy
OMD	Word2Vec	81.98
	Crawl GloVe	83.59
	Twitter GloVe	82.79
	FastText	**83.65**
HCR	Word2Vec	79.94
	Crawl GloVe	80.45
	Twitter GloVe	80.39
	FastText	**81.33**
STS-Gold	Word2Vec	82.69
	Crawl GloVe	84.12
	Twitter GloVe	**84.23**
	FastText	82.57

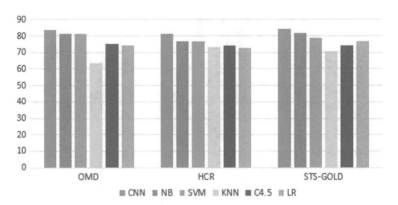

Fig. 5.3 Comparison of accuracy values between our best CNN version and baseline approaches

to the fact that the corpora used by this model originated from Twitter. On the other hand, Word2Vec seems not to be as effective as others models.

We further compare our method to five representative machine learning algorithms, namely Naïve Bayes, Support Vector Machine, k- Nearest Neighbor, Logistic Regression and C4.5. In [5], these classifiers have been evaluated using the same datasets as the current research, and their performance has been tabled. We use the best accuracy rates of CNN version for each dataset, in order to compare with the baseline algorithms. As represented in Fig. 5.3, the deep learning technique outperforms the other machine learning methods over all the datasets.

5.5 Conclusion and Future Work

Deep Learning and Big Data analytics are, nowadays, two burning issues of data science. Big Data analytics is important for organizations that need to extract information from huge amounts of data collected through social networking services. Such a social networking service is a microblogging service, with Twitter being its most representative platform. Through Twitter, millions of users exchange public messages daily, namely "tweets", expressing their opinion and feelings towards various issues. This makes Twitter one of the largest and most dynamic Big Data sets for data mining and sentiment analysis. Deep learning is a valuable tool for Big Data analytics. Deep Learning models have achieved remarkable results in a diversity of fields, including natural language processing and classification. Thus, in this paper, we employ a deep learning approach for Twitter sentiment classification, with the intention of evaluating different settings of word embeddings.

The development and use of word embeddings is an essential task in the classification process, as deep learning techniques require numeric values as input, and feeding the network with the proper input can boost the performance of algorithms. In this work, we examined four well-known pre-trained word embeddings, namely

Word2Vec, Crawl GloVe, Twitter Glove and FastText, that have been developed by reliable research teams and have been used extensively in deep learning projects. The results show that FastText provides more consistent results across datasets. Moreover, Twitter GloVe obtains very good accuracy rates despite its lower dimensionality.

Part of our future work is to study word embeddings trained from scratch on Twitter data using different parameters, as the training model, dimensional size, relevant vocabulary to test dataset etc., and evaluate their effect on classification performance. Moreover, another interesting research venue is concerned with the comparative analysis of different deep learning approaches regarding the algorithm structure, layers, filters, functions etc., in order to identify the proper network settings.

References

1. S. Sohangir, D. Wang, A. Pomeranets, T.M. Khoshgoftaar, Big Data: deep learning for financial sentiment analysis. J. Big Data **5**(1) 2018
2. V. Sahayak, V. Shete, A. Pathan, Sentiment analysis on twitter data. Int. J. Innov. Res. Adv. Eng. (IJIRAE) **2**(1), 178–183 (2015)
3. K.C. Tsai, L.L. Wang, Z. Han, Caching for mobile social networks with deep learning: twitter analysis for 2016 US election. IEEE Trans. Netw. Sci. Eng. (2018)
4. C. Troussas, A. Krouska, M. Virvou, Trends on sentiment analysis over social networks: pre-processing ramifications, stand-alone classifiers and ensemble averaging, in *Machine Learning Paradigms* (Cham, Springer, 2019), pp. 161–186
5. A. Krouska, C. Troussas, M. Virvou, Comparative evaluation of algorithms for sentiment analysis over social networking services. J. Univers. Comput. Sci. **23**(8), 755–768 (2017)
6. C. Troussas, A. Krouska, M. Virvou, Evaluation of ensemble-based sentiment classifiers for twitter data, in *2016 7th International Conference on Information, Intelligence, Systems & Applications (IISA)* (2016), pp. 1–6
7. H. Thakkar, D. Patel, Approaches for sentiment analysis on twitter: a state-of-art study (2015). arXiv:1512.01043
8. Q.T. Ain, M. Ali, A. Riaz, A. Noureen, M. Kamran, B. Hayat, A. Rehman, Sentiment analysis using deep learning techniques: a review. Int. J. Adv. Comput. Sci. Appl. **8**(6), 424 (2017)
9. T. Mikolov, K. Chen, G. Corrado, J. Dean, Efficient estimation of word representations in vector space (2013). arXiv:1301.3781
10. T. Mikolov, I. Sutskever, K. Chen, G.S. Corrado, J. Dean, Distributed representations of words and phrases and their compositionality, in *Advances in Neural Information Processing Systems* (2013), pp. 3111–3119
11. J. Pennington, R. Socher, C. Manning, Glove: global vectors for word representation, in *Proceedings of the 2014 Conference on Empirical Methods in Natural Language Processing (EMNLP)* (2014), pp. 1532–1543
12. P. Bojanowski, E. Grave, A. Joulin, T. Mikolov, Enriching word vectors with subword information (2016). arXiv:1607.04606
13. M. Cliche, BB_twtr at SemEval-2017 Task 4: twitter sentiment analysis with CNNs and LSTMs (2017). arXiv:1704.06125
14. H. Nguyen, M.L. Nguyen, A deep neural architecture for sentence-level sentiment classification in Twitter social networking, in *International Conference of the Pacific Association for Computational Linguistics* (2017), pp. 15–27
15. A. Severyn, A. Moschitti, Twitter sentiment analysis with deep convolutional neural networks. in *Proceedings of the 38th International ACM SIGIR Conference on Research and Development in Information Retrieval* (2015), pp. 959–962

16. A. Cocos, A.G. Fiks, A.J. Masino, Deep learning for pharmacovigilance: recurrent neural network architectures for labeling adverse drug reactions in Twitter posts. J. Am. Med. Inform. Assoc. **24**(4), 813–821 (2017)
17. T. Wu, S. Liu, J. Zhang, Y. Xiang, Twitter spam detection based on deep learning, in *Proceedings of the Australasian Computer Science Week Multiconference* (2017), p. 3
18. Z. Jianqiang, G. Xiaolin, Z. Xuejun, Deep convolution neural networks for Twitter sentiment analysis. IEEE Access **6**, 23253–23260 (2018)
19. J. Wehrmann, W. Becker, H.E. Cagnini, R.C. Barros, A character-based convolutional neural network for language-agnostic Twitter sentiment analysis, in *International Joint Conference on Neural Networks (IJCNN)* (2017), pp. 2384–2391
20. P. Vateekul, T. Koomsubha, A study of sentiment analysis using deep learning techniques on Thai Twitter data, in *2016 13th International Joint Conference on Computer Science and Software Engineering (JCSSE)* (2016), pp. 1–6
21. H. Saif, M. Fernandez, Y. He, H. Alani, Evaluation datasets for Twitter sentiment analysis: a survey and a new dataset, the STS-Gold, in *1st International Workshop on Emotion and Sentiment in Social and Expressive Media: Approaches and Perspectives from AI (ESSEM 2013)* (2013)
22. D.A. Shamma, L. Kennedy, E.F. Churchill, Tweet the debates: understanding community annotation of uncollected sources, in *Proceedings of the First SIGMM Workshop on Social Media* (2009), pp. 3–10
23. M. Speriosu, N. Sudan, S. Upadhyay, J. Baldridge, Twitter polarity classification with label propagation over lexical links and the follower graph, in *Proceedings of the First Workshop on Unsupervised Learning in NLP* (2011), pp. 53–63
24. A. Krouska, C. Troussas, M. Virvou, The effect of preprocessing techniques on twitter sentiment analysis, in *2016 7th International Conference on Information, Intelligence, Systems & Applications (IISA)* (2016), pp. 1–5
25. J. Zhang, C. Zong, Deep neural networks in machine translation: an overview. IEEE Intell. Syst. **30**(5), 16–25 (2015)

Chapter 6
A Good Defense Is a Strong DNN: Defending the IoT with Deep Neural Networks

Luke Holbrook and Miltiadis Alamaniotis

Abstract This chapter frames in the realm of Internet of Things (IoT) and provides a new deep learning solution for securing networks connecting IoT devices. In particular, it discusses a comprehensive solution to enhancing IoT defense in the form of a new protocol for IoT security. The need for IoT security solutions was revisited after the recent attacks on 120 million devices. In the current work, deep learning is studied for critical security applications by utilizing snapshots of network traffic taken from a set of nine real-world IoT devices. To that end, a set of learning tools such as Support Vector Machines (SVM), Random Forest and Deep Neural Network (DNN) are tested on a set of real-world data to detect anomalies in the IoT networks. Obtained results provided high accuracy for all tested algorithms. Notably, DNN exhibits the highest coefficient of determination among the tested models, thus, promoting the DNN as a more suitable solution in IoT security applications. Furthermore, the DNN's learning autonomy feature results in a time efficient real-world algorithm because it skips human intervention.

6.1 Introduction

Smart technologies in everyday life has driven the projected number of IoT devices to increase to over 46 billion by the year 2020 [1]. This influx of these "smart" devices has already shown major security breaches with Distributed Denial of Service (DDoS) attacks. Since IoT device security is being exploited at a high rate, immediate action is needed for our security [2]. Advancements in network security introduce a

---

L. Holbrook · M. Alamaniotis (✉)
Department of Electrical and Computer Engineering, University of Texas at San Antonio, San Antonio, USA
e-mail: Miltos.Alamaniotis@utsa.edu

L. Holbrook
e-mail: lholbrook003@gmail.com

G. A. Tsihrintzis and L. C. Jain (eds.), *Machine Learning Paradigms*, Learning and Analytics in Intelligent Systems 18, https://doi.org/10.1007/978-3-030-49724-8_6

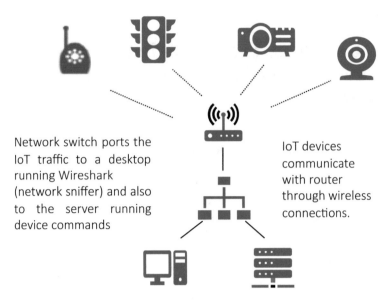

Network switch ports the IoT traffic to a desktop running Wireshark (network sniffer) and also to the server running device commands

IoT devices communicate with router through wireless connections.

Fig. 6.1 Design of experiment

promising unsupervised machine learning approach with the ability to detect anomalous network packets [3]. Although these approaches are proving valuable, this is still an untouched concept for IoT security, especially in regard to deep learning [4]. This chapter will investigate the use of deep learning for securing IoT networks.

Recent attacks have exploited commercial-off-the-shelf (COTS) IoT devices [5] and demonstrate an immediate need for robust defenses against theft of information and invasion of privacy. Other researchers have already analyzed the Mirai malware operations [6], but this chapter will focus on IoT network security measures against malware that executes at the server level. The devices tested are nine COTS IoT products vulnerable to malware attacks. Figure 6.1 illustrates the experimental design.

The assumption that continued popularity of IoT devices and Botnet attacks will be the driving factors behind security innovations and forms the basis of this chapter [7]. It is hypothesized that unsupervised deep learning models can be developed to autonomously perform anomaly detection to assist in securing IoT device networks. Specifically, DNNs are proposed as the superior algorithm for the detection of anomalous IoT network behavior. Data has been extracted from nine IoT devices to identify statistical features of normal IoT traffic, then train a DNN to learn that normal behavior. When an anomalous network packet is detected the deep learning neural network raises a red flag to indicate a malicious software attack. Extracting the network's statistical information from the cloud or server allows for a robust and practical IoT application. Installation of this research will rely on the DNN's efficiency to deploy the technology in a realistic commercial setting.

Typical research on traffic anomalies relies heavily on data from simulations for results, whereas this research uses real-world network traffic from IoT devices being attacked with two families of rampant network malware, Mirai and Gafgyt [8, 9]. The malware traffic datasets are available in [9]. This research delivers results that provide an objective solution to IoT network security, a challenging problem due to potentially large and dynamic systems. Results deliver a solution to malicious software attacks that is superior to other machine learning algorithms and is customizable per network being analyzed. Section 6.2 summarizes the state of the art in this area. Sections 6.3 and 6.4 deliver brief overviews of both IoT security and machine learning. Lastly, Sects. 6.5 and 6.6 discuss the experiment setup and the obtained results.

6.2   State of the Art in IoT Cyber Security

There is no objective protocol for IoT security, and data is plentiful. Since machine learning has been implemented in various ways, the research community has success-fully applied machine learning tools in classification of anomalous network traffic. However, application of machine learning to IoT network traffic has been largely overlooked up to this point. The current work contributes in the effectiveness of DNN for IoT network traffic anomaly detection and classification.

Doshi et al. [10] utilized machine learning for commercial IoT devices. Their research focuses solely on classic machine learning algorithms, such as K-nearest neighbor, linear kernel Support Vector Machine, Decision Tree, Random Forest and a Neural Network [10]. The researchers demonstrate that using IoT-specific network behaviors as features can result in a high accuracy for DDoS detection in IoT network traffic. Their results indicate that home gateway routers can automatically detect local sources of DDoS attacks using machine learning algorithms that do not require much computational processing and is protocol-agnostic traffic data [10]. Applying artificial neural networks to the IoT has been investigated by Canedo et al. [11]. Additionally, profiling IoT network behavior was studied in [12] where machine learning was used to apply a "semantic instance" to network behavior to determine a network-use trend. This research does not include deep learning and is inconclusive, probably caused by varying sizes in datasets which would introduce an additional error-prone parameter. The main result shows the success of supervised machine learning algorithms in stand-alone network environment. This chapter will focus on the unsupervised and autonomous nature of the DNN and how this compares to typical, well-studied machine learning algorithms.

Applying DNN to regular network security is not a new concept. Researchers at the Pacific Northwest National Lab (PNNL) have created a software tool named Safekit [13] which is an open source application for identifying internal threats on a network utilizing deep learning [13]. It uses a detailed analysis of an organization's computer network activity as a key component of early detection and mitigation of internal network threats. The researchers presented an online unsupervised deep learning approach to detect anomalous network activity from system logs in real time. Their model ranks anomaly scores based on individual user contributions. Behavior features are used for increased interpretability to aid analysts reviewing potential cases of insider threat. PNNL's research focus is not in IoT, leaving a gap in IoT network security research. Network security techniques utilizing RNN have been investigated in Du et al. [14]. DeepLog, a deep neural network model utilizing Long Short-Term Memory (LSTM), is used to model a system log file as a natural language sequence [14]. DeepLog is trained with "normal" log patterns and detects anomalies when patterns deviate from the norm. Additionally, DeepLog constructs workflows from the underlying system log so that once an anomaly is detected, users can diagnose the anomaly and take effective countermeasures. This research model provides results that determine a deep learning model is useful in monitoring network traffic and is extrapolated to the IoT.

Utilizing deep autoencoders in the detection of anomalous network behavior has also proven effective [9]. Meidan et al. have devised a method which extracts network traffic features and uses deep autoencoders to detect anomalies from IoT devices. To evaluate their method IoT devices were attacked experimentally with malware. Their evaluation showed the method's ability to accurately and instantly detect the attacks as they were being launched. This chapter differs from Meidan et al. since their research investigates deep autoencoders as opposed to DNN. Ultimately, this chapter fills the research gap of using DNN for IoT network security.

6.3 A Cause for Concern: IoT Cyber Security

6.3.1 Introduction to IoT Cyber Security

In February 2015, 2.2 million BMW vehicles were impacted by malware in their "Connected Drive" IoT software. This allowed for unwarranted remote unlocking of cars [15]. In July 2015, 1.4 million Chryslers had to be recalled due to a vulnerability in "Uconnect" IoT dashboard computers, allowing hackers to control dashboard functions, steering, transmission and brakes [16]. Various scenarios in which IoT devices are used and critical vulnerabilities of their services are given below [2]:

(1) "Public safety" organizations utilize many IoT devices, such as tablet computers in ambulances and military drones. IoT devices are used as tracking devices for persons of interest. The proven hacking of connected traffic lights can disrupt transportation and has the potential to cause accidents.
(2) "Digital health" may seem ideal to many who may not be able to leave the safety of their own home. This on the contrary may provide major security threats. Hospitals have already felt the effect of being taken hostage by virtual purveyors. Privacy concerns arise due to the sensitivity of the medical information being analyzed. Recently in India, over one billion user's biometric information was compromised due to the lack of proper security at the end point Application

Programming Interface (API) that connected the hand-held medical monitors to the government website [17].

(3) "Mobility" services introduce high number and availability of targets within the Internet of Vehicles and next generation of communications allowing self-driving cars. The attacks against these connected vehicles can severely impact society.

(4) "Manufacturing" can suffer from the reliance of IoT devices for industrial temperature and humidity monitoring. These sensors are utilized to rapidly manufacture products or provide services. Attacks to these can wreak havoc on productivity.

(5) "Smart cities" can be compromised by an attack on their infrastructure. This implies severe impacts on tourism experience, mobile payment methods and virtual reality.

(6) "Utilities" such as smart grid, metering and waste management can be affected by attacks against their IoT subsystems. With these systems down, swaths of the population will have blackouts. Researchers have shown the total debilitation of power grid transformers through network tampering [18].

(7) "Smart homes" can provide services include lighting, temperature, music and video. Here, IoT devices will consist of various sensors and electrical appliances, such as temperature sensor, fire sensor, humidity sensor, etc. An attacker can exploit IoT devices to act irrationally under the guise of normal use.

Figure 6.2 illustrates how IoT devices will be deployed in both low-priority and critical applications, such as military, emergency services, remote surgery in health-care, and self-driving cars [19–21]. In general, the attack on an entire IoT network will start with a single IoT device. These may be located in remote areas without any human supervision and may be stolen to modify their functions or steal their information to learn vulnerabilities [22]. The attacker may exploit those vulnerabilities to inject malicious code or to eavesdrop, which will result in full or partial loss of services or revenues for end users or networks. The recent Mirai botnet attack is a great example of the real-world repercussions from DDoS attacks [23]. User cyber awareness is another important element: typically default passwords are not changed and the attacker can easily enter IoT devices [5]. This implies that IoT devices must have strong authentication mechanisms and firewalls in place to prevent unauthorized access.

IoT malware is dominated by two botnet methodologies, Gafgyt and Mirai. In the following subsections, each malware will be discussed in detail to describe exactly how they work and how machine learning can be used to defend against them.

In summary IoT threats include service disruption, information theft, device theft, system manipulation and gaining unwarranted control of a system. Service can be disrupted by disabling IoT devices or carrying out a DDoS attack on the service platform.

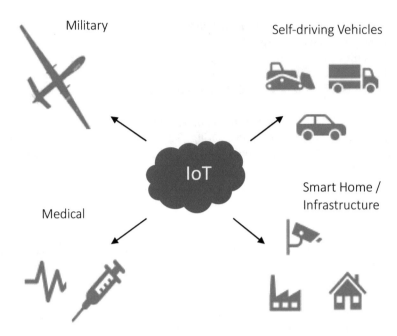

Fig. 6.2 Expanse of IoT

6.3.2 IoT Malware

IoT malware is dominated by two botnet methodologies, Gafgyt and Mirai. An understanding of the Mirai and Gafgyt malware are a necessity for enabling the user's protection. In the following subsections, each malware will be discussed in detail to describe exactly how they work and how machine learning can be used to defend against them.

6.3.2.1 Gafgyt Malware

Most embedded programs utilize a flavor of the Linux operating system with the application programs written in the C programming language. IoT devices tend to have lax security, which results from the simplicity of the device itself and lack of an objective protocol in place. The most prevalent malware currently analyzed is Gafgyt (also known as BASHLITE). This malware was first introduced in 2014 and infects devices by way of brute force potentially resulting in DDoS attacks [24]. This malware uses a predefined arsenal of usernames and passwords and attempts to access a local network by randomly choosing an IP address and attempting various username-password combinations until successful. Typically, the malware will hold TCP ports open by sending frames of junk information into the system and sometimes host-defined bash commands [25]. In August of 2016, of the identified botnet infections caused by Gafgyt, 96% were IoT devices (of which 95% were cameras and DVRs, 4% were home routers, and 1% were Linux servers) [25].

6.3.2.2   Mirai Malware

The Mirai malware is a virus that attacks networks by creating slave devices or "bots" out of infected devices to flood other networks with access requests in an attempt to force a reset and gain access to the network [26]. Bots enslaved by Mirai continuously scan the internet for the common IP addresses of IoT devices. Mirai utilizes a table of IP address ranges that it will not infect, including private networks and addresses allocated to the United States Postal Service and Department of Defense [26]. The malware then attempts to crack the security of the device by accessing a look-up-table of more than 60 commonly used manufacturer usernames and passwords. The only symptoms of an infection are device sluggishness and an increased use of bandwidth, which makes Mirai a highly volatile threat to users [27]. Rebooting the device is a simple way to clear the device of the malware, however, unless the username and password are immediately changed the device will quickly become re-infected. As of July 2018, thirteen versions of the Mirai malware were identified actively infecting Linux based IoT devices [7]. Instead of brute force attacks, several variants of Mirai adapt to device specific characteristics. Security seems to be an afterthought to many device manufacturers [5], a fact exploited by Gafgyt and Mirai.

6.4   Background of Machine Learning

Machine learning is the science of algorithm development for the purpose of performing predictive analysis utilizing trends in data [28]. The 21st century has been the period of machine learning application explosion: from playing chess against a virtual opponent to self-driving cars [6]. Based on the type of problem, the machine learning algorithms may be categorized into two broad groups: classification and regression models. Classification models are straight forward, their datasets are typically not time dependent and they apply a learned classifier to tested data. Whereas, a regression model's use is specified based on the use of a dataset with continuous, time-varying variables.

6.4.1   Support Vector Machine (SVM)

The optimization objective of Support Vector Machine (SVM) is to maximize the distance between adjacent margins between the separating hyperplane (decision boundary) and the training samples that are closest to this hyperplane [28], illustrated in Fig. 6.3.

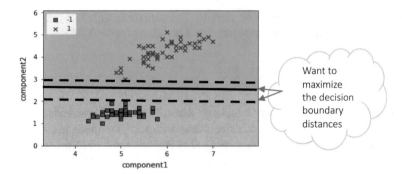

Fig. 6.3 Visualization of SVM

The margin maximization is mathematically analyzed below:

$$w_0 + w^T x_{pos} = 1. \tag{6.1}$$

$$w_0 + w^T x_{neg} = -1. \tag{6.2}$$

Subtracting (6.2) from (6.1), yields:

$$w^T (x_{pos} - x_{neg}) = 2. \tag{6.3}$$

Equation (6.3) is then normalized by the length of the vector **w**, defined as:

$$\|w\| = \sqrt{\sum_{j=1}^{m} w_j^2}. \tag{6.4}$$

This normalization allows for the expression defined by (6.5):

$$\frac{w^T (x_{pos} - x_{neg})}{\|w\|} = \frac{2}{\|w\|}. \tag{6.5}$$

The left side of (6.5) can then be interpreted as the distance between the parallel hyperplanes. This is the targeted margin for maximization. By maximizing the SVM function under the constraint that the samples are classified correctly, the following is concluded:

$$w_0 + w^T x^{(i)} \geq \text{ if } y^{(i)} = 1 \text{ for } i = 1 \ldots N. \tag{6.6}$$

$$w_0 + w^T x^{(i)} \leq \text{ if } y^{(i)} = -1 \text{ for } i = 1 \ldots N. \tag{6.7}$$

where N is the number of samples in the dataset. Equations (6.6) and (6.7) declare that all negative samples shall fall on one side of the first parallel hyperplane, whereas all the positive samples shall fall behind the second parallel hyperplane.

Wherever machine learning is utilized to make binary classifications, SVM is typically used. Examples can be found in facial detection, written word identification, and in bioinformatics [29].

6.4.2 Random Forest

Decision trees are the backbone of random forest and use a binary decision at each node that are stacked to develop a multi-layered "forest". The random forest algorithm is summarized in four steps from [28]:

1. Draw a random sample from the data of size n (randomly choose n samples from the training set with replacement)
2. Grow a decision tree from the sample. At each node:

    a. Randomly select d features without replacement
    b. Split the node into random subsets using the feature that provides the best split according to the objective function, for instance, maximizing the information gain

3. Repeat these steps k times
4. Aggregate the prediction by each tree to assign the class label by the majority vote.

Each decision node in Fig. 6.4 represents a pass through an activation function, where if the input data scores high correlation with the encountered feature, one or the other path will be chosen.

Since the user does not have to choose initial hyperparameter values and the algorithm does not need tuning, this algorithm is considered robust to noise from the individual decision trees that make up the forest. A general rule of thumb is the larger the number of trees, the better the performance of the random forest classifier at the expense of an increased computational cost.

The majority vote method allows for a weighted and deterministic identification of malicious network traffic and is identified as:

$$\hat{y} = \text{argmax} \sum_{j=1}^{m} w_j \chi_A \big(C_j(x) = i\big). \tag{6.8}$$

where w_j is a weight associated with the base classifier C_j, $\hat{y}$ is the predicted class label of the system, χ_A is the characteristic function and A is the set of unique class labels. For equal weights, (6.8) can be simplified to:

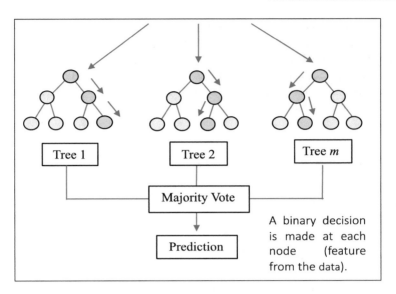

Fig. 6.4 Visualization of the Random Forest utilizing a majority vote technique

$$\hat{y} = \text{mode}\{C_1(x), C_2(x), \ldots, C_m(x)\}. \tag{6.9}$$

where the mode is the most frequent event or result in a set. As an example, the majority vote prediction of a binary classification task is shown in (6.10):

$$C(x) = \text{sign}\left[\sum_{j}^{m} C_j(x)\right]. \tag{6.10}$$

Majority vote will always give more accurate results for a binary model than when not used, assuming that all classifiers have an equal error rate [30]. Additionally, it is assumed that the classifiers are independent, and the error rates are not correlated.

Random forests are a popular model to use in various machine learning applications. Mainly appearing in banking, stock market, e-commerce and the medical field [31–34].

6.4.3 Deep Neural Network (DNN)

The discovery of layering and connecting multiple networks resulted in what is now called deep learning. Deep learning is observed to be subsets of Artificial Intelligence as shown in Fig. 6.5. In particular, this chapter will cover an application of deep learning known as Deep Neural Networks (DNN), or many connected layers of neural networks.

Fig. 6.5 The AI complex

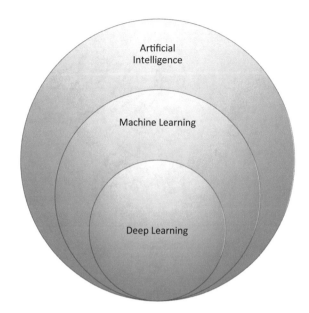

The process of forward propagation is used to calculate the output of the DNN and is characterized as follows [28]:

(1) Starting at the input layer, the model forward propagates the patterns of the training data through the network to generate an output.
(2) Based on the network's output, minimizing the error is calculated using a cost function
(3) Backpropagation is used to determine the error, finding its derivative with respect to each weight in the network, and updating the model.

Multiple epochs are used to evaluate the weights of the learning model. Once forward propagation finds the network output then backpropagation is applied to obtain the predicted class labels. Figure 6.6 illustrates forward and backpropagation.

The DNN in this research identifies an anomaly score from the feature vectors for a given IoT device's network. This model does not utilize a time-based approach, and instead compares all of the network data over one device simultaneously. The mathematical model of this DNN algorithm is largely based on Tuor et al. [13]. The model receives a series of T feature vectors for a device d and produces a series of T hidden state vectors. The final hidden state of the DNN is a function of the feature vectors as follows [13]:

$$z_{l,t}^d = a\left(W_l z_{l-1,t}^d + b_t\right). \tag{6.11}$$

Here a is a non-linear activation function, typically ReLU, tanh, or the logistic sigmoid. The tunable parameters are the weight matrices ($\mathbf{W}$) and bias vectors ($\mathbf{b}$).

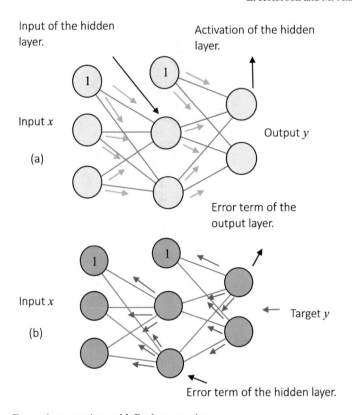

Fig. 6.6 **a** Forward propagation and **b** Backpropagation

6.5 Experiment

Raw events from system device logs are fed into the feature extraction system, which configures the data into identifiable feature vectors. A device's set of feature vectors are in turn inputted into a DNN, which trains to minimize the anomaly score through backpropagation [28].

In this research, the DNN is tasked with classifying test samples with the classification values gathered from minimizing the loss equation from the training data. Although the DNN basically learns from the device's "normal" behavior, the algorithm is unsupervised since no prior knowledge of the network is introduced. An anomaly is proportional to the prediction error with sufficiently anomalous behavior being flagged.

The IoT network traffic is captured and broken down into categorical network parameters per device listed in Table 6.1. Statistical parameters were collected for each stream aggregation feature over five distinct sample collection times. For each

Table 6.1 Parameters identified in research

Data features	Feature header	Definition of feature
Stream aggregation	H	– Source IP – Stats summarizing the recent traffic from this packet's host (IP)
	MI	– Source MAC-IP addresses – Stats summarizing the recent traffic from this packet's host (IP + MAC)
	HH	– Channel size – Stats summarizing changes to traffic going from this packet's host IP to the packet's destination IP
	HH_jit	– Channel size jitter – Stats summarizing changes in channel size of traffic leaving from this packet's host IP to the packet's destination IP
	HPHP	– Port – Stats summarizing the recent traffic leaving from this packet's host IP + port to the packet's destination IP + port – ex. 192.168.4.2:1242 –> 192.168.4.12:80
Statistics	Mean	– The mean deviation of the transmitted packet size
	STD	– The standard deviation of the packet size
	Number	– The number of items observed in recent history
	Magnitude	– The root squared sum of the two streams' means (host vs. destination addresses)
	Radius	– The root squared sum of the two streams' variances (host vs. destination addresses)
	Covariance	– An approximated covariance between two streams (host vs. destination)
	Correlation coefficient	– An approximated correlation coefficient between two streams
Time frame	L5–L0.01	– The decay factor Lambda used in the time window – Recent history of the stream is captured in these statistics

device d, for each time period, t, the parametric values and activity counts are exported into a 115-dimension numeric feature vector. The anomaly score is defined as the following:

$$\hat{a}_t^d = -\log P_\theta(x_t^d | h_t^d). \qquad (6.12)$$

here P is the probability, x is the input, and h is the associated hidden vector. This model will quantify the probability and assign an anomaly score to the current sample.

If the input is anomalous, then the algorithm is unlikely to assign the sample with a large probability density distribution, delivering a more narrow and higher loss value.

The anomaly scores start out as a maximum, where the model knows nothing about normal behavior and lowers over time when normal behavior patterns are learned. As each anomaly score is detected, an estimate of the exponentially weighted moving average of the mean and variance are computed and standardized. This is then used to properly place an anomaly score for device d at time t. The real-time method of DNN allows for the flexibility of immediately reacting to anomalous behavior.

6.5.1 Training and Test Data

The machine learning models are trained on statistical features extracted from benign network traffic data. The raw network traffic data is captured using port mirroring on a network switch. The IoT network traffic was collected immediately following the device's connection to the network. The network traffic collection process adheres to the following:

(1) Network traffic originated from a known IP address
(2) All network sources originated from the same sources of MAC and IP addresses
(3) Known data is transmitted between the source and destination IP addresses
(4) Destination TCP/UDP/IP port information is collected.

The same set of 23 features shown in Table 6.2 were extracted from each of five time-windows: 100 ms, 500 ms, 1.5 s, 10 s, and 1 min increments, totaling 115 features. These features can be computed very fast and incrementally to allow real time detection of malicious packets. Table 6.2 also lists the range of statistical values of the normal network traffic features over the longest sample collection time window (1 min). These features are useful for capturing source IP spoofing and other key malware exploits. For example, when a compromised IoT device spoofs an IP address, the features aggregated by the source MAC/IP (feature variable MI) and

Table 6.2 Statistic parameters collected from the data features

Statistics	HH	HH_jit	MI	H	HPHP
Mean	330.46	1.56E + 9	300.45	300.45	300.45
STD	145.71	5.66E + 17	2.12E + 4	2.12E + 4	71.57
Number	2.01	2.01	7.73	7.73	2.01
Magnitude	555.97	–	–	–	300.45
Radius	6.91E + 4	–	–	–	1.16E + 4
Covariance	5355.09	–	–	–	5372.24
Correlation coefficient	0.97	–	–	–	0.78
Number of features	7	3	3	3	7

IP/channel (feature variable HP) will immediately indicate a large anomaly score due to the behavior originating from the spoofed IP address [26].

6.5.2   Baselines of the Machine Learning Models

Baselines of the machine learning algorithms were determined manually. The training and test dataset sizes were tuned such that they represented different realistic environments [35]. For the DNN, the number of hidden layers were tested from 1 to 10. The hidden layer dimension was tested from 20 to 500, and various sizes of mini-batches were also tested to determine best fit for the tested malware. Additionally, the input vector to the DNN was tuned to be made compatible with the 115 features. For the SVM, a linear kernel was determined as the best fit option for classification and utilizes a random number generator when shuffling the data for probability estimates. For the random forest, the number of estimators was tested for optimization from 20 to 100. Furthermore, the percentage of the dataset the models were trained and tested on were experimented with where 5% of the dataset was used from training and the remaining 95% of the dataset was used for testing.

6.6   Results and Discussion

The results are divided between Gafgyt and the Mirai datasets. The results show that the DNN gives greater anomaly determination, compared to the supervised algorithms performing on the same data. The DNN also allows for a nearly autonomous solution that is robust enough to handle new threats as they emerge. In contrast, the other tested machine learning algorithms cannot immediately handle new types of network attacks. Since the DNN utilized mini-batches for the data processing, the confusion matrix will show a smaller sample size (318 test samples) compared to the SVM and random forest tests.

6.6.1   Results

The following analyses are broken up into their respective datasets. First the Gafgyt malware is discussed, followed by the Mirai malware.

6.6.1.1   Gafgyt Malware

The statistical performances of test data from the Gafgyt malware is shown in Table 6.3, and confusion matrices are shown in Fig. 6.7. In this chapter, the coeffi-

Table 6.3 Statistic results from the Gafgyt test data

Algorithm	SVM	RF	DNN
R^2	0.6995	0.8263	0.9998
MSE	0.0005	0.0003	0.0002
Accuracy	99%	99%	99%

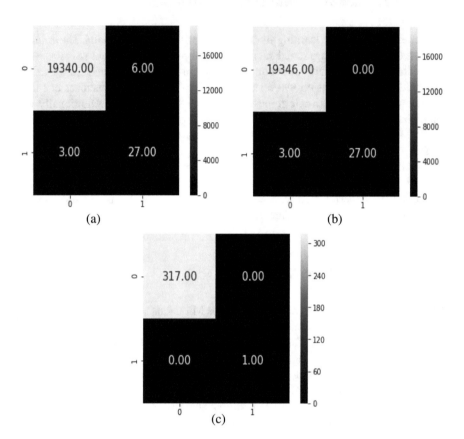

Fig. 6.7 Confusion matrices of the **a** SVM **b** Random Forest and **c** DNN from the Gafgyt malware test data

cient of determination is denoted R^2 and is the fraction of response variances of the feature values captured by the model. This value is defined by (6.13):

$$R^2 = 1 - \frac{SSE}{SST}. \tag{6.13}$$

here SSE is the sum of squares for the residuals, and SST is the total sum of squares. These values are the most common way of denoting results based on probabilistic outcomes. The mean square error (MSE) is a measure that expresses the minimized

Table 6.4 Statistic results from the Mirai test data

Algorithm	SVM	RF	DNN
R^2	0.6328	0.7689	0.9968
MSE	0.0006	0.0004	0.0001
Accuracy	99%	99%	99%

cost of fitting the machine learning models to the dataset. The mathematical formula of *MSE* is given by (6.14).

$$MSE = \frac{1}{n} \sum_{i=1}^{n} \left(y^{(i)} - \hat{y}^{(i)} \right)^2. \tag{6.14}$$

Analytically, the accuracy is defined by (6.15):

$$\text{Accuracy} = \frac{TN + TP}{TN + TP + FN + FP}. \tag{6.15}$$

where *TN* stand for True Negative, *TP* for True Positive, *FN* for False Negative and *FP* for False Positive.

The confusion matrices show that the SVM, random forest and DNN were each able to correctly classify 99% of the network traffic. This high accuracy further enforces the general notion of using machine learning algorithm to detect anomalous network behavior.

6.6.1.2 Mirai Malware

Detection of the Mirai and Gafgyt malwares largely showed similar results. The statistical performances are shown in Table 6.4 and the confusion matrices are shown in Fig. 6.8. Here the differences between the Mirai and the Gafgyt datasets are that the SVM and random forest each scored a lower coefficient of determinations and mean square errors. Again, all three of the algorithms scored 99% accuracy, despite decreasing statistical parameters.

6.6.2 *Discussion*

In this section we will dive into the implications based on the obtained results. Each of the machine learning models scored a high accuracy. This implies a heavy dependence on one of the features of the network traffic data. Based on the corresponding coefficient of determination values the DNN is the superior algorithm. Since the algorithm received a high value, this shows how the DNN was able to learn a correlated

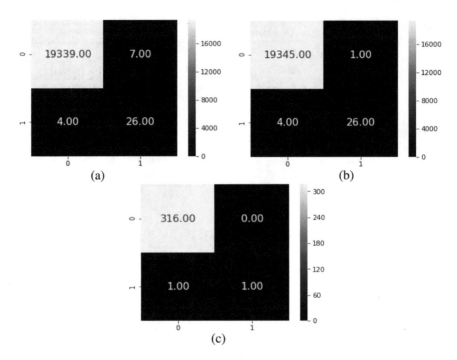

Fig. 6.8 Confusion matrices of the **a** SVM **b** Random Forest and **c** DNN from the Mirai malware test data

regression model and apply it to the tested data. This is suspected to have resulted from the 10-layers of neural networks the algorithm is built upon.

The DNN also had higher true negative rates, which means that the algorithm is superior for this network security application, with the notion we would rather be safe than sorry. The logistics for a deployable application requires an unsupervised approach due to environmental constraints such as network bandwidth and need for autonomy, both of which the DNN alleviate. For the supervised algorithms to persist under these constraints, additional techniques would have to be applied or an analyst would have to be present.

The channel jitter is observed to be the heaviest weighted feature by means of largest data value separations. This feature quantifies the variable network rates at which the traffic is sent. This means that the DNN learned the most from the established network channel.

Scalability is crucial for this research to be applied in a real-world environment. All three of these machine learning algorithms are scalable in their own respect, however the results from the DNN show that this model has the ability to provide the most value for its installation. Since large networks of IoT devices are prone to targeted DDoS attacks, the final application would require the ability to distinguish many devices [1]. With the DNN's proven performance on a single device, it would need supplementary work to distinguish n number of IoT devices in a network.

6.7   Conclusion

In this chapter we discussed the technical background and the importance of securing the IoT. Experimentation and results are walked through, depicting a solution to a lack of research in IoT networks. Future work and research areas are discussed in the following paragraphs.

Autonomy is the desired goal of the application of a machine learning model. To further robustness, continuous monitoring is required along with adaptive decision making. These challenges may be addressed with the assistance of Software Defined Networking (SDN) which can effectively handle the security threats to the IoT devices in dynamic and adaptive manner without any burden on the IoT devices. An SDN-based secure IoT framework called SoftThings has proposed the ability to detect abnormal behaviors and attacks as early as possible and mitigate as appropriate [36]. Machine learning was demonstrated to be used at the SDN controller to monitor and learn the behavior of IoT devices over time. Another option is utilizing a cloud deployable application. Since the cloud has the ability to remotely process and perform computations on large amounts of data, this is perfect for the DNN. Applications of a commercial cloud deployable application for IoT security is seen with Zingbox's Guardian IoT. Similar to this research, Zingbox's application employs deep learning to provide network monitoring of IoT devices.

Determining the heaviest feature will be an important key to running a robust algorithm. With the commercial sights on machine learning, a thorough analysis on the feature engineering will be useful to analyze how secure this application may be. At DEFCON 24 [37] researchers have shown that there are inherent weaknesses to the machine learning algorithms in use today. Additionally, researchers in [4] have exploited TensorFlow's machine learning algorithms using knowledge of the Linux operating system. The understanding of the proposed future work will be crucial in furthering the success of utilizing machine learning for network security.

References

1. M. Weldon, An Internet of Things blueprint for a smarter world (2015)
2. M. Liyanage, I. Ahmad, A. B. Abro, A. Gurtov, M. Ylianttila (Eds.), *A Comprehensive Guide to 5G Security* (Wiley, 2018), p. 640
3. J. Dromard, G. Roudiere, P. Owezarski, Online and scalable unsupervised network anomaly detection method. IEEE Trans. Netw. Serv. Manage. **14**, 34–47 (2017)
4. Q. Xiao, K. Li, D. Zhang W. Xu, Security Risks in Deep Learning Implementations (2017)
5. R. Sarang, Trending: IoT Malware Attacks of 2018 (2018). [Online] https://securingtomorrow. mcafee.com/consumer/mobile-and-iot-security/top-trending-iot-malware-attacks-of-2018/. Accessed on 2019
6. Anonymous, Is Artificial Intelligence a Detriment or a Benefit to Society? (2018). [Online] https://www.forbes.com/sites/forbestechcouncil/2018/06/19/is-artificial-intelligence-a-detriment-or-a-benefit-to-society/#1a9080f549a5
7. Anonymous, Mirai-Mirai on the wall. How Many Are You Now? (2018). [Online] https:// imgur.com/a/53f29O9

8. C. Kolias, G. Kambourakis, A. Stavrou and J. Voas, DDoS in the IoT: Mirai and other botnets, in *Computer* (2017)
9. Y. Meidan, M. Bohadana, Y. Mathov, Y. Mirsky, D. Breitenbacher, A. Shabtai, Y. Elovici, N-BaIoT: Network-based detection of IOT botnet attacks using deep autoencoders. IEEE Pervasive Comput. **17**, 12–22 (2018)
10. R. Doshi, N. Apthorpe, N. Feamster, Machine learning DDoS detection for consumer Internet of Things devices, in *IEEE Security and Privacy Workshop* (2018), pp. 29–35
11. J. Canedo, A. Skiellum, Using machine learning to secure IoT systems, in *Annual Conference on Privacy, Security and Trust* (2016)
12. S. Lee, S. Wi, E. Seo, J. Jung, T. Chung, ProFIoT: Abnormal behavior profiling of IoT devices based on a machine learning approach, in *IEEE International Telecommunication Networks and Applications Conference* (2017)
13. A. Tuor, S. Kaplan, B. Hutchinson, Deep learning for unsupervised insider threat detection in structured cybersecurity data streams (2017)
14. M. Du, F. Li, G. Zheng, V. Srikumar, DeepLog: Anomaly detection and diagnosis from system logs through deep learning, in *ACM SIGSAC Conference on Computer and Communications Security*, New York (2017)
15. T. Brewster, BMW update kills bug in 2.2 million cars that left doors wide open to hackers (2015). [Online] https://www.forbes.com/sites/thomasbrewster/2015/02/02/bmw-door-hacking/#435cfa346c92. Accessed on 2019
16. A. Greenberg, After jeep hack, Chrysler recalls 1.4 M vehicles for bug fix (2015). [Online] https://www.wired.com/2015/07/jeep-hack-chrysler-recalls-1-4m-vehicles-bug-fix/. Accessed on 2019
17. Z. Whittaker, A new data leak hits Aadhaar, India's national ID database (2018). [Online] https://www.zdnet.com/article/another-data-leak-hits-india-aadhaar-biometric-database/. Accessed on 2019
18. *CyberWar Threat.* [Film] (PBS, United States of America, 2015)
19. N. Suri, M. Tortonesi, Session 13: Military Internet of Things (IoT), autonomy, and things to come (2016)
20. A. Patel, R. Patel, M. Singh, F. Kazi, Vitality of robotics in healthcare industry: an Internet of Things (IoT) perspective, in *Internet of Things and Big Data Technologies for Next Generation Healthcare* (Springer, 2017), pp. 91–109
21. X. Krasniqi, E. Hajrizi, Use of IoT technology to drive the automotive industry from connected to full autonomous vehicles, in *IFAC Papers Online* (2016)
22. B. Barker, Robots take on tough tasks in transmission and distribution. Epri J. 1–3 (2019)
23. D. Su, Mirai DDoS attack: a wake-up call for the IoT industry (2016). [Online] https://www.nokia.com/blog/mirai-ddos-attack-wake-call-iot-industry/. Accessed on 2019
24. C. Cimpanu, There's a 120,000-strong IoT DDoS botnet lurking around (2016). [Online] https://news.softpedia.com/news/there-s-a-120-000-strong-iot-ddos-botnet-lurking-around-507773.shtml. Accessed on 2019
25. Anonymous, Attack of things (2016). [Online] https://www.netformation.com/our-pov/attack-of-things-2/
26. I. Zeifman, D. Bekerman, B. Herzberg, Breaking down Mirai: an IoT DDoS botnet analysis (2016). [Online] https://www.incapsula.com/blog/malware-analysis-mirai-ddos-botnet.html. Accessed on 2019
27. T. Moffitt, Source code for Mirai IoT malware released (2016). [Online] https://www.webroot.com/blog/2016/10/10/source-code-mirai-iot-malware-released/. Accessed on 2019
28. S. Raschka, V. Mirjalili, *Python Machine Learning* (Packt Publishing Ltd, Birmingham, 2017)
29. Anonymous, Real-life applications of SVM (Support Vector Machines) (2017). [Online] https://data-flair.training/blogs/applications-of-svm/
30. L. Lam, S. Suen, Application of majority voting to pattern recognition: an analysis of its behavior and performance, in *IEEE Transactions on Systems, Man and Cybernetics—Part A: Systems and Humans* (1997)

31. L. Khaidem, S. Saha, S. R. Dey, Predicting the direction of stock market prices using random forest (2016)
32. L. Min, X. Xu, Y. Tao X. Wang, An Improved Random Forest Method Based on RELIEFF for Medical Diagnosis, in *IEEE International Conference on Embedded and Ubiquitious Computing* (2017)
33. S. Akbari, F. Mardukhi, Classification of bank customers using the Random Forest algorithm. Int. J. Math. Comput. Sci. **29**, 828–832 (2014)
34. N. Donges, The Random Forest algorithm (2018). [Online] https://towardsdatascience.com/the-random-forest-algorithm-d457d499ffcd. Accessed on 2019
35. A. Galstev, A. Sukhov, Network attack detection at flow level, in *Smart Spaces and Next Generation Wired/Wireless Networking* (Springer, Berlin, 2011), pp. 326–334
36. S. Bhunia, M. Gurusamy, Dynamic attack detection and mitigation in IoT using SDN, in *IEEE International Telecommunication Networks and Applications Conference* (2017)
37. C. Chio, Machine duping: Pwning deep learning systems, in *DEFCON* (Las Vegas, 2016)

Part III
Deep Learning in the Medical Field

Chapter 7
Survey on Deep Learning Techniques for Medical Imaging Application Area

Shymaa Abou Arkoub, Amir Hajjam El Hassani, Fabrice Lauri, Mohammad Hajjar, Bassam Daya, Sophie Hecquet, and Sébastien Aubry

Abstract Unlike different sectors that people with specific backgrounds get interested in, the healthcare sector has always been the number one priority for all people all over the world. Thus, it has made people from different majors unite in requesting accurate healthcare services. The ever-increasing worldwide demand for early detection of diseases at many screening sites and hospitals had resulted in the need for new research avenues. As of that, researchers have employed deep learning algorithms in the medical industry due to its achieved state-of-art results across various domain's problems. These methods have emerged as the technique of choice for medical imaging interpretation, which solves previously intractable problems and enhances the performance level. This chapter highlights the primary deep learning

S. Abou Arkoub · A. Hajjam El Hassani (✉)
Nanomedicine Lab, University Bourgogne Franche-comté, UTBM, F-90010 Belfort, France
e-mail: amir.hajjam-el-hassani@utbm.fr

S. Abou Arkoub
e-mail: shymaa_arkoub@hotmail.com

F. Lauri
CIAD Lab, University Bourgogne Franche-comté, UTBM, F-90010 Belfort, France
e-mail: fabrice.lauri@utbm.fr

M. Hajjar · B. Daya
GRIT Lab, Lebanese University, Beirut, Lebanon
e-mail: mohammadhajjar@ul.edu.lb

B. Daya
e-mail: b_daya@hotmail.com

S. Hecquet · S. Aubry
Osteoarticular Radiology Department, CHU Jean Minjoz, Besançon, France
e-mail: sophiehecquet@yahoo.fr

S. Aubry
e-mail: sebastien.aubry@univ-fcomte.fr

G. A. Tsihrintzis and L. C. Jain (eds.), *Machine Learning Paradigms*, Learning and Analytics in Intelligent Systems 18, https://doi.org/10.1007/978-3-030-49724-8_7

techniques relevant to the medical imaging application area and provides fundamental knowledge of deep learning methods. Finally, this chapter ends by specifying
the current limitation and future research directions.

7.1 Introduction

Since the last four decades, medical technique imaging has been first introduced to
stun the world with its benefits and outstanding results. It has caused a revolution in
the medical aspect. Therapeutic techniques as tomography (CT), which is the first
medical imaging technique introduced by Godfrey Hounsfield, in 1979, magnetic
resonance imaging (MRI) and ultrasound in 1980, positron emission tomography
(PET) and X-rays, have scored an improvement in detecting, diagnosing and treating
diseases. Although it has excellent advantages, yet it depends on humans who are
limited by fatigue, errors, experience, and speed. These errors or delays may include
mistakes in diagnosing, causing harm to patients. That is why engineers and doctors
are trying to get assistance from computers, and here begins the story of how machine
learning interferes in this domain.

In the late 1990s, machine learning algorithms gained increased popularity in the
healthcare sector, especially in the medical imaging analysis field [1, 2]. These tools
were used to detect patterns from image data to predict and make a decision. However,
machine learning algorithms require task-related experts to extract essential features
from raw pixels before feeding it to machine learning models. Furthermore, these
methods cannot handle complex data (e.g., images) and suffer from critical issues
such as over-fitting, vanishing—exploding gradient, slow computation power, and
small dataset size. With passing time, the rapid increase of data volume and computation power along with the appearance of advances ml algorithms known as "deep
learning" solved the external constraint of ml algorithms.

Deep learning algorithms, a subfield of machine learning, emerged to overcome
the limitation and drastically enhance the performance of traditional machine learning
methods by consolidating feature extraction step within the learning process. This
end-to-end architecture is composed of many layers allowing it to automatically
extract hierarchical feature representations from raw data input with a high level of
abstraction and in a self-taught manner [3, 4]. The approaches of deep learning have
shown a record-breaking performance in various regions of medical imagining analysis like image processing (such as segmentation, registration, enhancement, reconstruction, etc.), computer-aided detection, computer-aided diagnosis, and prognosis.
This learning algorithm helps radiologists analyze image data rapidly and accurately,
allowing real-time disease detection and prevents them in time.

In terms of learning algorithms, deep neural networks categorized as supervised
learning that train on labeled samples (such as CNN, RNN, LSTM) and unsupervised
learning that train on unlabeled samples (like GAN, DBM, DBN, SAE, DAE, etc.).
Because of the structural representations of images and the correlated information
between neighboring pixels, CNN techniques are considered the method of choice

for understanding image content and have provided state-of-art accuracies across various image analysis applications [5]. However, unsupervised networks have also been effectively utilized in different medical tasks, for example, diagnosis of brain disease using Deep Auto-Encoder (DAE) [6].

The motive of this chapter is to introduce the key learning points of deep learning architecture that achieved state-of-art results in the medical imaging analysis domain. First, we gave a review of how traditional machine learning techniques advanced to deep learning algorithms. Then, we introduced a list of the most widely used deep learning algorithms in the medical sector, along with an explanation of their main building blocks. After that, we described how to apply MRI preprocessing steps to images before feeding them to DNN. Later, we surveyed recent studies of deep learning methods in medical imaging applications. Finally, we concluded current work limitations and future direction of deep learning in medical imaging applications.

7.2 From Machine Learning to Deep Learning

Although it is 77 years old, state of deep art learning recently has boomed solutions for various real-world domain problems in computer vision, language processing, sentiment analysis, and many more. Deep learning was born with a computer back in 1943. Deep Learning rose into prominently when Warren McCulloch and Walter Pitts [7] created a computational algorithm modeled on human neurons based on the linear threshold concept. Because of this idea, Donald Hebb wrote a book in 1949 about neuron pathways emphasizing a new concept, which is "neurons activated at the same time are linked together." During the same period, Rosenblatt [8] introduced the idea of a Perceptron in 1958. This idea states an algorithm with only two layers (the input node and output node) modeled on McCulloch-Pitts neuron for detecting patterns from data under study. Moving forward in 1960, Henry J. Kelley developed the basics of backpropagation, which is the backbone of the neural network training. Unfortunately, the backpropagation concept did not inject into the neural network domain before 1985. After that, Rumelhart et al. [9] meteorically demonstrated a re-discovery of backpropagation in ANN.

Fukushima et al. [10] was the first who introduced deep learning architecture in sophisticated vision fashion back in 1980. He added the idea of Necocognition, which is a multilayer hierarchal design and considered it as the first CNN. Each CNN consists of two layers: the convolution layers and the pooling layers. In 1989, LeCun et al. [11] was inspired by the work mentioned above of Kunihiko Fukushima, Rumelhart, Williams, and Hinton, and introduced the first practical system that combines both CNN and backpropagation to recognize handwritten digits. However, the training time for this system was impractically long, which takes around three days; also, training data used by the system was inefficient. Due to these and other problems, the next ten years came in as "AI winter" making the fate of the so-called Artificial Intelligence nothing but a hypothesis.

The thawing of this decade started in 1999 when computers began to gain faster processing with GPU, and data collection started to grow. These enhancements have caused revolutionary progress for neural networks. In 2012, Krizhevsky et al. [12] proposed a CNN model called AlexNet that won the ImageNet competition by a large margin. Further development was imposed in the following years using associated but deeper architectures including but not limited to large scale visual recognition (ILSVRC), face recognition (Facebook DeepFace, 2015), and AlphaGo algorithm (Google DeepMind, 2016). The promise of deep learning is back, and this time it was genuinely catching the imagination of the world and finally coming to par with and even overtaking its expectation.

7.3 Learning Algorithm

Machine learning is a data-driven model that learns to act like a human by feeding it the data without being explicitly programmed. There are two categories of machine learning, known as supervised and unsupervised learning types.

In supervised learning, each data sample is comprised of input (predictor) denoted as X, and the corresponding output (target/label) denoted as Y. Thus, the machine learning algorithm, which is based on loss function, will learn the weights that map the given input to the matched label from the training dataset. During testing, the model takes input data (x) only to evaluate its class from the test dataset. Supervised learning is referred to as a classification task when the output is discrete, and as regression, if the output is continuous.

In unsupervised Learning, each data sample consists of predicators only. Thus, during training, the model learns the hidden structure from un-label input data. Unsupervised Learning is referred to as clustering task when the output is discrete, and as dimensionality reduction, if the output is continuous.

7.4 ANN

ANN is a computer system that is roughly based on our brain neural nets. ANN is formed of interconnected layers, which are: one input layer collecting input signal from the previous layer, one hidden layer for training weights, and one output layer that form the prediction layer. Each layer is a set of basic units called neurons where its output is simultaneously the subsequent layer's input. Note that the input layer of the model is the underlying information layer that accepts your data (Fig. 7.1).

$$Output = \sigma(WT\sigma(WT \ldots \sigma(WT \times +b)) + b$$

A neuron is a mathematical computation unit inspired by the human nervous system. Neuron takes a set of inputs X from the previous layer connected to it,

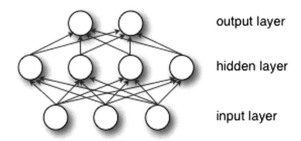

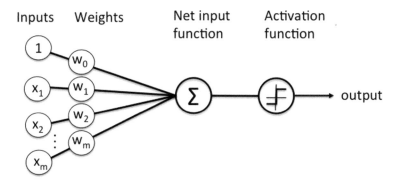

Fig. 7.1   ANN architecture

Fig. 7.2   Artificial neuron

multiplies these inputs with the corresponding set of weights W, and adds a bias B. Then, the output from this linear combination is applied to a non-linear transformation function known as activation function σ (Fig. 7.2).

$$\text{Output} = \sigma \, (W \cdot X + B)$$

7.4.1   Activation Function in ANN

The activation function is a mathematical equation added on every neuron in the network to map the input signal to a non-linear output value. It is used to make a mathematical decision on whether the neurons in the next layer should be activated (fire) or not. The role of the nonlinearity of the function is introduced to add a non-linear property to the network allowing it to learn and model complex tasks from high dimensional, unstructured, non-linear, and big datasets.

7.4.1.1 Types of Nonlinear Activation Function

A. Sigmoid Activation Function

S shaped curve range from 0 to 1 (Fig. 7.3).

$$F(x) = 1/1 + \exp(x)$$

Major advantages of sigmoid function:

- Easy to apply.
- Smooth gradient preventing jumps in output values.
- Normalized output since its output value range between zero and one.

Major disadvantages of sigmoid function:

- Gradient saturation–> introduces vanishing gradient problem.
- Slow convergence –> expensive computation.
- Hard optimizations since the sigmoid are not zero centered.

B. Tanh activation function

S shaped curve range from −1 to 1 (Fig. 7.4).

$$F(x) = 1 - \exp(-2x)/1 + \exp(-2x)$$

Major advantages of Tanh function

- Easy optimization since it is zero centered.
- Normalized output.
- Smooth gradient.

Fig. 7.3 Sigmoid activation function

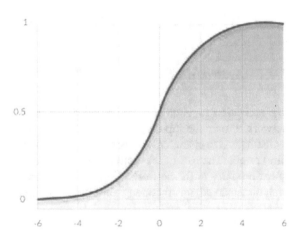

Fig. 7.4 Tanh activation function

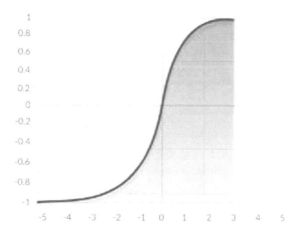

Major disadvantages of Tanh function

- Gradient saturation –> introduces vanishing gradient problem.
- Slow convergence –> computationally expensive.

C. Softmax activation function

Range from 0 to 1

$$F(x) = \exp(Xi) \sum \exp(Xi).$$

Major advantages of softmax activation function

- Used for multi-classification task.

D. Relu activation function

Range from 0 to +infinity (Fig. 7.5).

$$R(x) = \max(0, x)$$

Major advantages of Relu function

- Unsaturated gradient–> solves the problem of vanishing gradient.
- Fast convergence.
- Nonlinearity–> allows efficient backpropagation.

Major disadvantages of Relu function

- Dying Relu problem results in dead neurons when the input is below zero, which prevents networks from performing efficient backpropagation.

Fig. 7.5 Relu activation
function

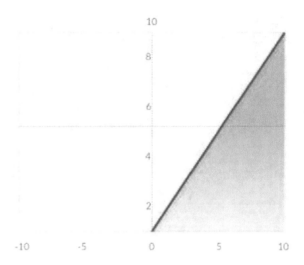

E. Leaky Relu activation function

Range from −infinity to +infinity (Fig. 7.6).

$$F(x) = \max(0.1 \times x, x)$$

Major advantages of Leaky Relu function

- Prevents dying Relu problem.
- Fast convergence.
- Prevents vanishing gradient problem.

Major disadvantages of Leaky Relu function

- Inconsistence prediction due to the inefficiency of negative input results.

Fig. 7.6 Leaky Relu
activation function

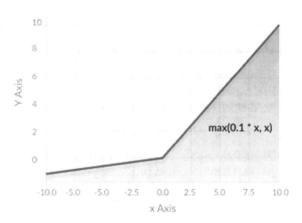

Fig. 7.7 Swish activation
function

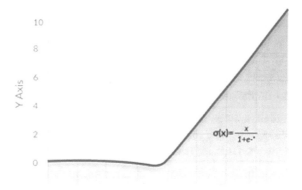

F. Parametric Relu

Range from –infinity to +infinity

$$F(x) = \max(a * x, x)$$

Major advantages of Parametric Relu function

- Allows negative input values to be learned.
- Prevents dying Relu problem.
- Fast convergence.
- Prevents vanishing gradient problem.

Major disadvantages of Parametric Relu function

- Different performance, depending on the task.

G. Swish

New activation function discovered by Google [13], which performs better than Relu
with the same level of computational efficiency (Fig. 7.7).

$$F(x) = x / \exp(-x)$$

7.4.2 Training Process

The training process is a must to allow the designed model to learn from the provided
data. The whole training process may be summarized as follow:

7.4.2.1 Model Initialization

Determining a proper initialization strategy is a critical step for the training process to achieve higher accuracy and higher convergence.

There are three kinds of parameters initializations which are:

A. Initialize parameter to zero.

Initializing weights and biases to zero is not a good practice as all weights will receive the same update regardless of how many iterations we backpropagate through the network. Thus, the neuron won't be able to learn because of this symmetric property.

B. Initialize parameter randomly using uniform or standard normal distribution.

This technique is more efficient than the zero-initialization one because it breaks symmetry and allows the neuron to learn. The problem with this strategy is that when the weights are initialized with large or small values and applied to activation function like sigmoid or tanh, gradient saturation will occur, which will result in slow learning.

C. Initialize parameter using Glorot et al. [14] or He et al. [15] initialization techniques.

These methods avoid slow convergence, give lower training error rates, and solve vanishing and exploding gradient issues by initializing weights to an appropriate value.

Xavier initialization used for tanh activation function:

$$\sqrt{\frac{1}{\text{size}^{(l-1)}}}$$

$$W^l = \text{np.random.randn}(\text{size}_l, \text{size}_{l-l}) * \text{np.sqrt}\left(\frac{1}{\text{size}_{l-l}}\right)$$

Kaiming initialization used for Relu activation function:

$$\sqrt{\frac{2}{\text{size}^{(l-1)}}}$$

$$W^l = \text{np.random.randn}(\text{size}_l, \text{size}_{l-l}) * \text{np.sqrt}\left(\frac{2}{\text{size}_{l-l}}\right)$$

7.4.2.2 Feed Forward

The natural step after randomly initializing weights is to check the model performance, which is done by passing the inputs (data understudy) through the network layers (hidden layers) and then calculating the actual output straightforwardly.

7.4.2.3 Loss Function

The loss function is a function of weights and biases, which is the difference between the ground truths (target value) given to the network and the value outputted from ANN. This cost must be carried on over the whole dataset. Thus, the loss function must be summed up across all samples present in the dataset to give the total error. Generally speaking, machine learning algorithm goals lies in minimizing loss function to reach a value close to zero.

7.4.2.4 Gradient Descent

Gradient Descent (Fig. 7.8) is a mathematical concept that leads to the position of global minima or local minima of a function. It deals with the partial derivatives of the loss function concerning the model's weights to find the direction in which the cost function decreases more. Gradient Descent shows how much each weight contributes to the total error. As an example, imagine a person standing on the top of a mountain and trying to find a way down; ideally, the person will take a small step toward the direction of steepest descent. The person will iteratively hike down with a small step at a time until he reaches the bottom. Gradient descent operates in a similar way where it finds the minima of a function starting by a position (the current value of weights and biases). Then, it calculates the gradient in a negative direction. Finally, it takes a small step toward that direction to define a new model with new weights (W'), which leads to a minor error that classifies points better. By repeating this process, the minima would eventually be reached.

The gradient of the cost function is given by:

$$y \frac{\partial C}{\partial w^l_{jk}}$$

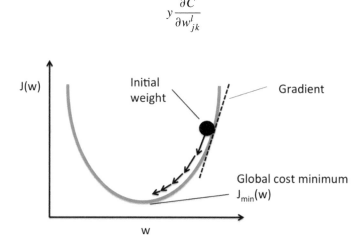

Fig. 7.8 Gradient descent process

where y represents the step size.

In machine learning, the step size is often referred to as learning rates. The learning rate should neither be too large or otherwise; the minima will be lost, nor too small or otherwise; the training process will be too slow. It should be within a range of 10^{-1} to 10^{3}.

A. Types of gradient descent

i.   Batch Gradient Descent:

All training samples are processed per iteration of computing gradient descent. However, it is computationally expensive for the large training dataset.

ii. Stochastic Gradient Descent:

This strategy processes a single training sample per iteration. Hence, the parameters get updated per iteration. The major disadvantage of this approach is when dealing with a large data size, which results in an expensive computational process when the number of iteration is large.

iii. Mini-batch Gradient Descent:

This approach is faster than the previous ones, as it processes a batch of training samples per iteration, allowing for fast convergence.

B. Gradient descent optimization algorithms

- Momentum
- Nesterov accelerated gradient
- Adagrad
- Adadelta
- RMSprop
- Adam
- AdaMax
- Nadam
- AMSGrad.

C. Gradient descent problems

One of the problems with the training process of deep architecture networks is the vanishing and exploding gradient explained by Pascanu et al. [16]. This problem arises when the gradient back propagates through a series of matrix multiplication using the chain rule from the output layer and progressing intensely toward the initial layer. This issue summarized as:

"When the gradient becomes negligible (<1), subtracting it from the weight matrix makes no changes. Hence, the model stops learning, which is known as the vanishing gradient issue. While if the gradient value becomes too large (>1), subtracting it from the weight matrix results in a high value. Hence, the model will crash, which is known as the exploding gradient issue."

Without clipping   With clipping

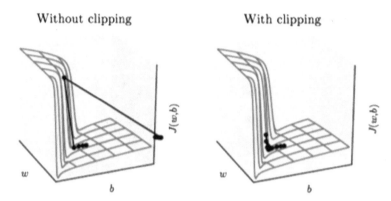

Fig. 7.9 Gradient clipping

D. Solution of gradient issue.

i. Gradient clipping for exploding gradient descent:

This concept is used to clip (rescale) the gradient value if it goes beyond a specific threshold, thus preventing the weights from reaching undesired value and keeps the network stable (Fig. 7.9).

$$g \leftarrow \eta g / \|g\|.$$

ii. Using relu instead of tanh/sigmoid activation functions

Tanh ranges from -1 to 1 and sigmoid range from 0 to 1, the derivative of these functions saturated at the extreme giving zero value as x-axis increases (Figs. 7.10, 7.11).

Fig. 7.10 Derivative of sigmoid function

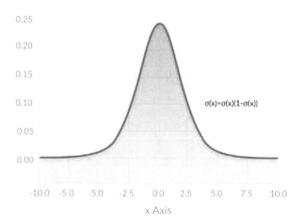

Fig. 7.11 Derivative of Tanh
function

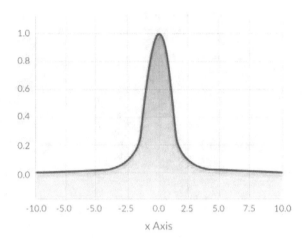

As shown from the previous graphs, as the value on the x-axis approaches the extreme of the function, its derivative becomes near zero and thus vanishes as we go backward deeper into the network. Hence, the gradient will be multiplied over and over with a value (<1). Thus, the backpropagated gradients in each previous layer will become smaller and smaller, until reaching almost zero when entering the first hidden layer. Therefore, the weights of the first layer remain constant. Relu activation (Fig. 7.12) solves this problem by giving a derivative of 1, making gradient stay the same as we move back through the network.

iii.   Weight initialization

Weight should be randomly initializing to break symmetry and should not be too large or too small to avoid vanishing/exploding gradient problems.

Fig. 7.12 Derivative of Relu
function

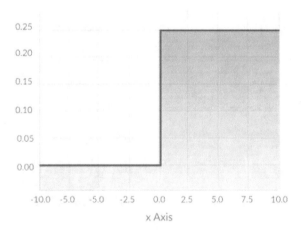

iv.   Applying regularization for exploding gradient descent

Applying L1 and L2 penalty to weights of large value to prevent them from exceeding the value of 1.

7.4.2.5   Backpropagation

Backpropagation is the process of propagating total error backward through the network by calculating the gradient descent concerning neurons input at each layer. This method is applied to optimize weights, biases, and ultimately to reduce the total error. Gradient descent exploits the chain rule mathematical concept to compute the derivative on the composition of function (loss function) over any variable in the nested equations from node to output. Therefore, the chain rule is a mathematical formula used to find the derivative of a series of nested equations.

It's written using Leibniz's notation, as shown in (Fig. 7.13).

If you have a variable A applied to a function of x, as:

$$A = f(x)$$

and another function g applied to f(x), as:

$$B = g(f(x)),$$

then derivative of B with respect to x is represented as a multiplication between the partial derivative of B with respect to A and of A with respect to x.

$$\frac{dB}{dx} = \frac{dB}{dA} \times \frac{dA}{dx}$$

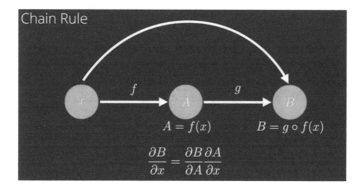

Fig. 7.13 Representation of the chain rule concept

7.4.2.6 Update Weights

New weight $=$ old weight—Derivative Rate * learning rate

7.4.2.7 Iterates Until Convergence

Repeat steps 2–6 until reaching parameters values that give low training error rates, speed up convergence, and increase the accuracy of the model.

7.5 DNN

Deep learning is a neural network with multiple hidden layers, which essentially allows the vibrant representation of features and reduces the need for manual feature engineering. This ability enables deep learning algorithms to achieve state-of-art results across various domains problems.

7.5.1 Supervised Deep Learning

7.5.1.1 Convolution Neural Network

CNN is a type of neural network where both consist of the same building blocks and learning concept: they are made up of a set of layers where one layer feeds its output to the next layer. Each layer comprises a collection of neurons with learnable weights and biases. The weights of each neuron get updated per iteration of backpropagation and gradient descent optimization techniques. These techniques used to reduce the error between the outputted model's signal and the ground truth label. Note that the process of updating model parameters is known as training. So what is the difference between CNN and ANN?

Recall the definition of neural network architecture from the previous section: each neuron per layer is fully connected to all neurons in the previous layer, where the neurons within the same layer are independent and do not share parameters. Also, to carry on a successful training process, we should make sure that the format of the input data is binarized (manual feature engineering) for ANN to recognize the prioritize patterns. This dense architecture does not work well when it comes up to extensive raw pixel data input. For example, imagine a case of colored images with resolution equals to 250×250, which will lead to $250 \times 250 \times 3 = 187,500$ weights. To learn from this vast number of inputs, one should use a multi-layered design with hundreds if not thousands of neurons. Thus, using a Multi-layered design will kill the model's performance and introduce over-fitting.

CNN architecture mimics the human visual cortex allowing it to encode information from the underlying image's content where every neuron in one layer is attached to a small region of neurons in the previous one. Thus, the number of calculated weights is minimized. This property reduces memory storage requirements and increases the algorithm's efficiency. CNN achieves state-of-art results across various image domain tasks such as classification, detection, segmentation, making it the technique of choice in vision-related applications. This model is designed to extract local spatial features from raw pixel images using a topology of learning stages, which consist of the convolution layer, nonlinear function, and subsampling layer.

This hierarchal-like architecture allows CNN to automatically learn low and mid features on the layer that are closer to the input, which in turn are combined to form abstract representation to extract high-level features at the model's deeper hidden layers. Deep CNN gains its popularity due to the availability of both big datasets and high computational power resources in addition to the enhancement in the representation capacity of CNN to deal with more complex problems as model depth increased. The significant motivations of CNN are automatic feature extraction, hierarchal learning process, weight sharing, sparse connections, and its ability to preserve the spatial relationship.

Basic component of CNN

The building blocks of CNN are made up of three types of learning stages: convolution layer for automatic feature extraction, pooling layer for subsampling, and the fully connected layer used to map the extracted feature into a final output class. The full ConvNet architecture is composed of several stacks from these three layers. This arrangement plays a critical role in achieving enhanced performance.

A. Convolution Layer

Convolution layer is the core of CNN architecture. This layer is composed of a set of convolutional kernels (each kernel represents one neuron), where each kernel represents a small spatially grid of learnable weights. The kernel, with its fixed weights, is applied at each and every small region of an image (the small image position that has the same width and height as that of the filter is known as a respective field) to detect the feature of interest anywhere in the image. That causes CNN to be highly effective in the area of image processing. This process is known as parameter sharing.

Two fundamental mathematical functions are defined in this layer to perform automatic feature extraction, which are linear (convolution function) and nonlinear (activation function) operations. Convolution function works as an operation (Fig. 7.14) of two grid of numbers where the first array of numbers called kernel is applied across the second grid, which is the respective field of the input image. The element-wise dot product is computed between the filter and the image respective field and then summed up to append the outputted value to the associated location in the feature map. Then recursively, the filter shifted, as defined by stride length, over the width and height of the image to cover it, to fill the elements of the feature map. Moreover, if different types of features are presented within the same layer, a collection

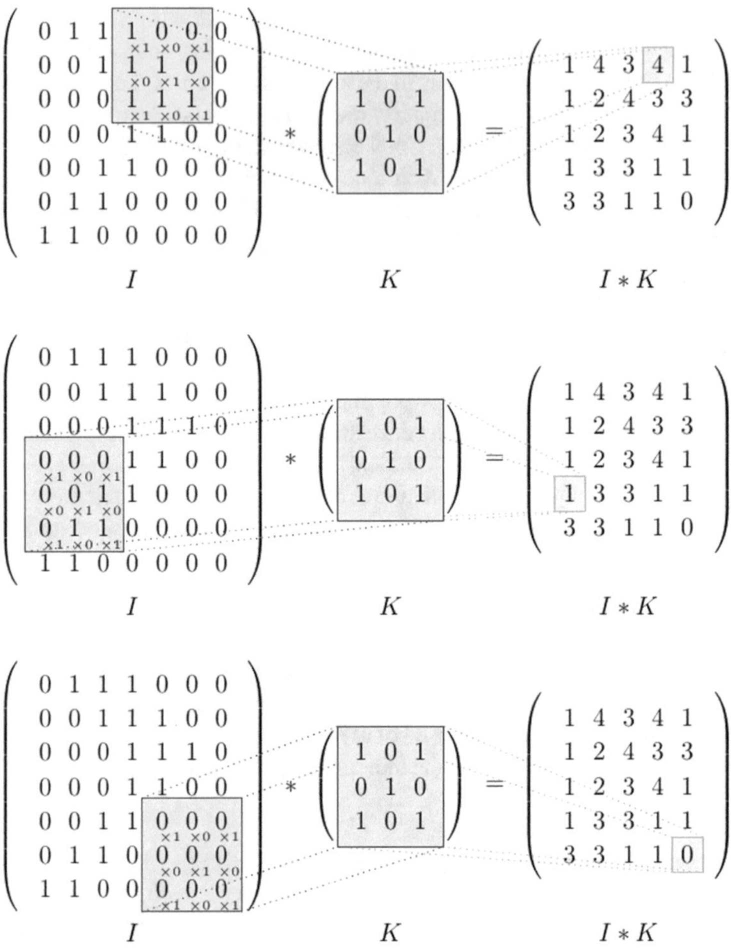

$$
\begin{pmatrix}
0 & 1 & 1 & 1 & 0 & 0 & 0 \\
0 & 0 & 1 & 1 & 1 & 0 & 0 \\
0 & 0 & 0 & 1 & 1 & 1 & 0 \\
0 & 0 & 0 & 1 & 1 & 0 & 0 \\
0 & 0 & 1 & 1 & 0 & 0 & 0 \\
0 & 1 & 1 & 0 & 0 & 0 & 0 \\
1 & 1 & 0 & 0 & 0 & 0 & 0
\end{pmatrix} \quad * \quad
\begin{pmatrix}
1 & 0 & 1 \\
0 & 1 & 0 \\
1 & 0 & 1
\end{pmatrix} \quad = \quad
\begin{pmatrix}
1 & 4 & 3 & 4 & 1 \\
1 & 2 & 4 & 3 & 3 \\
1 & 2 & 3 & 4 & 1 \\
1 & 3 & 3 & 1 & 1 \\
3 & 3 & 1 & 1 & 0
\end{pmatrix}
$$

$$\qquad\qquad I \qquad\qquad\qquad\qquad K \qquad\qquad\qquad I * K$$

$$
\begin{pmatrix}
0 & 1 & 1 & 1 & 0 & 0 & 0 \\
0 & 0 & 1 & 1 & 1 & 0 & 0 \\
0 & 0 & 0 & 1 & 1 & 1 & 0 \\
0 & 0 & 0 & 1 & 1 & 0 & 0 \\
0 & 0 & 1 & 1 & 0 & 0 & 0 \\
0 & 1 & 1 & 0 & 0 & 0 & 0 \\
1 & 1 & 0 & 0 & 0 & 0 & 0
\end{pmatrix} \quad * \quad
\begin{pmatrix}
1 & 0 & 1 \\
0 & 1 & 0 \\
1 & 0 & 1
\end{pmatrix} \quad = \quad
\begin{pmatrix}
1 & 4 & 3 & 4 & 1 \\
1 & 2 & 4 & 3 & 3 \\
1 & 2 & 3 & 4 & 1 \\
1 & 3 & 3 & 1 & 1 \\
3 & 3 & 1 & 1 & 0
\end{pmatrix}
$$

$$\qquad\qquad I \qquad\qquad\qquad\qquad K \qquad\qquad\qquad I * K$$

$$
\begin{pmatrix}
0 & 1 & 1 & 1 & 0 & 0 & 0 \\
0 & 0 & 1 & 1 & 1 & 0 & 0 \\
0 & 0 & 0 & 1 & 1 & 1 & 0 \\
0 & 0 & 0 & 1 & 1 & 0 & 0 \\
0 & 0 & 1 & 1 & 0 & 0 & 0 \\
0 & 1 & 1 & 0 & 0 & 0 & 0 \\
1 & 1 & 0 & 0 & 0 & 0 & 0
\end{pmatrix} \quad * \quad
\begin{pmatrix}
1 & 0 & 1 \\
0 & 1 & 0 \\
1 & 0 & 1
\end{pmatrix} \quad = \quad
\begin{pmatrix}
1 & 4 & 3 & 4 & 1 \\
1 & 2 & 4 & 3 & 3 \\
1 & 2 & 3 & 4 & 1 \\
1 & 3 & 3 & 1 & 1 \\
3 & 3 & 1 & 1 & 0
\end{pmatrix}
$$

$$\qquad\qquad I \qquad\qquad\qquad\qquad K \qquad\qquad\qquad I * K$$

Fig. 7.14 Convolution operation

of kernels is required to represent these different characteristics, thus resulting in several feature maps. The map represents the activated region at every spatial image position where the filter sees the feature of interest such as line, edge, and curve. This activation map becomes the input of the next layer.

The output S of the convolution operation (convolution operation represented by symbol '*') between spatial image locality I and kernel K is defined as follows.

$$S = (I * K)$$

During the convolution operation, filters may extend outside the image preventing the convolution layer from capturing information at the edge regions of the input image. Therefore, the zero-padding technique is used to give the filter more space to move. This operation works by adding a border of zero values around the image edges. Now the convolution layer gets a contribution from every region in the image.

Because images are far from linearity and because the outputted values from the convolution function are linear, non-linear activation function is added directly after the convolution operation to introduce non-linearity properties to the activation map. The popular activation function used presently in CNN is Relu.

B. Pooling Layer.

Feature maps are sensitive to the location of features in the input image. Thus, different feature maps present for any small change that occurs in the input image (Image changes such as clipping, rotation, and scaling), making the CNN model far from automatic features extraction ability. To address this issue, the experts inserted a new layer after the convolution layer, especially after the activation function. This layer is used to downsample the feature map to a smaller version (by reducing the height and width of the activation map). Downsampling preserves the essential features that come from the activation map elements and remove the useless detailed elements by applying pooling operation (filter) across the feature map. This layer is useful as any movement of features in the input images will result in the same pooling map; this property of the pooling layer is known as local translation invariance. The pooling filter size is smaller than that of the feature map. Usually, pooling operation represents a filter of 2×2 dimension with a stride of length equal to 2.

In summary, the key advantages of pooling layer are:

- Reducing computational complexity
- Extracting low-level feature
- Minimizing the variance.

Presently the most two used pooling methods (Fig. 7.15) are:

- Average pooling: takes the average of feature map elements that are covered by the pooling filter
- Max pooling: takes the most activated element of feature map elements that are covered by the pooling filter.

Fig. 7.15 Pooling operation

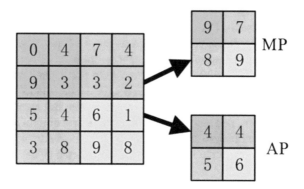

C.   Fully connected layer.

The output from the last convolution or pooling layer is flattened into a one-dimensional array and fed to one or more dense layers. So, every neuron in the fully connected layer connects learnable weights links to every neuron in the preceding layer to compute the probability score for the classification task.

D.   Dropout layer.

Overfitting is a significant problem that affects any neural network where the model's weights are tuned well on the training data but fail to generalize on the test data. Dropout is a technique that fights this issue by randomly setting some of the activation nodes in the dropout layer to zero. This algorithm takes a parameter that represents the probability of the number of nodes that should be dropped at a particular epoch.

E.   Batch normalization.

Batch normalization is a normalizing technique discovered by Ioffe and Szegedy from Google Brain to address the problem of internal covariance shift in feature maps. This issue arises from the change in data distribution within hidden layers, which leads to vanishing gradient issues. Hence, slowing down convergence and preventing neuron from learning. This algorithm solves this problem by unifying images to zero mean and unit variance. This layer is inserted in the hidden layer before the Relu activation function.

Transfer learning with CNN

As mentioned above, deep learning algorithms require a huge dataset to run effectively and to achieve high accuracy. In domains such as medical imaging analysis, it is rare to have sufficient dataset size; hence, training deep learning algorithms from scratch is useless. To come across this problem, scientists introduced a machine learning technique known as Transfer Learning. Transfer learning involves using a pre-trained model as a feature extractor or weight initializer for the new CNN model. Depending on the new dataset size and how similar this new dataset is to the original dataset, two significant types of transfer learning are introduced, which are:

A.   Original model to be used as a fixed feature extractor:

Recall the difference between machine learning and deep learning where DNN suppresses ANN by its automatic feature extraction ability. The fixed feature extractor technique holds the same idea as the ANN process, where in ANN, we manually extracted features from input data before feeding them to the model. However, in the case of transfer learning, the heavy lifting calculations of feature extraction are done by the pre-trained deep model. So, for the pre-trained model to act as a fixed feature extractor on the new model, we remove the last layer "fully connected layer" from the pre-trained model. Then, we treat the rest of its layers (that were already learned) as a fixed feature extractor for the target model by connecting the pre-trained layer to the input of the new model fully connected layer.

B.   Fine tuning the pre-trained model:

Unlike the previous strategy, the fine-tuning technique modifies some or all learned weights of the pre-trained model. This technique is used when the distribution of the new dataset is a bit different than the original dataset. Keeping in mind that the earlier layers of the deep learning model hold generic feature representation (such as edges, lines, colors, etc.), while later layers carry dataset-specific features. So, it's possible to fine-tune some or all high-level portions of the network which are specific to the classification task of interest.

The selected transfer learning type is based on the similarities between the base dataset and the new one along with the size of the target dataset. Here are some regular, dependable guidelines for exploring the four unique situations:

i.   Target dataset is smaller than and similar to the original dataset:

The fixed feature extractor type is the best fit in this case since the new dataset has the same data distribution as the original one. Because of this similarity, both the pre-trained model and the target model will have relevant high-level features representation. However, using the fine-tuning type in this scenario would be referred to as a bad practice since the model would overfit because of insufficient data size. Below are the steps used to perform transfer learning as a feature extractor on the new model.

- Remove the fully connected layer from the pre-trained model
- Add a new fully connected layer with output nodes matches the number of targeted classes
- Randomly initialize weights of the fully connected layer in the targeted model
- Freeze the weights of the pre-trained model
- Train the new fully connected layer using the target dataset.

ii.   Target dataset is large and similar to the original dataset:

Because of the efficient target dataset size, the model won't overfit if fine-tuning is used in this scenario. Following the below steps to fine-tune the targeted network.

- Remove the dense layer from the pre-trained model
- Add a new fully-connected layer with output nodes matches the number of targeted classes
- Randomly initialize the weights of the new fully connected layer
- Unfreeze the weights of the pre-trained model
- Retrain the full network from the target dataset.

iii.   Target dataset is smaller than and different from the original dataset:

Using fine-tuning alone will result in overfitting problems, and using fixed feature extractor only is not accurate since the later hidden layers will carry features from a different distribution that is specific to the targeted dataset. Instead, the network will work on a combination of both transfer learning types by using the lower layers of the pre-trained model as a feature extractor and fine-tuning the rest of the model. This schema is implemented as follow:

- Take the earlier layers from the pre-trained model
- Add a new fully-connected layer to the selected layers
- Randomly initialize the weights of the new fully connected layer
- Train the new fully connected layer from the target dataset.

iv. Target dataset is large and different from the original dataset:

Training the model from scratch is feasible because the amount of data is efficient. However, it's advantageous to initialize the weights from the pre-trained model and adjust them so that the training process becomes faster. The implementation steps to be performed in this scenario are the same as those listed in the third case.

Transfer learning is commonly used in the medical imaging community because of insufficient annotated dataset size. Initially, the medical imaging analysis domain was centered on an unsupervised pre-trained model such as SAE and RBMs. Authors in [17, 18] leveraged the transfer learning algorithm and applied a pre-trained SAE and DBN on magnetic resonance imaging (MRI) to classify patients as normal or with Alzheimer's disease (AD).

In the more recent papers, a big jump towards a pre-trained supervised model like CNN is observed. Shin et al. [19] evaluate the influence of transfer learning in a CT scan dataset to detect abdominal lymph node (LN) and interstitial lung disease (ILD). The authors fine-tuned a CNN model that was pre-trained on a large-scale labeled dataset known as ImageNet [20]. This approach achieves the state-of-art result in detecting LN with 86% sensitivity with three false-positive rates per each patient. Ravishankar et al. [21] examine the process of transferring learned weights from a pre-trained CNN that was modeled on ImageNet dataset to kidney recognition task using ultrasound images. They concluded that the increase in the degree of transfer learning positively affects CNN performance. Tajbakhsh et al. [22] compared the performance of the pre-trained CNN model against a model trained from scratch using medical imaging data. The study was carried on four different medical imaging applications (involving Polyp detection, pulmonary embolism detection, colonoscopy frame classification, and Intima-media boundary segmentation) with three different medical imaging modalities. They found that deep tuning of the CNN model outperforms prepared CNN from scratch.

Few researchers focused on examining the type of transfer learning (feature extractor and fine-tuning) that gives the best outcomes when applying it to the medical imaging dataset. Under this study, two researchers (Antony and Kim) provided conflicting results. Antony et al. [23] proposed a comparison of knee osteoarthritis (OA) severity classification results between fine-tuning a pre-trained CNN model and using this model as a feature extractor. In this approach, fine-tuning outperformed feature extraction and achieved 57.6% classification accuracy over 53.4%. In Kim et al. [24] study, the feature extraction strategy overtook fine-tuning on the cytopathology images by recording 70.5% over 69.1%. Recently, researchers [25, 26] showed the power of fine-tuning in classifying diseases from medical images.

The authors utilized GoogleNet Inception v3 CNN architecture [27] that was pretrained on ImageNet [20] dataset with 1.28 million images. Their results achieved a human expert level of performance. These state-of-art results have not yet been accomplished by just utilizing pre-trained systems as feature extractors.

7.5.1.2   Recurrent Neural Network

In reality, sequential data is all around us; for example, audio represents a sequence of sound waves, while the text is made up of a series of either characters or words, etc. Tradition feedforward model cannot handle this type of data due to a set of model's properties limitations, which are demonstrated by the following scenarios:

- Feedforward networks only accept fixed-length vector as an input; this is a problem for the task of sequence modeling in general because, in sequential modeling, the number of input words that are used to predicate the next word has to be dynamic.
- To handle the issue presented in the first point, scientists introduced a small fixed window size with a limited number of words. This solution arises another problem in this field where long-term dependency in the input data cannot be modeled due to the limited history given to this network.
- To address the problem presented in the second point, the researchers proposed another approach known as "bag of words." This approach represents the entire sequence as a set of counts. In using counts, all sequential information is wholly taken off, for example, imagine the case of two sentences with the same bag of words representation but with entirely different semantic meaning.
- The issue that emerges in the third point was resolved by extending the size of the window using the solution presented in the first point to adopt more words. In this strategy, each input of the sequence is connected to the network with a separate weight. This approach prevents the learned weights from one sequence at one point to transfer across the other sequence.

These issues showed that traditional neural networks are not well suited to handle sequential data and furthered motivate a particular set of designed criteria for sequence modeling problems. So, the newly designed model needs to be able to:

- Handle variable-length sequence
- Track long term dependency
- Maintain information order
- Share parameters across the sequence.

Deep sequential modeling or RNN (Fig. 7.16) is a neural model designed in a way that allows it to address the problems of sequential processing of data due to its capacity to generate text. The configuration layer used by the RNN model allows for information persistence. RNN topology has been utilized in various text analysis applications such as speech recognition [28], image caption generation [29], machine translation [30], and many other MLP tasks.

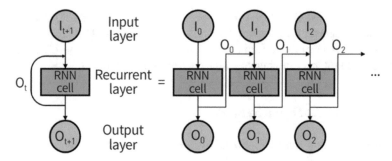

Fig. 7.16 RNN architecture

Each output from hidden layers in the RNN model is not only connected to the inputs of the next layer but also back to its input to update its internal state at time t, which we call ht. Therefore, the hidden state is maintained by applying a recurrence relation at every time step as a function of weights depending on the previous state ht-1 and the current input xt.

$$ht = \sigma(xt + Rht - 1).$$

The same parameters (weights and R) are used at each time step, which allows parameter sharing in the context of sequence modeling.

Depending on the computed internal state at each time step and on a separate weight matrix, RNN generates the predictive output yt using the following formula.

$$yt = W * ht$$

This model suffers from the gradient issues [31] (vanishing gradient or exploding gradient) because of the many weight matrix multiplications and the repeated use of the activation function derivatives that have to be computed while propagating the error back to the input through time. To this end, a few specific memory units have been built, the soonest and most well-known unit is the Long Short-Term Memory (LSTM) [32] and the Gated Recurrent Units (GRUs) [33].

Although RNN gained its popularity in the text analysis field, RNNs are progressively used in imaging analysis applications and especially in medical imaging analysis space. Shin et al. [34] trained the RNN model using deep CNN features to depict the context of an identified sickness. Stollenga et al. [35] used Multi-Dimensional Recurrent NNs to segment 3D-Brain images in a pixel-wise manner, where every pixel is doled out to a class. The authors showed that RNN outperforms CNN in image segmentation tasks since CNN only sees a small local context of the pixels while RNN perceives the whole spatiotemporal context of each pixel. Chen et al. [36] proposed a DL framework built from a fully convolutional network (FCN) and a recurrent neural network (RNN) to segment neuronal and fungal from 3D images.

7.5.2 Unsupervised Learning

7.5.2.1 Auto-Encoders (AEs) and Stacked Auto-Encoders (SAEs)

Auto-encoders [37] is a type of unsupervised neural network that consists of two main processes: encoding (or downscaling) and decoding (or upscaling) (Fig. 7.17).

During encoding, the model learns to extract a compressed features representation from the input data by limiting the reconstruction error between the input and the output values. At this stage, the number of nodes in the input layers are reduced to a smaller amount in the hidden layers.

As for decoding, the model uses the extracted representation from the hidden layers to restore the input information (this process called reconstruction). At this stage, the number of nodes in the hidden layers increased to reach the same number of input nodes at the output layer. During the decoding process, auto-encoders use a cost function that penalizes the encoder when the input and the output values are different.

The primary vital advantages of auto-encoders are:

- This encoder reduces the need for annotated data since it works in an unsupervised manner.
- This encoder reduces the input dimensionality because the number of hidden nodes is smaller than the inputs; this forces the model to keep only valuable input features that are useful for reconstruction and thus reject the noises from the information.
- This encoder generates new data that is similar to the training sample, and this is useful in domains with small amounts of data.

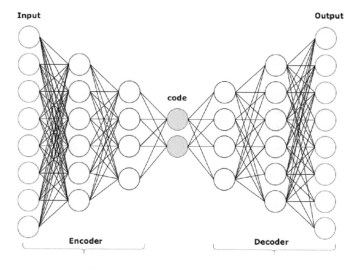

Fig. 7.17 Auto-encoder architecture

The simple architecture of AE exceptionally restricts its power in reconstructing complex input data. For this end, multiple auto-encoders are stacking together to form a robust design known as SAE, where the trained hidden layer from the first encoders is used as input for the next AE and so on [38].

SAE represents a deep learning model as the first AEs will learn low-level features, and the later AE will combine these features to form high-level representations. During the training process, simple stacking AE architecture may overfit by just replicating the input data representation to the output layers since it has various ways to represent it. For the model to learn valuable representation and thus enhance the discovery process, new optimization techniques must be employed, which are de-noising and sparse auto-encoders.

The De-noising auto-encoder [39] add noise (such as Gaussian noise, salt-and-pepper-noise, others) to the inputs which force the model to reconstruct back the noise-free inputs. Another constraint is the sparse auto-encoder [40], where a non-sparsity penalty is added to the cost function, which deactivates some neurons when a specific threshold reached.

Medical applications surveyed this designed model with supervised learning to make predictions. Kallenberg et al. [41] trained auto-encoders from unlabeled data to extract feature hierarchy. They used the learned features to train a supervised model to classify the severity of mammographic texture and segment breast density.

7.5.2.2 Restricted Boltzmann Machines (RBMs) and Deep Belief Networks (DBNs)

Restricted Boltzmann Machine [42, 43] is a bidirectional graphical model which consists of visible layers that hold input data and hidden layers that hold the features representations. RBM architecture assumes no connections within the layers themselves. However, the relationships are built between the invisible and visible nodes in a bidirectional manner, so from input vector x, features representation h obtained, and vice versa. In light of this symmetric connection's property, RBM works as AE in the context of reconstructing input samples from hidden representations. Like AE, RBM can also be stacked with one another to construct a deep architecture called DBN [44]. The top two layers form a bidirectional design while the lower layers form a directional model.

7.5.2.3 Variational Auto-Encoders

Variational auto-encoder (VAE) [45] is a generative model that learn probability distribution from input data. VAE consists of Bayesian inference encoder-decoder and two loss functions (a reconstruction loss function and the Kullback–Leibler divergence) [46]. The encoder encodes the input vector into a latent representation space while the decoder takes the latent representation to estimate the probability distribution of the data.

7.6   MRI Preprocessing

Before feeding the MRI images into the model, some preprocessing steps must be applied to the input image. These steps are listed as followed:

7.6.1   Inter-series Sorting

It happens that within the same series, we find images of different cuts that must be separated. This can be done automatically using the histogram image analysis algorithm, which allows us to detect discontinuities between histograms of images within the same series.

7.6.2   Registration

Sometimes the patient moves or their body is not at the center of the MRI scanner, which results in images that are offset from one another. At that point, the image should initially be adjusted to a typical direction.

7.6.3   Normalization

For the same patient, with the same protocol, on the same machine under the same clinical conditions, two MR shots can have different light intensities. These differences are accentuated by the change of machine model, change of contrast agent, etc. For better convergence, it is necessary to standardize images. Many MRI segmentation models have used the Nyul et al. [47] algorithm. However, to standardize the images, it is first necessary to label each pixel of the MRI. In fact, since somebody parts like muscles do not have the same luminous value as fatty tissues or bones, it is absurd to apply the same standardization to them. However, a pixel segmentation on MRI is only possible in areas where the task is easy (Ex: MRI of the brain). Therefore, this algorithm will not perform well in some areas that research might be interested in because there is virtually no segmentation solution to such areas.

7.6.4   Correction of the Bias Field

MRI images are biased in the high-frequency domain, which causes the intensity to vary across the same tissue. This phenomenon is known as bias field distortion, and it is related to several factors such as (external interference, intracellular mechanism, material defect, etc.). This issue tends to degrade the quality of the images and can harm the convergence of the neural network. The image noise is a quiet problem that prevents us from getting correct and precise results. That is why solving this problem will enhance the outcomes. Perspective and retrospective methods have been introduced to deal with this issue [48].

Prospective methods deal with intensity corruption as a systematic error of the MRI acquisition process. It emphasizes minimizing the bias field from the devices. It depends on several techniques, which are the Phantom, Multi-coil, and unique sequences.

Regarding the Phantom method, an initiative method is applied using oil or water for a phantom image and smoothing out the images using median filtering. However, this method left a 30% inhomogeneity rate, which is quite high. This rate is a significant drawback that can't correct patient-induced inhomogeneity. Therefore, another approach is delivered by scientists and physicists. This procedure depends on carefully recording the orientation and position of the phantom and the coils [49]. It mainly focuses on geometrically transforming the estimated inhomogeneity field to any acquired image using mathematical techniques and tactics. These mathematical operations are polynomials [50], curves [51], and integration based on the Biot-Savart law [52]. These operations miniature the inhomogeneity field and then fit it to the phantom image. To decrease the dependency of the inhomogeneity field on shaping the outcome on the scanner, Collewet et al. [53] proposed a mathematical model based on the T1 signal equation, which uses a phantom image, flip angle mapping, and reference objects. However, the benefits of this solution are limited, which requires specific imaging conditions and sensitivity to input parameters.

Another prospective method is Multi-coil Imaging. Multi-coil Imaging is a method that depends on taking more images via different coils. There are two types of coils: surface coils which have a high SNR but severe bias field and volume coils, which is quite the opposite. In 1988 Brey & Narayana studied and understood the characteristics of each coil. They introduced a way to combine the surface and the body coil to get a moderated SNR and bias-free image. Brey & Narayan depends on splitting the filtered surface coil image and the body coil image, and then smoothing is occurred to get the required, resulting image. However, there are two major drawbacks to this method. These drawbacks can be summarized as having long image acquisition time and difficulty in separating the body coil image entirely from the surface coil one. So now the primary aim of the scientist is how to solve or minimize these disadvantages. To solve the time issue, Lai and Fang in 2003 registered a low-resolution body image to the full resolution surface coil image. Then the spline surface modeled the bias field used for correction. Regarding the separation problem, Fan in 2003 suggests integrating one body coil with multiple surface coil images, and thus the bias can be obtained for estimation.

The final method of the prospective is unique sequences. This method mainly focuses on the hardware part to solve and ensure the design of the prospective process. It can be briefly described as a way of estimating the spatial distribution of the flip angle and then used the obtained result to measure the intensity inhomogeneity. It depends on a mathematical operation which requires the acquisition of two images where the second image is the double nominal flip angle of the first. This method has another technique that mainly relied on sensitivity encoding by multiple receiver coils. Although this method minimizes the inhomogeneity artifact, its main purpose is to speeds up the scanning process.

The retrospective method is more efficient than the perspective. It is relatively general, where it usually uses information from acquired images in which anatomical information integrates with the intensity inhomogeneity information. It relies on several approaches, such as filtering-based method, surface fitting-based methods, segmentation methods, and histogram-based methods, where each technique has its strategies.

First, the filtering Based Method is a way that recognizes the bias field as a low-frequency piece that can be extracted from the high-frequency part of acquired images using a low-pass filter. The significant advantages of this technique are that it is cheap and straightforward to apply. The filtering method has two categories derived under it. They are the Homomorphic filtering and the Homomorphic Unsharp Masking. The homomorphic filtering method is used on log-transformed image intensities, while the Homomorphic unsharp Masking is used to correct intensity inhomogeneity in MRI.

Another retrospective method is Surface fitting, where intensity and Gradient approaches lie under it. Surface fitting methods can be summarized in fitting parametric surface to a sequence of image features that hold bias field information. As mentioned above, surface fitting relied either on the intensity-based method or gradient-based method. Intensity-Based mainly depends on fitting the least-squares parametric surface that comes in the forms of thin plates splines to intensities of a set of pixels. This method can be applied using either manual selection, automated iteration, or Gaussian primary tissue model. While the gradient-based process counted on the assumption that appropriately large homogeneous spaces are equally distributed over the whole image. This approach was tested using both least-squares fitting and independent tissue segmentation in the polynomial surface. This strategy did not use all the information to find the inhomogeneity field. Other researchers introduced a similar approach, where the finite element surface model was used to calculate and minimize the difference between homogeneous area derivatives and the surface model. Another way depends on integrating the derivative in the homogeneous areas to find the bias field. However, these methods work only if homogenous regions of the image are large and typical.

In addition to that, there are the segmentation Based Methods, which are considered as one of the retrospective types. Segmentation Based method makes intensity inhomogeneity correction easy and applicable. It mainly relies on ML, MAP, Fuzzy c-means, and Nonparametric. ML, MAP is used to approximate the image probability distribution by parametric models using either maximum likelihood (ML) or maximum a posteriori probability (Map) criterion. It mainly uses finite Gaussian mixture models to integrate the inhomogeneity. This model uses the expectation-maximization (EM) algorithm [54, 55] to estimate its parameter.

S. Abou Arkoub et al.

Moreover, Fuzzy C-Mean is the second segmentation-based approach. Its principal property is the soft classification model. This property eliminates the explicitly modeling of mixed classes and thus enhances the results. Authors in [56] proposed adaptive fuzzy c-means where they multiply the class centroid by location function and thus approximates the intensity inhomogeneity. Finally, the nonparametric approach, the last segmentation method, uses nonparametric max-shift or mean-shift clustering to ensure the targeted purpose. It doesn't require any prior knowledge of the intensity probability distribution, which is quite an improvement. It mainly relies on iterative minimization of class square error.

Also, as briefly discussed above, the retrospective also relies on Histogram-Based Methods. Histogram-Based Methods focus directly on image intensity histograms and require little or no initialization with optional prior Knowledge on the intensity probability of the imaged structure. There are three approaches that Histogram-Based depends on which are: High-Frequency Maximization, Information Minimization, and Histogram Matching. High-Frequency Maximization is an automatic process that does not needs any prior knowledge and operated to nearly all MR images. Its target is to find the smooth multiplicative field [57]. While Information minimization removes intensity inhomogeneity using a constrained minimization of image information estimated by image entropy. This approach hangs on the assumption that intensity inhomogeneity corruption brings additional information to the inhomogeneity-free image. Finally, the last Histogram-Based Methods is Histogram Matching. Histogram Matching cuts the image into small sub volumes taking into consideration having stable intensity inhomogeneity. It is a finite Gaussian mixture with seven parameters initialized from the global histogram of the image. However, it is less general than information minimization methods.

7.7 Deep Learning Applications in Medical Imagining

Medical imaging is an imperative segment with an enormous number of utilizations within various clinical events. It is not only limited to clinical diagnosis but also extended to the region of arranging, culmination, and evaluation of surgical and radiotherapeutic strategies. Many imaging modalities (such as CT scan, MRI, X-ray, PET, etc.) are used by physicians to view the internal body structure, to detect abnormalities, and to provide an accurate diagnosis of the case. However, interpretation of complex data (images) by humans is constrained by its subjectivity, significant variations across translator, fatigue, and big image data amount that surpass processing power.

Furthermore, recall that radiologist's role is not linked only to analyze medical imaging, but also to other responsibilities which will be hard on the radiologist to allocate time to as the workload for the image analysis is increasing. Interestingly, a computer vision technique called deep learning alleviated the previously described problems enabling the radiologist to perform more valuable medical tasks rapidly and more accurately.

7.7.1 Classification

In medical imaging, classification refers to assigning labels to imaging data from predefined categories. Depending on the classification task, the output could be binary (e.g., distinguish normal from affected lesion) or multiple classes (e.g., identify the severity of the understudy disease, discrimination between various conditions, etc.).

7.7.1.1 Classification Using Supervised Learning

Many research papers investigated the usefulness of adopting the CNN model to classify chest X-ray into benign or malignant. Rajkumar et al. [58] introduced the activity of both the concepts of transfer learning (pre-training and fine-tuning) and data augmentation to solve the problem of small quantities of radiological image data. They adopted GoogLeNet [59] CNN of a 22-convolution layer model that was trained on 1.2 million non-radiological images to classify view orientation of chest radiographs from 1850 chest X-ray augmented to 150,000 observation. Rajpurkar et al. [60] developed the so-called chestnut, a modified pre-trained DenseNet [61] with 121 layer convolution neural networks trained on 112,120 frontal views of chest X-ray images from ChestX-ray14 dataset [62], these images are categorized into 14 thoracic diseases including pneumonia. First, this model was designed to output a binary label from X-ray images for the pneumonia detection task. The process of the model was analyzed and then compared using the test dataset of 420 frontal chest X-rays against four radiologists with 4, 7, 25, and 28 years of experience respectively, the model achieves higher F1 score (0.435) incomparable to the radiologist averaged F1 score (0.387). Then the model was extended to classify the presence/absence of 14 thoracic diseases, the performance of this model outperforms the previous state of art results done by Wang et al. [62], Yao et al. [63] with a margin more significant than 0.05 AUROC.

The deep neural network has been broadly utilized for analyzing brain images to classify Alzheimer's disease from MRI images. Hosseini-Asl et al. [64] designed a model to classify Alzheimer's disease from 3D MRI brain images; they divided their work into three steps refers as preprocessing, features extraction, and AD classification. During the preprocessing phase, they applied spatial normalization and skull stripping on MRI images. For the features extraction phase, they used 3D Convolutional Autoencoder (3D-CAE) pre-trained on the CADDementia dataset to extract local patterns from 3D MRI images. The features learned by this phase are then fed to the next stage, where the authors employed a 3D Deeply Supervised Adaptive CNN (3D-DSA-CNN) to categorizing patients into healthy brains, MCI, or AD from the ADNI database. This structure achieved a higher accuracy (99%) than another state of art predictor. Korolev et al. [65] proposed a model that can classify Alzheimer's disease (AD) from 3D MRI images with a deep focus on bypassing the caveats that were raised by previous studies on AD classification task using deep learning design. They overcame the two well-known problems of MRI classification, which are:

- Use pretraining techniques to train an in-depth network on a small neuroimaging dataset. Note that a large data size is required to train an in-depth network from scratch.
- Use end-to-end models for automatic feature extraction to avoid the delay that was introduced by handcrafted feature generation and extraction process.

The design of the deep 3D convolutional neural nets (VoxCNN and ResNet) approaches derived from plain (VGG) and residual (VoxResNet) 3D CNN architecture, respectively, were trained on Alzheimer's Disease Neuroimaging Initiative (ADNI) dataset for a task of classifying AD from MRI brain images. Although previous studies such as that done by [64] achieve higher accuracy, this end-to-end approach is the most straightforward and thus take less time to compute.

7.7.1.2 Classification Using Unsupervised Learning

Unsupervised learning algorithms are progressively used to discriminate malignant and benign organs from image-based information. Chen et al. [66] proposed a deep convolutional auto-encoder framework to assist the classification of lung nodules from unlabeled CT scan images. This study was introduced in an unsupervised manner because of the insufficient annotated dataset, where this model can effectively learn image features from small labeled data size.

Cheng et al. [67] exploited stacked denoising auto-encoder (SDAE) on two different modalities (ultrasound and CT) to avoid significant image processing errors and inaccurate feature representation that may lead to classification bias. The primary vital properties of using SDAE over traditional CADx are automatic feature extraction and noise tolerance, which allow the model to deal with noise property that comes from different imaging modalities.

Plis et al. [68] demonstrated deep belief networks (DBN) in the application of brain imaging and neurosciences to learn outstanding feature representation and discover the spatial relationship in functional and structural neuroimaging data (fMRI).

Suk et al. [6] used Deep Auto-Encoder (DAE) to extract non-linear hierarchal-like functional relationships across various brain regions to discriminate patient's fMRI images between normal or mild cognitive impairment.

7.7.2 Detection

Detection task refers to localizing the region from medical imaging that contributes to the disease, which is done by highlighting voxel positions that are involved with the disease. Radiologists manually scan the images looking for a specific anatomic signature that differentiates one organ structure from the other. However, human interpretation is time-consuming and suffers from variation, subjectivity, and fatigue.

Therefore, recent researches have revealed a deep learning-based method to come over these issues.

7.7.2.1 Detection Using Supervised Learning

CNN model has been adopted in detection applications because of its precise detection performance. Dou et al. [69] proposed an automatic novel method based on a 3D convolutional neural network to detect CMBs from magnetic resonance imaging (MRI). The 3D model was employed in this study because of its availability in capturing more high-level feature representation with minimum computational power and thus achieved high detection accuracy. Shin et al. [70] evaluated five different CNN architectures for detecting thoracoabdominal lymph node (LN) and interstitial lung disease (ILD) from CT scans imaging to examine the usefulness of transfer learning parameters from a pre-trained model to computer-aided detection problems.

7.7.2.2 Detection Using Unsupervised Learning

Few researchers perform deep unsupervised learning on computer-aided detection. Xu et al. [71] leverage Stacked Sparse Autoencoder (SSAE) to localize nuclei from histopathological images for the application of breast cancer detection. SSAE extract high-level features using only pixels intensities. The learned feature is then fed to a classifier to classify each image patch as nuclear or non-nuclear categories.

7.7.3 Segmentation

Segmentation is the process of partitioning imaging data into different organs and other substructures by drawing outlines around the objects of interest, which permits quantitative examination of clinical parameters to identify both volume and shape. From image information such as intensity and texture, segmentation task maps a set of voxels that make up the interior of targeted objects to the corresponding classes. Recent advances in deep learning frameworks have been translated to medical imaging applications, which show a massive improvement in the performance of image semantic segmentation tasks.

7.7.3.1 Segmentation Using Supervised Learning

Many studies have successfully applied CNN models on different medical modifies to segment organs of interest and use the obtained results for further application tasks. Brain segmentation is a critical step for cancer diagnosis; this step should be done with high accuracy to build a successful diagnosis process. Zhao et al. [72]

proposed a unified framework for tumor segmentation that consists of both CNN and RNN models in which CNN trained on multimodal brain MRI image patches while RNN trained on image slices and use the learned parameters from CNN. Then, both models were fine-tuning using image slices. Moeskops et al. [73] present automatic segmentation on MRI brain images by employing three CNN models, each with different convolution kernel size. This multi-scale architecture was proposed to assign different classes to each tissue where the model with small patch size capture local texture while that with larger patch size digest spatial features. Overall, the method achieved promising results, averaging from 0.82 to 0.91 on five different datasets.

Other research areas focused on prostate segmentation. Yang et al. [74] addressed the issue of losing local information in hand-crafted feature descriptors by proposing an RNN-based model. Their study is comprised of three learning stages, starting by serializing ultrasound images into dynamic sequences. Then, the multi-view fusion strategy is introduced to integrate predicated shapes from different views. Finally, sequential data is feed to the RNN model for automatic prostate segmentation. The proposed architecture outperforms several advanced methods in dealing with prostate incompleteness boundaries. Milletari et al. [75] trained end-to-end fully convolutional neural networks on 3-dimensional MRI prostate images from the PROMISE2012 dataset to predict prostate segmentation.

7.7.3.2 Segmentation Using Unsupervised Learning

Unsupervised deep learning techniques have been widely used for medical imaging segmentation applications because of their availability on working on training samples that lack annotations. Avendi et al. [76] proposed an automatic segmentation method based on stacked auto-encoders to segment the right ventricle (RV) from cardiac MRI. The model has trained on 16 cardiac MRI images from the MICCAI 2012 RV Segmentation Challenge database and achieved a dice ratio averaging to 82.5%.

Bi et al. [77] used the generative adversarial network (GAN) model as a pre-processing step to overcome the limitation of rare annotated imaging data. GAN algorithm is employed because of its inherent ability in generating realistic feature representation so that ROI features can be captured at a different level of complexity. The extracted feature is then fed to a fully convolutional network (FCNs) to segment the region of interest from underlying medical images.

Su et al. [78] applied sparse reconstruction and stacked denoising autoencoder (DAE) for cell segmentation applications.

7.7.4 Registration

Registration, known as spatial alignment, is an integration task of separate but complementary images data modalities, which is done by transforming coordinate from one image to another. The objective of the registration task is to locate the optimal transformation that best aligns objects of interest from the input images. This process provides clinicians with various information from different resources. With the fast developments of deep learning methodologies, researchers have started applying these frameworks in the field of registrations to get the best registration performance results. These advances have improved clinical event tracks such as understanding the etiology of illnesses, improving medical procedure arranging and execution, generally identifying unnoticed medical issue flag, and mapping functionalities of the brain.

7.7.4.1 Registration Using Supervised Learning

CNN-based techniques could be used as automatic registration tasks to address the limitation of traditional registration methods such as intensity variation, image noise, slow computation, small capture region, and large-scale images, etc.

To locate the pose and location of implanted organs, Miao et al. [79] leveraged 5 CNN models on synthetic X-ray images by performing three-dimensional models of a knee implant, a hand implant, and a trans-esophageal probe to 2D X-ray registration. Each CNN model was trained on a separate local region and connected in a hierarchal manner to separate the intricate enlistment into less complicated sub-issues to be adapted independently. This approach achieved a higher registration rate of 79–99% and faster registration speed (took 0.1 s) than traditional intensity-based registration methods.

Yang et al. [80] introduced quicksilver, a model built from a stacked CNN layer in an encoder-decoder manner. The author used the deformation diffeomorphic metric mapping (LDDMM) registration model to predict transformation parameters from image patches. This architecture reduces the computation time to 1500 × for 2D and to 66 × for 3D.

Cheng et al. [81] used deep learning techniques to study the similarity metric for multimodal medical images. These images consist of two different imaging modalities. The authors generate a shared feature representation by training two deep learning techniques (stacked denoising autoencoder and denoising autoencoder) on image data from the first and the second imaging modality, respectively. Then, these feature representations were used to initialize the weights of a deep classifier that was trained on the aligned image data from both the first and the second modalities to generate a similarity metric. Finally, the image registration method used the generated similarity metric to map voxels from the first image data to the second.

A similar strategy was proposed by Simonovsky et al. [82]. The only difference is that the authors used a different deep method (CNN) to estimate the similarity costs between two patches of different modalities and to optimize the transformation parameters of the registration task. The authors prepared the input data by aligning images from different modalities with one another to feed them into a CNN model, which consists of a set of volumetric convolutional layers and ReLU non-linearity.

7.8 Conclusion

Deep learning algorithm has made remarkable progress in computer vision-related tasks. However, some challenges associated with the use of these techniques in medical imaging applications still exist.

A recurring theme in analyzing medical imaging using deep learning techniques is because of the lack of public datasets in general, and the scarcity of high quality annotate images in particular. Annotating dataset is difficult, time-consuming, and requires domain experts only because of the high sensitivity needed for healthcare data. Another vital challenge is the data imbalance between the targeted classes understudy, which is an extreme case when it comes to studying rare diseases.

A workaround has been done by researchers to come over these obstacles, which are:

- Using transfer learning procedures to transfer learning weights from pre-trained model to your own DL model.
- Using data augmentation to produce more training samples by making a different variation on images (such as translation, rotation, cropping, flipping, etc....). Note that data augmentation may result in the risk of overfitting.
- Using generative models (such as VAEs and GANS) to generate synthetic medical data.
- Sharing data among various service providers. Potential issues of sharing data are the heterogeneity of image quality and the disclosure of patient's privacy and their legal rights in protecting their personal information.

Deep learning methods have achieved astonishing improvements across a wide variety of healthcare applications; however, the growth is yet to begin. Regarding the limitation of the current ML and DL techniques, future directions and researching areas have been opened by the investors. These directions are listed in the following sections.

The core of DL training process is known as hyperparameter (e.g., number of hidden neurons, dropout probability, learning rate, number of epochs), where multi-tuning operations for these parameters are required to reach the desired performance of DL model. Active research is being performed to handle hyperparameter tuning dynamically and hence find the optimal ones. Domhan et al. [83] proposed a method that can automatically detect the optimal parameters based on the drawn curve from the first few iterations. This process can lead to bad performance since it depends on a few initial data points. Shen et al. [104] solved this issue by introducing an intelligence system that finds the direction and magnitude of the desired parameters.

As mentioned previously, annotating data is challenging in DL training process. The quick proposed solutions for this problem will not achieve excellent performance for different application scenarios. Note that from a decade PACS system had been routinely utilized in most hospitals to annotate millions of images with their free-text reports, which is a time-consuming process. Thus, by using deep learning methods in text-mining tasks, these reports could be used to generate an accurate annotation for images. However, these methods are sophisticated and require human experts to be utilized; this problem leads to opening important research areas to provide smooth and efficient solutions.

Furthermore, the availability of labeled data does not mean that these data comprise free-noise. Label noise is a constraining component in training DL models, which leads to lousy classification performance. This issue is still an open challenge that should be addressed in the near future.

Another new application was reported by Nie et al. [84, 85], who proposed the idea of generating brain CT scans from MRI images using the GAN algorithm. This idea is helpful as it avoids patients from being exposed to radiation and reduces cost.

References

1. S. Wang, R.M. Summers, Machine learning and radiology. Med. Image Anal. (2012)
2. M. de Bruijne, Machine learning approaches in medical image analysis: from detection to diagnosis. Med. Image Anal. (2016)
3. Y. Bengio, *Learning Deep Architectures for AI: Foundations and Trends in Machine Learning* (Now, Boston, 2009), 127 pp
4. Y. LeCun, Y. Bengio, G. Hinton, Deep learning. Nature **521**, 436–44 (2015)
5. D.C. Ciresan, A. Giusti, L.M. Gambardella, J. Schmidhuber, Mitosis detection in breast cancer histology images using deep neural networks, in *Proceeding of Medical of Image Computing and Computer-Assisted Intervention* (2013)
6. H.I. Suk, C.Y. Wee, S.W. Lee, D. Shen, State-space model with deep learning for functional dynamics estimation in resting-state fMRI. NeuroImage **129**, 292–307 (2016)
7. W.S. McCulloch, W. Pitts, A logical calculus of the ideas immanent in nervous activity. Bull. Math. Biol. **5**(4), 115–133 (1943)
8. F. Rosenblatt, The perceptron: a probabilistic model for information storage and organization in the brain. Psychol. Rev. **65**(6), 365–386 (1958)
9. D.E. Rumelhart, G.E. Hinton, R.J. Williams, Learning representations by back-propagating errors. Nature **323**, 533–536 (1986)
10. K. Fukushima, S. Miyake, Neocognitron: A Self-organizing Neural Network Model for a Mechanism of Visual Pattern Recognition in Competition and Cooperation in Neural Nets (Springer, Berlin, Germany, 1982), pp. 267–285
11. Y. LeCun et al., Backpropagation applied to handwritten zip code recognition. Neural Comput. **1**(4), 541–551 (1989)
12. A. Krizhevsky, I. Sutskever, G.E. Hinton, ImageNet classification with deep convolutional neural networks, in *Proceedings of Advance Neural Information Processing System* (2012), pp. 1097–1105
13. P. Ramachandran, B. Zoph, Q.V. Le, Swish: a self-gated activation function (2017)
14. X. Glorot, Y. Bengio, Understanding the difficulty of training deep feedforward neural networks. Aistats **9**, 249–256 (2010)

15. K. He, X. Zhang, S. Ren, J. Sun, Delving deep into rectifiers: surpassing human-level performance on imagenet classification, in *Proceedings of the IEEE International Conference on Computer Vision* (2015), pp. 1026–1034
16. R. Pascanu, T. Mikolov, Y. Bengio, On the difficulty of training recurrent neural networks, in *Proceedings of 30th International Conference on Machine Learning* (2013), pp. 1310–1318
17. T. Brosch, R. Tam, Manifold learning of brain MRIs by deep learning, in *Medical Image Computing and Computer-Assisted Intervention*, vol. 8150. Lecture Notes on Computer Science (2013), pp. 633–640
18. H.-I. Suk, S.-W. Lee, D. Shen, Hierarchical feature representation and multimodal fusion with deep learning for AD/MCI diagnosis. NeuroImage **101**, 569–582 (2014)
19. Shin HC, Roth HR, GAO M, Lu L, Xu Z, et al., Deep convolutional neural networks for computeraided detection: CNN architectures, Dataset characteristics and transfer learning, IEEE Trans. Med. Imaging **35**, 1285–98 (2016)
20. O. Russakovsky et al., Imagenet large scale visual recognition challenge. Int. J. Comput. Vis. **115**, 211–252 (2015)
21. H. Ravishankar et al., Understanding the mechanisms of deep transfer learning for medical images, in *Proceeding of International Workshop Large-Scale Annotation Biomedical Data* and *Expert Label Synthesis* (2016), pp. 188–196
22. N. Tajbakhsh et al., Convolutional neural networks for medical image analysis: full training or fine tuning? IEEE Trans. Med. Imag. **35**(5), 1299–1312 (2016)
23. J. Antony, K. McGuinness, N.E.O Connor, K. Moran, Quantifying radio-graphic knee osteoarthritis severity using deep convolutional neural networks. arxiv: 1609.02469 (2016)
24. J. Kim, V.D. Calhoun, E. Shim, J.H. Lee, Deep neural network with weight sparsity control and pre-training extracts hierarchical features and enhances classification performance: evidence from whole-brain resting-state functional connectivity patterns of schizophrenia. Neuroimage **124**, 127–146 (2016)
25. A. Esteva, B. Kuprel, R.A. Novoa, J. Ko, S.M. Swetter, H.M. Blau, S. Thrun, Dermatologist-level classification of skin cancer with deep neural networks. Nature **542**, 115–118 (2017)
26. V. Gulshan, L. Peng, M. Coram, M.C. Stumpe, D. Wu, A. Narayanaswamy, S. Enugopalan, K. Widner, T. Madams, J. Cuadros, R. Kim, R. Raman, P.C. Nelson, J.L. Mega, D.R. Ebster, Development and validation of a deep learning algorithm for detection of diabetic retinopathy in retinal fundus photographs. JAMA **316**, 2402–2410 (2016)
27. C. Szegedy, V. Vanhoucke, S. Ioffe, J. Shlens, Z. Wojna, Rethinking the inception architecture for computer vision (2015). https://arxiv.org/abs/1512.00567
28. H. Sak, A. Senior, F. Beaufays, Long short-term memory recurrent neural network architectures for large scale acoustic modeling, in *Proceedings of Interspeech* (2014)
29. A. Karpathy, L. Fei-Fei, Deep visual-semantic alignments for generating image descriptions, in *Proceedings of Conference on Computer* Vision and *Pattern Recognition* (2015), pp. 3128–3137
30. I. Sutskever, O. Vinyals, Q. V. Le, Sequence to sequence learning with neural networks. Tech. rep. NIPS (2014)
31. Y. Bengio, P. Simard, P. Frasconi, Learning long-term dependencies with gradient descent is difficult. IEEE Trans. Neural Netw. **5**, 157–166 (1994)
32. S. Hochreiter, J. Schmidhuber, Long short-term memory. Neural Comput. **9**(8), 1735–1780 (1997)
33. K. Cho, B. Van Merrienboer, C. Gulcehre, D. Bahdanau, F. Bougares, H. Schwenk, Y. Bengio, Learning phrase representations using rnn encoder-decoder for statistical machine translation. arXiv:1406.1078 (2014)
34. H.-C. Shin, K. Roberts, L. Lu, D. Demner-Fushman, J. Yao, R.M. Summers, Learning to read chest X-rays: recurrent neural cascade model for automated image annotation, in *Proceedings of IEEE Conference on Computer Visual Pattern Recognition* (2016), pp. 2497–2506
35. M.F. Stollenga, W. Byeon, M. Liwicki, J. Schmidhuber, Parallel multi-dimensional LSTM, with application to fast biomedical volumetric image segmentation, in *Advances in Neural Information Processing Systems* (2015), pp. 2998–3006

36. J. Chen, L. Yang, Y. Zhang, M. Alber, D.Z. Chen, Combining fully convolutional and recurrent neural networks for 3D biomedical image segmentation, in Proceedings of Advance Neural Information Processing System (2016), pp. 3036–3044
37. H. Bourlard, Y. Kamp, Auto-association by multilayer perceptrons and singular value decomposition. Biol. Cybern. **59**, 291–94 (1988)
38. Y. Bengio, P. Lamblin, D. Popovici, H. Larochelle, Greedy layer-wise training of deep networks, in *Proceedings of the 19th Conference on Neural Information Processing Systems* (NIPS 2006), ed by B. Scholkopf, J.C Platt, T. Hoffmann (2007), pp. 153–60
39. P. Vincent, H. Larochelle, I. Lajoie, Y. Bengio, P.-A. Manzagol, Stacked denoising autoencoders: learning useful representations in a deep network with a local denoising criterion. J. Mach. Learn. Res. **11**(12), 3371–3408 (2010)
40. C. Poultney, S. Chopra, Y.L. Cun, et al., Efficient learning of sparse representations with an energy-based model, in *Proceedings of Advances in Neural Information Processing Systems* (2007), pp. 1137–1144
41. M. Kallenberg et al., Unsupervised deep learning applied to breast density segmentation and mammographic risk scoring. IEEE Trans. Med. Imag. **35**(5), 1322–1331 (2016)
42. G. Hinton, A practical guide to training restricted Boltzmann machines. Momentum **9**(1), 926 (2010)
43. P. Smolensky, Information processing in dynamical systems: Foundations of harmony theory, in *Parallel Distributed Processing: Explorations in the Microstructure of Cognition: Foundations*, vol. 1 (MIT Press, Cambridge, MA) (1986), pp. 194–281
44. G.E. Hinton, S. Osindero, Y.-W. Teh, A fast learning algorithm for deep belief nets. Neural Comput. **18**(7), 1527–1554 (2006)
45. D.P. Kingma, M. Welling, Auto-encoding variational Bayes (2013)
46. J. Ker, L. Wang, J. Rao, T. Lim, Deep Learning Applications in Medical Image Analysis. IEEE Access **6**, 9375–9389 (2018)
47. L. G. Nyul, J. K. Udupa, X. Zhang, New variants of a method of MRI scale standardization. IEEE Trans. Med. Imaging **19**(2), 143–150 (2000)
48. U. Vovk, F. Pernus, B. Likar, A review of methods for correction of intensity inhomogeneity in MRI, IEEE Trans. Med. Imaging **26**(3) (2007)
49. D.A. Wicks, G.J. Barker, P.S. Tofts, Correction of intensity nonuniformity in MR images of any orientation. Magn. Reson. Image. **11**, 183–196 (1993)
50. M. Tincher, C.R. Meyer, R. Gupta, D.M. Williams, Polynomial modeling and reduction of RF body coil spatial inhomogeneity in MRI. IEEE Trans. Med. Imag. **12**(2), 361–365 (1993)
51. B.R. Condon, J. Patterson, D. Wyper, A. Jenkins, D.M. Hadley, Image non-uniformity in magnetic resonance imaging: its magnitude and methods for its correction. Br. J. Radiol. **60**, 83–87 (1987)
52. E.R. McVeigh, M.J. Bronskill, R.M. Henkelman, Phase and sensitivity of receiver coils in magnetic resonance imaging. Med. Phys. **13**, 806–814 (1986)
53. G. Collewet, A. Davenel, C. Toussaint, S. Akoka, Correction of intensity nonuniformity in spin-echo T(1)-weighted images. Magn. Reson. Image. **20**, 365–373 (2002)
54. W.M. Wells, W.E.L. Grimson, R. Kikinis and F.A. Jolesz, Adaptive segmentation of MRI data. IEEE Trans. Med. Imag. **15**(8), 429–442 (1996)
55. A.H. Andersen, Z. Zhang, M.J. Avison, D.M. Gash, Automated segmentation of multispectral brain MR images. J. Neurosci. Meth. **122**, 13–23 (2002)
56. D.L. Pham, J.L. Prince, An adaptive fuzzy C-means algorithm for the image segmentation in the presence of intensity inhomogeneities. Pattern Recognit. Lett. **20**, 57–68 (1998)
57. J.G. Sled, A.P. Zijdenbos, A.C. Evans, A nonparametric method for automatic correction of intensity nonuniformity in MRI data. IEEE Trans. Med. Imag. **17**(1), 87–97 (1998)
58. A. Rajkumar, S. Lingam, A.G. Taylor, M. Blum, J. Mongan, High-throughput classification of radiographs using deep convolutional neural networks. J. Digit. Image. **30**(1), 95–101 (2017)
59. C. Szegedy et al., Going deeper with convolutions, in *Proceedings of IEEE Conference on Computer Visual Pattern Recognition* (2015), pp. 1–9

60. P. Rajpurkar et al., CheXNet: Radiologist-level pneumonia detection on chest X-rays with deep learning (2017)
61. G. Huang, Z. Liu, L. Van Der Maaten, K. Q. Weinberger, Densely connected convolutional networks (2016)
62. X. Wang, Y. Peng, L. Lu, Z. Lu, M. Bagheri, R. M. Summers, ChestX-ray8: hospital-scale chest X-ray database and benchmarks on weakly-supervised classification and localization of common thorax diseases (2017)
63. L. Yao, E. Poblenz, D. Dagunts, B. Covington, D. Bernard, K. Lyman, Learning to diagnose from scratch by exploiting dependencies among labels. arXiv preprint arXiv: 1710.10501 (2017)
64. E. Hosseini-Asl et al., Alzheimer's disease diagnostics by a 3D deeply supervised adaptable convolutional network. Front Biosci. **23**, 584–596 (2018)
65. S. Korolev, A. Safiullin, M. Belyaev, Y. Dodonova, Residual and plain convolutional neural networks for 3D brain MRI classification (2017)
66. M. Chen, X. Shi, Y. Zhang, D. Wu, M. Guizani, Deep features learning for medical image analysis with convolutional autoencoder neural network. IEEE Trans. Big Data (2017)
67. J.Z. Cheng, D. Ni, Y.H. Chou, J. Qin, C.M. Tiu, Y.C. Chang et al, Computer-aided diagnosis with deep learning architecture: applications to breast lesions in US images and pulmonary nodules in CT scans. Sci. Rep. (2016)
68. S.M. Plis, D.R. Hjelm, R. Salakhutdinov, E.A. Allen, H.J. Bockholt, J. D. Long, H.J. Johnson, J.S. Paulsen, J.A. Turner, V.D. Calhoun, Deep learning for neuroimaging: a validation study. Front Neurosci (2014)
69. Q. Dou, H. Chen, L. Yu, L. Zhao, J. Qin, D. Wang, et al, Automatic detection of cerebral microbleeds from MR images via 3D convolutional neural networks. IEEE Trans. Med. Imaging (2016)
70. H.-C. Shin et al., Deep convolutional neural networks for computer-aided detection: CNN architectures dataset characteristics and transfer learning. IEEE Trans. Med. Imag. **35**(5), 1285–1298 (2016)
71. J. Xu, L. Xiang, Q. Liu, H. Gilmore, J. Wu, J. Tang, et al, Stacked Sparse Autoencoder (SSAE) for nuclei detection on breast cancer histopathology images. IEEE Trans. Med. Imaging (2016)
72. X. Zhao, Y. Wu, G. Song, Z. Li, Y. Zhang, Y. Fan, A deep learning model integrating FCNNs and CRFs for brain tumor segmentation. Med. Image Anal. (2018)
73. P. Moeskops, M.A. Viergever, A.M. Mendrik, L.S. de Vries, M.J.N.L. Benders, I. Išgum, Automatic segmentation of MR brain images with a convolutional neural network. IEEE Trans. Med. Imag. **35**(5), 1252–1261 (2016)
74. X. Yang, L. Yu, L. Wu, Y. Wang, D. Ni, J. Qin, et al, Fine-grained recurrent neural networks for automatic prostate segmentation in ultrasound images, in Thirty-first AAAI Conference on Artificial Intelligence (AAAI-17), ed by S.S. Singh, S. Markovitch (2017 Feb 4–9, San Francisco, CA; AAAI Press, Palo Alto (CA)
75. F. Milletari, N. Navab, S. Ahmadi, V-net: fully convolutional neural networks for volumetric medical image segmentation, in *Proceedings of 4th international conference* on *3D Vision* (3DV) (2016), pp. 565–571
76. M.R. Avendi, A. Kheradvar, H. Jafarkhani, Automatic segmentation of the right ventricle from cardiac MRI using a learning-based approach. Magn. Reson. Med. **78**(6) (2017)
77. L. Bi, D. Feng, J. Kim, Dual-path adversarial learning for fully convolutional network (FCN)-based medical image segmentation. Vis. Comput. **34**(6–8) (2018)
78. H. Su, F. Xing, X. Kong, Y. Xie, S. Zhang, L. Yang, Robust Cell Detection and Segmentation in Histopathological Images Using Sparse Reconstruction and Stacked Denoising Autoencoders. Lecture Notes in Computer Science (Springer, Cham, 2018), p. 9351
79. S. Miao, Z.J. Wang, R. Liao, A CNN regression approach for real-time 2D/3D registration. IEEE Trans. Med. Imag. **35**(5), 1352–1363 (2016)
80. X. Yang, R. Kwitt, M. Styner, M. Niethammer, Quicksilver: fast predictive image registration—a deep learning approach. Neuroimage **158**, 378–396 (2017)

81. X. Cheng, L. Zhang, Y. Zheng, Deep similarity learning for multimodal medical images. Comput. Methods Biomech. Biomed. Eng. Imag. Vis. **6**(3), 248–252 (2016)
82. M. Simonovsky, B. Gutiérrez-Becker, D. Mateus, N. Navab, N. Komodakis, A deep metric for multimodal registration, in *International Conference on Medical Image Computing and Computer-Assisted Intervention* (Springer, 2016), pp. 10–18
83. T. Domhan, J.T. Springenberg, F. Hutter, Speeding up automatic hyperparameter optimization of deep neural networks by extrapolation of learning curves, in *Proceedings of the 24th International Conference on Artificial Intelligence*, ed by Q. Yang, M.J. Wooldridge (2015)
84. D. Nie et al., Medical image synthesis with context-aware generative adversarial networks, in *Proceedings of International Conference on Medical Imaging and Computer-Assisted Intervention* (2017), pp. 417–425
85. D. Nie, X. Cao, Y. Gao, L. Wang, D. Shen, Estimating CT image from MRI data using 3D fully convolutional networks, in *Proceedings of the 4th International Workshop on Large-Scale Annotation of Biomedical Data and Expert Label Synthesis* (2016), pp. 170–178

Chapter 8
Deep Learning Methods in Electroencephalography

Krzysztof Kotowski, Katarzyna Stapor, and Jeremi Ochab

Abstract The volume, variability and high level of noise in electroencephalographic (EEG) recordings of the electrical brain activity make them difficult to approach with standard machine learning techniques. Deep learning methods, especially artificial neural networks inspired by the structure of the brain itself are better suited for the domain because of their end-to-end approach. They have already shown outstanding performance in computer vision and they are increasingly popular in the EEG domain. In this chapter, the state-of-the-art architectures and approaches to classification, segmentation, and enhancement of EEG recordings are described in applications to brain-computer interfaces, medical diagnostics and emotion recognition. In the experimental part, the complete pipeline of deep learning for EEG is presented on the example of the detection of erroneous responses in the Eriksen flanker task with results showing advantages over a traditional machine learning approach. Additionally, the refined list of public EEG data sources suitable for deep learning and guidelines for future applications are given.

8.1 Introduction

The human brain contains around 86 billion neurons [1], these nerve cells are a source of the electric activity that is the biological basis of learning, memory, behavior, perception, and consciousness. Revealing how billions of electric impulses map into

K. Kotowski · K. Stapor (✉)
Institute of Informatics, Silesian University of Technology, Gliwice, Poland
e-mail: katarzyna.stapor@polsl.pl

K. Kotowski
e-mail: krzysztof.kotowski@polsl.pl

J. Ochab
Marian Smoluchowski Institute of Physics and Mark Kac Center for Complex Systems Research, Jagiellonian University, Kraków, Poland
e-mail: jeremi.ochab@uj.edu.pl

G. A. Tsihrintzis and L. C. Jain (eds.), *Machine Learning Paradigms*, Learning and Analytics in Intelligent Systems 18, https://doi.org/10.1007/978-3-030-49724-8_8

191

our mind is the ultimate goal of neuroscience. The advantages in this domain may help treat psychological and neurological disorders, boost our cognitive skills, speed up communication between people and computers and support the new ideas for technological advancements. The complexity of the human brain makes it extremely hard to decode. Thousands of tedious experiments supported by the technological development of measuring devices and computer science provided only basic principles of neuroscience. The deep artificial neural networks (inspired by the brain itself) seem to be the next significant step towards exploring massive amounts of data produced each second by our biological neural networks.

8.1.1 A Short Introduction to EEG

The electrical activity of the brain may be measured non-invasively by means of electroencephalography (EEG, from the ancient Greek word "*encephalo*" which means "within the head"). This method was introduced in 1924 by German neurologist Hans Berger [2]. The EEG devices consist of a cap with a set of electrodes (like presented in Fig. 8.1), a signal amplifier and a recorder. Their prices span from a few hundred dollars for low-cost personal devices like Emotiv EPOC + used in one of our studies [3] to tens of thousands dollars for research-grade systems like Brain-Products actiChamp (BrainProducts Inc., Gilching, Germany) used in the experiment described in this chapter. The problem with measuring the brain activity using the EEG system consisting of several electrodes placed on the scalp is like using a glass against a wall to eavesdrop on billions of conversations at a noisy cocktail party—it is impossible to precisely locate the source of the signal. The alternative non-invasive

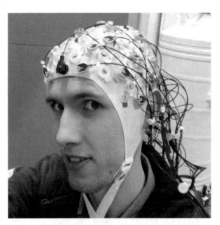

BrainProducts actiCAP   Emotiv EPOC+

Fig. 8.1 On the left: the BrainProducts actiCAP EEG cap with 64 electrodes. On the right: the Emotiv EPOC + wireless headset with 14 electrodes. *Source* flic.kr/p/diA5hh

method that gives this opportunity is the functional magnetic resonance imaging (fMRI) which infers the brain activity from blood oxygenation level observed in the high-resolution brain scans. However, the fMRI scanners are more expensive (millions of dollars), very heavy and have a low temporal resolution of around one second per scan (comparing to thousands of samples per second in EEG). Thus, EEG stays the most popular and most accessible method of brain monitoring. Currently, only the EEG-based solutions are portable and cheap enough to enable brain reading in practical applications like brain-computer interfaces (BCI). Other non-invasive methods similar to EEG is magnetoencephalography (MEG) measuring magnetic fields induced by electrical brain activity. The invasive version of EEG is called electrocorticography (ECoG) with the electrodes placed directly on the surface of the brain.

A typical EEG signal sample is a 2D matrix of real values representing potentials recorded by each electrode (channel) at consecutive timestamps (defined by the sampling frequency of the device). Physiological signals, especially EEG, introduce problems with the low signal-to-noise ratio. While the order of the typical signal is from several to tens of microvolts (μV), the noise is of the order of 1 μV and the artifacts may reach several tens of microvolts (e.g., eye blinks, muscle motions, electrode pops). The manual analysis of the EEG time series is challenging even for experts in the domain. The computerized process is necessary to extract relevant information from this data. The simplest and frequently used method is to reduce the dimensionality by computing power spectral density (PSD) of the signal (like the one presented in Fig. 8.2). This spectrum of the signal may be divided into specific frequency bands called delta (1–4 Hz), theta (4–8 Hz), alpha (8–12 Hz), beta (13–30 Hz) and gamma (30–70 Hz) brain waves. The magnitude of neural oscillations in each band was shown to be the indicator of the specific state of mind, e.g., alpha waves can be detected from the occipital lobe during relaxed wakefulness and can be increased by closing the eyes (Fig. 8.2). This analysis is usually used in studies about emotion, stress, fatigue, concentration, sleep disorders and depth of anaesthesia. Brain waves are much easier to interpret than a raw EEG signal and are used as an input to simple classifiers.

A great number of traditional machine learning and pattern recognition algorithms have been applied to the EEG data [4], most frequently Support Vector Machines (SVMs), Fisher's Linear Discriminant Analysis (LDA) and Bayesian classifiers built on top of the features calculated with Power Spectral Density, Common Spatial Patterns [5] and xDAWN [6] designed specifically for EEG. However, the recent development of deep learning and massive datasets for EEG analysis (described in Sect. 8.2.1) strongly attracts the attention of researchers.

If the 2D EEG samples are time-locked to the moments of specific events, i.e., stimulus presentation, they are called epochs and they are used in event-related potential (ERP) analysis [7] in the time domain. The EEG reflects thousands of simultaneously ongoing brain processes, so the specific component of brain response reflecting a single event or stimulus may be indistinguishable in the EEG recording of a single epoch. The experimenter must collect many epochs for each class of the stimuli and average them in order to attenuate irrelevant activity in ERP and notice

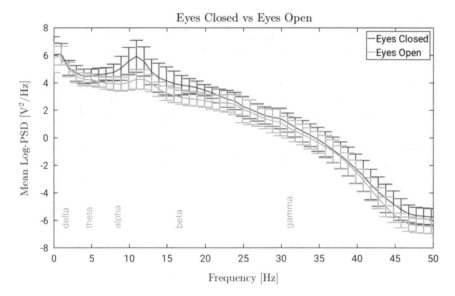

Fig. 8.2 The power spectral density (PSD) of EEG signal for two conditions: eyes closed (red) and eyes opened (green). The brain waves ranges are marked with grey vertical lines and the name of the band. *Source* sapienlabs.org/eyes-open-eyes-closed-and-variability-in-the-eeg/

the differences between waveforms for each class. However, both classic machine learning methods [8] and deep learning methods [9] are capable of classifying ERPs by single trials, without averaging, which makes them suitable for real-time BCIs. The ERP waveforms consist of underlying components (like presented in Fig. 8.3). One of the most popular ERP components in practical applications is P300 (positive potential peaking around 300 ms after the stimulus onset on the electrodes covering the parietal lobe), used in mind typing BCIs since 1988 [10], and is now heavily explored by means of deep learning [11–13]. In the experimental Sect. 8.3 of this chapter, the error-related negativity (ERN) component is described in more detail and analyzed using deep learning methods. Thus, the chapter as a whole is oriented more to ERP analysis.

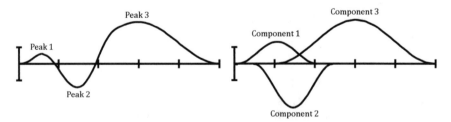

Fig. 8.3 On the left: the example of an ERP waveform with marked peaks. On the right: components underlying this ERP waveform (the example adapted from [7])

8.2   Literature Review

Deep learning methods are now extremely popular in computer vision applications and are becoming the state-of-the-art in EEG analysis. This is indicated by a number of recent review articles [4, 9, 14, 15] reporting hundreds of related articles published in the domain since 2010, and the quadratic-like manner of increase in the number of articles each year.

8.2.1   Public Datasets

A large base of publicly available datasets is a crucial element of a wide adaptation of deep learning solutions in any domain. It provides training data, states a comparison point for other researchers and supports reproducibility of the results. Deep learning shows state-of-the-art performance in the computer vision domain mainly because of high accessibility of massive datasets like ImageNet [16]. Good quality EEG data are much harder and more expensive to obtain than images, fortunately, there is a growing set of public resources. The refined list of public datasets suitable for deep learning (big enough, diverse and clearly structured) is presented in Table 8.1.

Besides, there are multiple public data banks containing smaller EEG sets of recordings that are not necessarily useful for deep learning purposes but may be useful for supplementing other datasets, including PhysioBank [34], BNCI Horizon 2020 [35], UCSD HeadIT [32], UCI Machine Learning Repository [36], BCI competitions [37], Kaggle platform [38], OpenNeuro platform [39], Open Science Framework [40], Scientific Data journal [41] and others [42, 43].

8.2.2   Preprocessing Methods

EEG data in its nature are very noisy. They are highly affected by all other physiological and environmental signals, such as heartbeats, eye blinks and movements, breathing, swallowing and other movements of the experimental subject, sweating and changing skin potentials, electrode displacements, cable movements and interference with external electrical and magnetic fields. Formerly, the EEG signals were manually inspected and filtered by human experts. Recently, the automatic identification and removal of EEG artifacts is an extensively studied topic in the literature [44].

Filtering in the frequency domain is the most common and effective preprocessing method in the research practice. Filtering of the low frequencies of up to 0.1 Hz removes the DC component and helps to reduce slow drifts of the potentials caused by varying electrode impedances. Filtering of frequencies up to 4 Hz reduces the impact of motion artifacts, heartbeat, and eye movements however, at the expense of losing delta band important, i.e. in sleep analysis. The fastest oscillations generated

Table 8.1 The refined list of public EEG datasets suitable for deep learning

Database name (Year) [Ref]	Dataset size	Dataset purpose
NEDC TUH EEG Seizure (2018, v1.5.0) [17]	642 subjects, 5610 recordings (920 h)	Epileptic seizure detection
TUH EEG Artifact Corpus (2018, v1.0.0) [17]	213 subjects, 310 recordings	Automatic EEG artifacts detection
CHB-MIT (2009) [18]	23 subjects, 664 recordings (844 h)	Epileptic seizure detection
Niuewland et al. (2018) [19]	356 subjects, 160 trials each, 9 different laboratories	Prediction of the upcoming words
DEAP (2012) [20]	32 subjects, over 40 min each	Emotion recognition
DREAMER (2018) [21]	23 subjects, around 60 min each	Emotion recognition, low-cost devices
MAHNOB-HCI (2012) [22]	27 subjects, around 30 min each	Emotion recognition
SEED (2018) [23]	15 subjects, 45 recordings (around 48 min) each	Emotion recognition
MASS (2014) [24]	200 subjects, whole night recording each	Sleep stages analysis
The CAP Sleep Database (2001) [25]	108 subjects, whole night recording each	Sleep disorders scoring
BCI2000 Motor Movement/Imagery (2004) [26]	109 subjects, 1500 one- and two-minute recordings each	Motor movements/imagery classification
Kaya et al. (2018) [27]	13 subjects, 75 recordings (60 h)	Motor imagery classification
Cho et al. (2017) [28]	52 subjects, 36 min each	Motor imagery classification
Cao et al. (2018) [29]	27 subjects, 62 90-min recordings	Driver's fatigue level
WAY-EEG-GAL (2014) [30]	12 subjects, 3936 trials each	Decoding sensation, intention and action
Babayan et al. (2019) [31]	216 subjects, 16 min each	Resting state—eyes closed and eyes opened classification
IMAGENET of The Brain (2018) [32]	1 subject, 70'060 3-s trials	Perceived image classification, low-cost devices
MNIST of Brain Digits (2015) [33]	1 subject, 1'207'293 2-s trials	Perceived image classification, low-cost devices

in the human brain are gamma waves of frequencies up to 70 Hz, so usually low-pass filtering of higher frequencies helps to remove any irrelevant high-frequency noise. Usually, the zero-phase forward-backward digital Butterworth filters are used because they do not change the phase and do not shift the signals. When experiments are not conducted in an electrically-shielded room, the notch filter of the 50 Hz (or 60 Hz) is used to reduce the influence of the electrical devices. Alternatively, the

50 Hz (or 60 Hz) harmonics may be filtered more accurately with the multi-taper regression technique for identifying significant sinusoidal artifacts [45].

Baseline correction of the ERP epochs may be an alternative or extension to low-frequencies filtering. It is a very simple operation of subtracting the mean value of samples from the pre-stimulus period (usually 100 ms or 200 ms right before the stimulus presentation) from all the samples in the epoch.

Common average referencing (CAR) is the operation of subtracting from each channel the sample-wise average of all the channels. This simple operation has been proven to minimize the effects of reference-site activity and accurately estimate the scalp topography of the measured electrical fields [46].

Eye blinks filtering may be achieved by extracting independent components (ICA) [47] of the EEG signal, detecting the proper components manually by looking for characteristic patterns in topographical maps (Fig. 8.4), and removing the components from the original signal. An automatic blinks correction requires additional electrooculography recordings (EOG) measuring the corneo-retinal standing potential that exists between the front and the back of the human eye. The blinks are easily detected in EOG and may be removed from EEG data using several methods. The simplest one is to remove signal fragments affected by blinks. The more complex one involves extracting independent components (ICA) of the EEG signal and removing components highly correlated with EOG. However, the latter method is computationally expensive and may not be suitable for real-time applications.

Automatic artifact detection and removal using deep learning is one of the future directions in EEG preprocessing. For example, the eye blinks detection mentioned in the previous paragraph may be accomplished with the deep learning model trained on a proper dataset or in the preparation part of the experiment [48]. The artifact detection using deep learning is targeted by the recently published TUH EEG Artifact Corpus [17] which includes chewing, shivering, electrode pops, muscle artifacts, and background events.

Lack of preprocessing is also an option in case of deep learning. One of the great advantages of deep learning is that it can learn from raw, unfiltered data. It may go

Fig. 8.4 Characteristic patterns of topographical maps for independent components connected to eye blinks

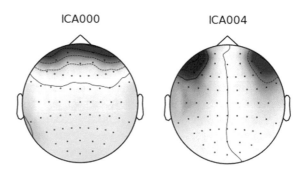

against intuition, but sometimes, it even achieves better results without any preprocessing [14, 49]. This suggests that preprocessing efforts in deep learning solutions should be limited to methods removing obvious artifacts or clearly preventing
overfitting.

8.2.3 Input Representation

In deep learning for EEG, there are 3 dominant input formulation strategies [14]:

1. using features extracted with traditional methods,
2. using a raw matrix of signal values,
3. using EEG signals transformed into the form of images.

The first approach is the most prevalent one in the literature because it allows
the use of much simpler deep learning models. The majority of papers about deep
learning in simple EEG applications use power spectral density (PSD) or statistics of
the signal (mean, variance, skewness, and kurtosis, median or standard deviation) as
the input to the network. More complex methods use wavelet transform, short-time
Fourier transform (STFT) or common spatial patterns (CSP).

However, independent feature extraction methods are useful when designing
simple neural networks but may be suboptimal while using deep models and large
datasets. The popularity of deep learning is rising because deep neural networks are
capable of dealing with high-dimensional data without any additional effort from
the researcher. They are able to extract proper features themselves in the process
of training. Although, some basic preprocessing (e.g., 4 Hz low-pass filter for eye
movements [43] or removing eye blinks using EOG [50]) may prevent overfitting in
some specific tasks.

Additionally, the high popularity of convolutional neural networks (CNNs) in
image processing has led to the idea of reusing these architectures in EEG by
converting brain signals into images. There are multiple types of these transfor-
mations:

- simple 2D signal matrix—where each pixel corresponds to the potential value on
  a different channel and a different number of the sample [48]. This representation
  is equivalent to the raw signal but is treated as an image by CNN.
- 3D signal grid—where 2D signal matrices for consecutive time frames are stacked
  into 3D volume [49].
- spectral topography maps—where electrode locations in 3D are projected onto
  a 2D plane, brain waves power (alpha, beta, gamma, etc.) is calculated for each
  electrode over the time frame of the trial, and the 2D plane is filled with linearly
  interpolated values between electrodes positions [51].
- frequency maps—where the standard visualizations of Fourier transform energy
  maps [52] or short-time Fourier transform [53] are used.

8.2.4 Data Augmentation

Data augmentation is one of the most effective methods of combating overfitting and increasing the amount of training data. It is extremely useful for highly imbalanced datasets containing rare events like seizures or erroneous responses in the Flanker task described in Sect. 8.3.1.2. In the computer vision domain, it is easy to assess the impact and suitability of different types of augmentations (rotations, flips, crops, deformations) by visual inspection of the image. EEG data is much harder to inspect, even when using image-like representation. Thus, all the augmentation techniques must be supported by the knowledge about brain functioning. EEG augmentation is a field of emerging interest with only three detailed studies on the topic [54–56] showing very promising results. Even the simplest approach of adding Gaussian noise to the EEG samples drastically increased the accuracy of emotion recognition on the SEED and MAHNOB-HCI datasets [55]. The second reported method is to generate Fourier transform surrogates for the underrepresented class of EEG signals, it increased accuracy up to 24% on the CAP Sleep Database [56]. The most complex approach is to use deep generative models, like the generative adversarial network (GAN), trained on the massive real EEG datasets in order to produce artificial samples. This approach was used in the motor imagery BCI competition improving the testing accuracy from 83% to 86%. Other papers use augmentations by adding different types of noise, overlapping time windows, sampling original recordings with different phase or swapping the hemispheres; but only a few papers report its impact on the resulting accuracy [15].

8.2.5 Architectures

Deep learning is inherently connected with artificial neural networks. In the past few years, the different types of architectures have abounded, especially in the domain of computer vision. Traditional fully-connected artificial neural networks, multi-layer perceptrons (MLPs), Restricted Boltzmann Machines (RBMs) and Deep Belief Networks (DBNs) have been replaced with more specialized models and layers. The most popular architecture in the EEG domain is, similarly as in the computer vision domain, the convolutional neural network (CNN). It is used in more than 40% of articles about deep learning for EEG and is the most frequently adopted architecture regardless of the type of experimental task [14, 15]. The design of CNN allows encoding spatial and temporal relationships among the EEG data in the form of a set of trainable convolutional filters stacked in multiple layers of the network. Only the last two or three layers constitute a traditional fully-connected (dense) classifier. Additionally, multiple supporting layers have been designed to prevent overfitting, speed up the training and decrease the number of parameters. The most popular ones are pooling (mainly max pooling, less frequently average pooling), dropout and batch

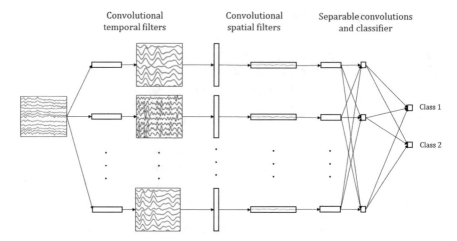

Fig. 8.5 Overall visualization of the EEGNet architecture. Lines denote the connectivity between inputs and outputs (feature maps). The network starts with a temporal convolution to learn temporal filters, then uses a depthwise convolution connected to each feature map individually, to learn spatial filters. The separable convolution is a combination of a depthwise convolution, which learns a temporal summary for each feature map individually, followed by a pointwise convolution, which learns how to optimally mix the feature maps together (the figure inspired by [57])

normalization layers. All of them have been used in state-of-the-art deep architectures dedicated to EEG classification, namely EEGNet [57] presented in Fig. 8.5, and DeepConvNet from [49]. As reported by their authors, the use of Dropout and Batch-Normalization significantly increased accuracy. Interestingly, an additional significant increase in accuracy in both models was achieved by replacing the very common (70% of the articles in EEG domain [14]) Rectified Linear Units (ReLU) activations with Exponential Linear Units (ELU).

The EEG recordings are, before all, the time series. Thus, the Recurrent Neural Networks (RNNs) designed for finding patterns in data series like speech or text have been applied to the EEG domain as well. Majority of the recent RNN architectures use Long short-term memory (LSTM) cells to effectively explore the temporal dependencies in EEG signals. Previously used Gated recurrent Units (GRUs) have been shown to be strictly worse than LSTMs [58]. The RNNs give promising results in emotion recognition and seizure detection [14] but are currently less suitable than CNNs for motor imagery [59], also no studies using RNNs for ERP analysis were reported. The increasing number of RNNs EEG papers meets the needs for more thorough research and better input formulation in this type of architecture. RNNs are frequently combined with CNNs into the hybrid architectures where CNNs are outputting features into LSTM modules. These hybrid architectures are state-of-the-art in emotion recognition [60].

The high popularity of deep autoencoders (AEs) and generative architectures like variational autoencoders (VAEs) and generative adversarial networks (GANs) starts to spread over the EEG domain. They support unsupervised learning and feature

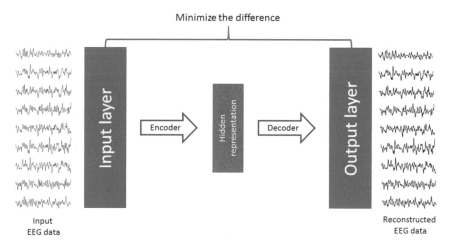

Fig. 8.6 The simplified structure of a standard autoencoder. Encoder and decoder steps may be realized with fully-connected or convolutional layers. The loss of the autoencoder is defined as the difference between the input and output of the model

extraction by learning how to reduce the dimensionality of the data without losing relevant information. The encoder part of the autoencoder compresses the input data to the hidden representation of much lower dimensionality and the decoder part tries to reconstruct the original input minimizing the loss of the model. The simplified structure of the autoencoder is presented in Fig. 8.6. This approach is extremely useful in reconstructing missing or corrupted EEG fragments [61, 62] and extracting features from the hidden representation [63]. Generative models are based on the autoencoder idea, but they introduce additional mechanisms (discriminator network in GANs and "reparametrization trick" in VAEs) and loss functions to regularize the hidden representation space. Then, this regularized representation retransformed by the decoder can be used to generate novel samples mimicking the characteristics of original data as mentioned in Sect. 8.2.4. Also, GANs may serve as the high-quality upsampling method of the EEG data [64]. The autoencoder-based architecture called U-Net [65] shows spectacular results in segmentation of images, however, to the best of our knowledge it has not been applied to EEG yet which may be a promising path to follow.

8.2.6 Features Visualization

One of the greatest disadvantages of deep learning is low interpretability of the extracted features. There are recent efforts for better understanding feature relevance attribution in the computer vision community resulting with algorithms like DeepLIFT [66, 67]. The authors of DeepConvNet and EEGNet applied it successfully to EEG data like presented in Fig. 8.7. Besides the diagnostic and explanatory

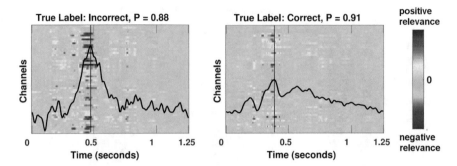

Fig. 8.7 Single-trial EEG feature relevance for a cross-subject trained EEGNet-4,2 model, using DeepLIFT, for the one test subject of the ERN dataset. On the left: correctly predicted the trial of incorrect feedback. On the right: correctly predicted trial of correct feedback. Both along with its predicted probability P. The black waveform denotes the average ERP for true label epochs with peak marked with a thin vertical line. Figures adapted from [57]

functions, this method is a powerful tool for data mining and exploration of the EEG data. Conclusions from this kind of analysis may serve as the foundations for new neuroscientific theories.

8.2.7 Applications

Deep learning methods for EEG have been proposed for a wide range of applications in medicine, psychology, automotive and entertainment. A thorough summary with references and types of networks used in each application is given in [14, 15]. Table 8.2 is only a brief listing of all the reported applications in our proposed division into five main groups: medicine, supportive BCI, entertainment BCI, automotive, and psychology.

Table 8.2 Applications of deep learning in EEG grouped into five domains

Domain	Specific applications
Medicine	Seizures detection and prediction. Monitoring the depth of anaesthesia. Diagnosis of: epilepsy, Alzheimer's disease, dementia, ischemic stroke, schizophrenia, sleep and consciousness disorders, ADHD and autism spectrum disorders
Supportive BCI	Text and mouse input. Silent speech decoding. Interface for grasping and lifting using the robotic arm. Motor imagery. Controlling an electric wheelchair. Person identification
Entertainment BCI	Gaming input devices. Monitoring mental states. Sleep scoring
Automotive	Monitoring driver's fatigue and drowsiness
Psychology	Emotion and mood recognition. ERP detection. Supporting new theories

8.3 Practical Example—Eriksen Flanker Task

This section introduces a basic pipeline of deep learning applied to EEG, from the design of the experiment, through data preparation, to the training and validation of the model. It is not meant to achieve the best possible results but to present the approach from the practical point of view and to assess the effectiveness of one of the state-of-the-art models.

8.3.1 Materials

In this particular experiment, we were interested in classifying errors and correct responses in a modified Eriksen flanker task [69]. The task consists of presenting a target binary symbol (e.g., one of a set of two letters—as in the original formulation by the Eriksens—or an arrow pointing in one of two directions [68], as done in our example), which is surrounded by additional four symbols—the so-called flankers—two on each side. The flankers are either all the same as the central target symbol (the congruent condition) or they are all different from it (incongruent). The subject is asked to respond what is the target symbol.

This cognitive task is designed to test among others inhibition (suppressing the response suggested by context) and selective attention. Thanks to inducing a significant number of errors, it can also be used to study error monitoring. Specifically, in EEG studies one can detect so-called "error-sensitive" ERP components, and in particular, the error-related negativity (ERN) [69]. The ERN is a negative deflection in EEG recording that occurs after committing an error in performance task; in the reaction-time tasks, it peaks 40–100 ms after response with maximal amplitude at fronto-central recording sites of the scalp [70, 71]. Existence of such an event-related potential strongly suggests that deep learning might be able to detect errors in human performance from EEG only.

8.3.1.1 Participants

There were 80 participants: 39 men and 41 women, 23.1 ± 3.4 y.o. during acquisition (mean and standard deviation). 4 subjects were excluded due to various reasons described later on. The eligibility criteria were: right-handedness and no use of neuropharmacological substances. All of the subjects provided their written informed consent and were financially reimbursed for their time. The study was approved by the Bioethics Committee of the Jagiellonian University, Kraków, Poland.

8.3.1.2 Procedures

The study protocol included a 6-min resting state recording (3-min with eyes open followed by 3-min with eyes closed), and the Eriksen flanker task. The resting state recordings are not used here. The flanker task consisted of presenting on-screen the four stimuli types (left and right, congruent and incongruent) 384 times, in random order and in equal proportions. A fixation cross was presented between consecutive stimuli. The stimulus duration was set to 250 ms, and the time interval between stimuli was randomly chosen from: 800, 1000, 1200, and 1400 ms. Responses after more than 800 ms were not recorded. Stimuli were presented with E-Prime 2.0 (Psychology Software Tools, Pittsburgh, PA) on a 60 Hz screen. The time course of the experiment is presented in Fig. 8.8.

The participants were asked to respond as fast and as accurately as possible; they were to press button 1 with their index finger if the on-screen central arrow pointed left, or press button 2 with their middle finger if the central arrow pointed right. The answers were registered using PST Chronos response and stimulus device. The flanker test began with a 20-stimuli training with onscreen feedback about response correctness.

The whole task took on average 11 min. 3 subjects were excluded because of too many errors or late responses. For others, response time was 408 ± 52 ms (median and median deviation, after excluding over 8% too late responses). In total, there were $83.6 \pm 9.0\%$ of correct responses and $9.3 \pm 5.3\%$ of errors. The numbers of correct, incorrect and late answers for each subject are presented in Fig. 8.9.

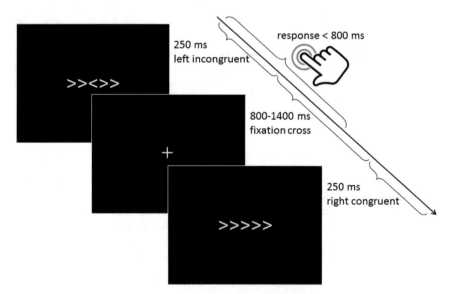

Fig. 8.8 A course of the flanker task with two different stimuli

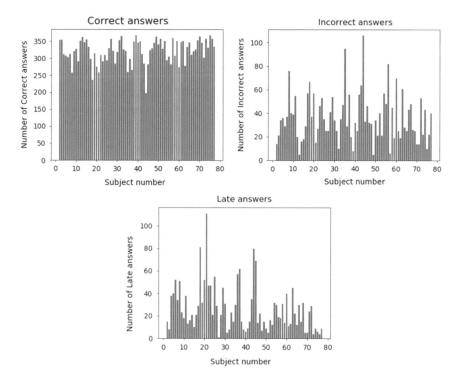

Fig. 8.9 The numbers of correct, incorrect and late responses for subjects in our Flanker experiment

8.3.1.3   EEG Registration

The data were registered with BrainVision Recorder software at 2500 Hz sampling with actiCHamp (Brainproducts GmbH, Gilching, Germany) amplifier, 64 active electrodes (with active shielding on) placed according to 10–10 system, Cz as a reference electrode, and passive EOG electrodes (a bipolar montage with one electrode below and one in the corner of the eye, allowing recording of both horizontal and vertical eye movements). The impedance of electrodes was maintained below 10 kΩ (5.9 ± 5.1 kΩ on average, with 2% of channels above 20 kΩ). One pilot recording has been removed from further analysis due to lower sampling.

8.3.1.4   Methods

Data import and filtering were done in Matlab R2016a in EEGLAB 13.5.4b [70]. All further procedures were implemented in Python 3.6.2 using the NumPy 1.16.2 package for vector operations, the MNE 0.17.1 package [72] for processing and visualization of EEG data, and Tensorflow GPU 1.12 with Keras 2.2.4 for neural networks.

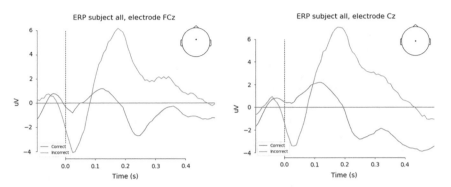

Fig. 8.10 Grand averages over all the participants for incorrect and correct answers, after all the preprocessing procedures. On the left: electrode FCz. On the right: electrode Cz

8.3.1.5 Preprocessing

All signals were initially bandpass filtered in the range from 1 to 1245 Hz using the default two-way least-squares FIR filter. Further, to avoid undesirable notch filtering, the 50 Hz AC line noise and its harmonics were reduced using the CleanLine plugin. Individual bad channels detected during manual analysis were interpolated with a spherical spline algorithm [73]. EEG was re-referenced to the common average reference (CAR). To speed up the computations and decrease the dimensionality of the training data the EEG was downsampled from 2500 to 125 Hz sampling rate. Eye blinks filtering was applied where possible by removing ICA independent components highly correlated with EOG. If the EOG filtering failed for any reason the epochs of peak-to-peak potentials over 125 μV at any electrode were discarded. It was important as the event related-negativity (ERN) is the strongest at fronto-central electrode sites [71] and frontal electrodes are highly affected by blinks. The same article [71] was an argument for limiting the subset of analysed channels to 14 most relevant ones (FCz, Fz, FC1, FC2, FC3, FC4, Cz, CPz, C1, C2, CP1, CP2, Pz, POz) in order to speed up the training and decrease the complexity of the network. In the end, the epochs were extracted using the range from 100 ms before to 500 ms after the response of the participant, and 100 ms baseline corrected. The ERP waveforms for incorrect and correct answers averaged, after the described preprocessing, over all the participants are presented in Fig. 8.10 for FCz and Cz electrodes. Before feeding the network, epochs were standardized according to the global mean and standard deviation.

8.3.1.6 Architecture and Input Formulation

The EEGNet was selected for implementation because of its superior performance on the ERP datasets [57] and easily accessible implementation on Github (github. com/vlawhern/arl-eegmodels). As recommended by the authors of the EEGNet architecture, the input to the network is represented by a simple 2D signal matrix with 14 channels as consecutive rows and 75 samples as consecutive columns. The channels were ordered exactly as listed in the previous section. The architecture was parameterized to EEGNet-8,2 version with $F_1 = 8$ temporal filters of 64 kernel length (authors found that setting it to be half the sampling rate worked well), $D = 2$ spatial filters, $F_2 = 16$ pointwise filters, $p = 0.5$ dropout rate and $N = 2$ output classes corresponding to correct or incorrect answer. The loss of the network was set to be the categorical cross-entropy. The F1-score metric was used to assess the model. The adaptive learning rate Adam optimizer was used for gradient-based optimization.

The training was performed in a cross-subject manner. 15% randomly selected correct and 15% randomly selected incorrect samples from each subject were left as the test set. Other 70% correct and 70% incorrect answers from each subject were added to the training set. The rest was merged into the validation set. The Flanker task is characterized unbalanced classes with much more correct than incorrect answers. This is a common problem of many datasets for deep learning. It was solved here by simply adding additional samples augmented with Gaussian noise (zero mean, 0.05 standard deviation) to balance training, validation and test sets. That means expected accuracy of random classifier is 50%. In the end, the training set included 35'927 epochs and validation and test sets both had 7'654 epochs.

8.3.1.7   Training and Results

The EEGNet network was trained for 50 epochs witch batch size 16. The training time was around 1 h on a mediocre PC with i5-4210U 1.7 GHz CPU, NVIDIA GeForce 840 M 2 GB GPU and 12 GB RAM memory. The training loss achieved 0.2914 with the corresponding training F1-score of 0.9333. Validation loss and F1-score for the best model were 0.3137 and 0.9230.

Exactly like in [57], the results of the network were compared with the state-of-the-art traditional machine learning approach employing a combination of xDAWN spatial filtering, Riemannian geometry, channel subset selection, and L_1 feature regularization (xDAWN + RG) which achieved the best result in Kaggle BCI competition (github.com/alexandrebarachant/bci-challenge-ner-2015) in 2015. The comparison of results for xDAWN + RG and EEGNet on the testing and whole dataset are presented in a form of confusion matrices in Fig. 8.11. The corresponding overall accuracies and F1-scores are presented below the figures. EEGNet achieved slightly better result than xDAWN + RG. It suggests that deep learning allows to easily achieve results of the best traditional machine learning methods and may be applied to classification of erroneous responses and other types of ERPs. The classification time of the whole dataset was two times lower for EEGNet (5.98 s) than for xDAWN + RG (11.92 s). The classification time of order of milliseconds per epoch allows using deep learning in real-time applications.

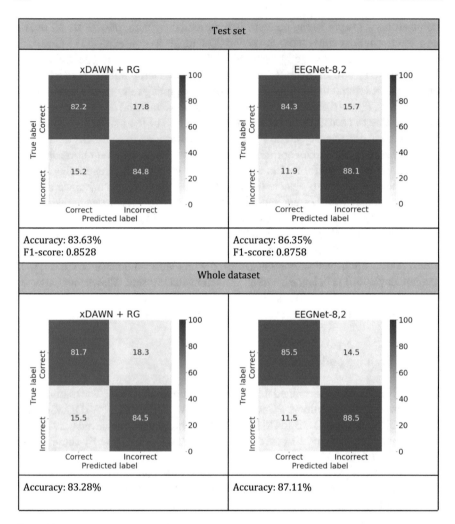

Fig. 8.11 Confusion matrices and overall scores for xDAWN + RG method (on the left) and EEGNet (on the right) on the testing set and over all samples (training, validation and test ones)

8.4 Summary

Deep learning has been shown as a state-of-the-art approach to EEG analysis wherever enough data is accessible. Multiple deep learning methods and architectures, including CNNs, RNNs, DBMs, and hybrid types allow to successfully address various EEG tasks, including medical and BCI applications, achieving comparable or better results than traditional algorithms and traditional machine learning methods. The possibilities of deep learning in EEG do not end at classification problem. Many of them were inspired or adapted from the computer vision domain. Autoencoders and generative models are used in unsupervised learning, feature extraction, data enhancement, artifacts filtering and artificial data generation. However, there is still only a little research in these areas. The popularity of deep learning lays in the straightforward usage of neural networks with ready-to-use public repositories and libraries with code and data. Growing public datasets allow training of deeper models while no need for tedious preprocessing and selecting proper feature extraction methods reduces the development and research time to minimum. One of the greatest disadvantages of deep learning—the lack of features interpretability has been addressed by feature visualization methods. They may be a useful tool for EEG exploration and supporting new neuroscience theories.

The experimental part of the chapter has shown the simplicity and effectiveness of using the deep convolutional neural network (namely EEGNet) to the classification of event-related potentials of erroneous responses. The results were slightly better than the state-of-the-art traditional machine learning method. This supports the conclusions from the thorough literature review. However, many aspects of deep learning for EEG still need much more research. Including the better input formulations (especially for RNNs), the impact of preprocessing on the results, specialized layers for EEG and adapting more novel methods from the computer vision domain. The future directions of development may be dictated by a rising popularity of low-cost wearable EEG devices characterized with lower quality of signal and exposed to many artifacts.

Acknowledgments This work was supported partially by statutory funds for young researchers of Institute of Informatics, Silesian University of Technology, Gliwice, Poland (BKM/560/RAU2/2018) and partially by National Science Center of Poland grant no. 2015/17/D/ST2/03492 (JO). We would like to thank Magda Gawłowska for designing the EEG experiment.

References

1. F.A.C. Azevedo et al., Equal numbers of neuronal and nonneuronal cells make the human brain an isometrically scaled-up primate brain. J. Comp. Neurol. **513**(5), 532–41 (2009)
2. L.F. Haas, Hans Berger (1873–1941), Richard Caton (1842–1926), and electroencephalography. J. Neurol. Neurosurg. Psychiatry **74**(1), 9–9 (2003)
3. K. Kotowski, K. Stapor, J. Leski, M. Kotas, Validation of emotiv EPOC + for extracting ERP correlates of emotional face processing. Biocybern. Biomed. Eng. **38**(4), 773–781 (2018)
4. F. Lotte et al., A review of classification algorithms for EEG-based brain–computer interfaces: a 10 year update. J. Neural Eng. **15**(3), 031005 (2018)
5. Z.J. Koles, M.S. Lazar, S.Z. Zhou, Spatial patterns underlying population differences in the background EEG. Brain Topogr. **2**(4), 275–284 (1990)
6. B. Rivet, A. Souloumiac, V. Attina, G. Gibert, xDAWN algorithm to enhance evoked potentials: application to brain-computer interface. IEEE Trans. Biomed. Eng. **56**(8), 2035–2043 (2009)
7. S.J. Luck, Ten simple rules for designing and interpreting ERP experiments, in *Event-related Potentials: A Methods Handbook* (The MIT Press, 2004), pp. 17–32
8. H. Cecotti, A.J. Ries, Best practice for single-trial detection of event-related potentials: Application to brain-computer interfaces. Int. J. Psychophysiol. **111**, 156–169 (2017)

9. F.A. Heilmeyer, R.T. Schirrmeister, L.D.J. Fiederer, M. Völker, J. Behncke, T. Ball, A large-scale evaluation framework for EEG deep learning architectures, in *2018 IEEE International Conference on Systems, Man, and Cybernetics (SMC)* (2018), pp. 1039–1045

10. L.A. Farwell, E. Donchin, Talking off the top of your head: toward a mental prosthesis utilizing event-related brain potentials. Electroencephalogr. Clin. Neurophysiol. **70**(6), 510–523 (1988)

11. R. Joshi, P. Goel, M. Sur, H.A. Murthy, Single trial P300 classification using convolutional LSTM and deep learning ensembles method, in *Intelligent Human Computer Interaction*, vol. 11278, ed. by U.S. Tiwary (Springer, Cham, 2018), pp. 3–15

12. M. Liu, W. Wu, Z. Gu, Z. Yu, F. Qi, Y. Li, Deep learning based on batch normalization for P300 signal detection. Neurocomputing **275**, 288–297 (2018)

13. H. Shan, Y. Liu, T. Stefanov, A simple convolutional neural network for accurate P300 detection and character spelling in brain computer interface, in *Proceedings of the Twenty-Seventh International Joint Conference on Artificial Intelligence* (Stockholm, Sweden, 2018), pp. 1604–1610

14. A. Craik, Y. He, J.L. Contreras-Vidal, Deep learning for electroencephalogram (EEG) classification tasks: a review. J. Neural Eng. **16**(3), 031001 (2019)

15. Y. Roy, H. Banville, I. Albuquerque, A. Gramfort, T. H. Falk, J. Faubert, Deep learning-based electroencephalography analysis: a systematic review. J. Neural Eng. (2019)

16. O. Russakovsky et al., ImageNet large scale visual recognition challenge. Int. J. Comput. Vis. (IJCV) **115**(3), 211–252 (2015)

17. V. Shah et al., The temple university hospital seizure detection corpus. Front Neuroinf. **12**, 83 (2018)

18. A.H. Shoeb, Application of machine learning to epileptic seizure onset detection and treatment. Thesis, Massachusetts Institute of Technology (2009)

19. M.S. Nieuwland et al., Large-scale replication study reveals a limit on probabilistic prediction in language comprehension. eLife **7**, e33468 (2018)

20. S. Koelstra et al., DEAP: a database for emotion analysis; using physiological signals. IEEE Trans. Affect. Comput. **3**(1), 18–31 (2012)

21. S. Katsigiannis, N. Ramzan, DREAMER: a database for emotion recognition through EEG and ECG signals from wireless low-cost off-the-shelf devices. IEEE J. Biomed Health Inf. **22**(1), 98–107 (2018)

22. M. Soleymani, J. Lichtenauer, T. Pun, M. Pantic, A multimodal database for affect recognition and implicit tagging. IEEE Trans. Affect. Comput. **3**(1), 42–55 (2012)

23. W. Zheng, W. Liu, Y. Lu, B. Lu, A. Cichocki, EmotionMeter: a multimodal framework for recognizing human emotions. IEEE Trans. Cybern. 1–13 (2018)

24. C. O'Reilly, N. Gosselin, J. Carrier, T. Nielsen, Montreal archive of sleep studies: an open-access resource for instrument benchmarking and exploratory research. J. Sleep Res. **23**(6), 628–635 (2014)

25. M.G. Terzano et al., Atlas, rules, and recording techniques for the scoring of cyclic alternating pattern (CAP) in human sleep. Sleep Med. **2**(6), 537–553 (2001)

26. G. Schalk, D.J. McFarland, T. Hinterberger, N. Birbaumer, J.R. Wolpaw, BCI2000: a general-purpose brain-computer interface (BCI) system. IEEE Trans. Biomed. Eng. **51**(6), 1034–1043 (2004)

27. M. Kaya, M.K. Binli, E. Ozbay, H. Yanar, Y. Mishchenko, A large electroencephalographic motor imagery dataset for electroencephalographic brain computer interfaces. Sci. Data **5**, 180211 (2018)

28. H. Cho, M. Ahn, S. Ahn, M. Kwon, S.C. Jun, EEG datasets for motor imagery brain–computer interface. GigaScience **6**(7), gix034 (2017)

29. Z. Cao, C.-H. Chuang, J.-K. King, C.-T. Lin, Multi-channel EEG recordings during a sustained-attention driving task. Sci. Data **6**(1), 19 (2019)

30. M.D. Luciw, E. Jarocka, B.B. Edin, Multi-channel EEG recordings during 3,936 grasp and lift trials with varying weight and friction. Sci. Data **1**, 140047 (2014)

31. A. Babayan et al., A mind-brain-body dataset of MRI, EEG, cognition, emotion, and peripheral physiology in young and old adults. Sci. Data **6**, 180308 (2019)

32. D. Vivancos, IMAGENET of the brain (2018). [Online] http://www.mindbigdata.com/opendb/imagenet.html. Accessed 08 July 2019
33. D. Vivancos, The MNIST of brain digits (2015). [Online] http://mindbigdata.com/opendb/index.html. Accessed 08 July 2019
34. Ary L. Goldberger et al., PhysioBank, PhysioToolkit, and PhysioNet. Circulation **101**(23), e215–e220 (2000)
35. BNCI Horizon 2020. [Online] http://bnci-horizon-2020.eu/database/data-sets. Accessed 08 July 2019
36. UCI Machine Learning Repository. [Online] https://archive.ics.uci.edu/ml/index.php. Accessed 08 July 2019
37. BCI Competitions. [Online] http://www.bbci.de/competition/. Accessed 08 July 2019
38. Kaggle EEG competitions. [Online] https://www.kaggle.com/datasets?search=eeg. Accessed 08 July 2019
39. OpenNeuro public datasets. [Online] https://openneuro.org/public/datasets. Accessed 08 July 2019
40. Open Science Framework. [Online] https://osf.io/search/?q=eeg. Accessed 08 July 2019
41. Scientific Data Journal EEG datasets. [Online] https://www.nature.com/search?q=eeg&journal=sdata. Accessed 08 July 2019
42. EEG/ ERP data available for free public download. [Online] https://sccn.ucsd.edu/~arno/fam2data/publicly_available_EEG_data.html. Accessed 08 July 2019
43. brainsignals.de. [Online] http://www.brainsignals.de/. Accessed 08 July 2019
44. M.K. Islam, A. Rastegarnia, Z. Yang, Methods for artifact detection and removal from scalp EEG: a review. Neurophys. Clinique/Clin. Neurophys. **46**(4), 287–305 (2016)
45. P. Mitra, H. Bokil, *Observed Brain Dynamics* (Oxford University Press, New York, 2007)
46. J. Dien, Issues in the application of the average reference: review, critiques, and recommendations. Behav. Res. Methods Instrum. Comput. **30**(1), 34–43 (1998)
47. S. Makeig, A.J. Bell, T.-P. Jung, T.J. Sejnowski, Independent component analysis of electroencephalographic data, in *Proceedings of the 8th International Conference on Neural Information Processing Systems* (Denver, Colorado, 1995), pp. 145–151
48. B. Yang, K. Duan, C. Fan, C. Hu, J. Wang, Automatic ocular artifacts removal in EEG using deep learning. Biomed. Signal Process. Control **43**, 148–158 (2018)
49. R.T. Schirrmeister et al., Deep learning with convolutional neural networks for EEG decoding and visualization: convolutional neural networks in EEG analysis. Hum. Brain Mapp. **38**(11), 5391–5420 (2017)
50. R. Manor, A.B. Geva, Convolutional neural network for multi-category rapid serial visual presentation BCI. Front. Comput. Neurosci. **9**, 146 (2015)
51. P. Bashivan, I. Rish, M. Yeasin, N. Codella, Learning representations from EEG with deep recurrent-convolutional neural networks. arXiv:1511.06448 [cs] (2015)
52. W. Abbas, N.A. Khan, DeepMI: deep learning for multiclass motor imagery classification, in *2018 40th Annual International Conference of the IEEE Engineering in Medicine and Biology Society (EMBC)* (2018), pp. 219–222
53. Y.R. Tabar, U. Halici, A novel deep learning approach for classification of EEG motor imagery signals. J. Neural Eng. **14**(1), 016003 (2016)
54. Q. Zhang, Y. Liu, Improving brain computer interface performance by data augmentation with conditional deep convolutional generative adversarial networks. CoRR **abs/1806.07108** (2018)
55. F. Wang, S. Zhong, J. Peng, J. Jiang, Y. Liu, Data augmentation for EEG-based emotion recognition with deep convolutional neural networks, in *MultiMedia Modeling* (2018), pp. 82–93
56. J.T. Schwabedal, J.C. Snyder, A. Cakmak, S. Nemati, G.D. Clifford, Addressing class imbalance in classification problems of noisy signals by using fourier transform surrogates. arXiv preprint arXiv:1806.08675 (2018)
57. V.J. Lawhern, A.J. Solon, N.R. Waytowich, S.M. Gordon, C.P. Hung, B.J. Lance, EEGNet: a compact convolutional neural network for EEG-based brain–computer interfaces. J. Neural Eng. **15**(5), 056013 (2018)

58. D. Britz, A. Goldie, M.-T. Luong, Q. Le, Massive exploration of neural machine translation architectures, in *Proceedings of the 2017 Conference on Empirical Methods in Natural Language Processing* (Copenhagen, Denmark, 2017), pp. 1442–1451

59. Z. Wang, L. Cao, Z. Zhang, X. Gong, Y. Sun, H. Wang, Short time Fourier transformation and deep neural networks for motor imagery brain computer interface recognition. Concurr. Comput. Pract. Exp. **30**(23), e4413 (2018)

60. B.H. Kim, S. Jo, Deep physiological affect network for the recognition of human emotions. IEEE Trans. Affect. Comput. 1–1 (2018)

61. J. Li, Z. Struzik, L. Zhang, A. Cichocki, Feature learning from incomplete EEG with denoising autoencoder. Neurocomputing **165**, 23–31 (2015)

62. Y. Jia, C. Zhou, M. Motani, Spatio-temporal autoencoder for feature learning in patient data with missing observations, in *2017 IEEE International Conference on Bioinformatics and Biomedicine (BIBM)* (Kansas City, MO, 2017), pp. 886–890

63. M. Dai, D. Zheng, R. Na, S. Wang, S. Zhang, EEG classification of motor imagery using a novel deep learning framework. Sensors **19**(3), 551 (2019)

64. I.A. Corley, Y. Huang, Deep EEG super-resolution: upsampling EEG spatial resolution with generative adversarial networks, in *2018 IEEE EMBS International Conference on Biomedical Health Informatics (BHI)* (2018), pp. 100–103

65. O. Ronneberger, P. Fischer, T. Brox, U-Net: convolutional networks for biomedical image segmentation, in *Medical Image Computing and Computer-Assisted Intervention—MICCAI 2015* (2015), pp. 234–241

66. A. Shrikumar, P. Greenside, A. Kundaje, Learning important features through propagating activation differences, in *Proceedings of the 34th International Conference on Machine Learning* **70**, 3145–3153 (2017)

67. M. Ancona, E. Ceolini, C. Öztireli, M. Gross, Towards better understanding of gradient-based attribution methods for deep neural networks. arXiv preprint arXiv:1711.06104 (2017)

68. K.R. Ridderinkhof, G.P. Band, G.D. Logan, A study of adaptive behavior: Effects of age and irrelevant information on the ability to inhibit one's actions. Acta Physiol. (Oxf) **101**(2–3), 315–337 (1999)

69. S.J. Luck, *An Introduction to the Event-Related Potential Technique* (MIT Press, 2014)

70. M. Falkenstein, J. Hohnsbein, J. Hoormann, L. Blanke, Effects of crossmodal divided attention on late ERP components. II. Error processing in choice reaction tasks. Electroencephalogr. Clin. Neurophysiol. **78**(6), 447–455 (1991)

71. S. Hoffmann, M. Falkenstein, Predictive information processing in the brain: errors and response monitoring. Int. J. Psychophys. **83**(2), 208–212 (2012)

72. A. Gramfort et al., MEG and EEG data analysis with MNE-Python. Front. Neurosci. **7**, 267 (2013)

73. R. Srinivasan, P.L. Nunez, D.M. Tucker, R.B. Silberstein, P.J. Cadusch, Spatial sampling and filtering of EEG with spline laplacians to estimate cortical potentials. Brain Topogr. **8**(4), 355–366 (1996)

Part IV
Deep Learning in Systems Control

Chapter 9
The Implementation and the Design of a Hybriddigital PI Control Strategy Based on MISO Adaptive Neural Network Fuzzy Inference System Models–A MIMO Centrifugal Chiller Case Study

Roxana-Elena Tudoroiu, Mohammed Zaheeruddin, Sorin Mihai Radu, Dumitru Dan Burdescu, and Nicolae Tudoroiu

Abstract This paper continues the presentation of new contributions to improve significantly the results disseminated in the recent paper presented for FedCSIS' 2018 International Conference on single-input single-output autoregressive moving average with exogenous input models in terms of accuracy, simplicity and real time fast implementation. The improvement is performed by developing accurate multi-inputs single-output adaptive neuro-fuzzy inference system models, suitable to be tailored also for modelling the nonlinear dynamics of any process or control systems. Compared to the conference paper the novelty is a further investigation on design and implementation of hybrid closed-loop control strategy structures consisting of the proposed improved models in combination with digital PI controllers that enhance significantly the tracking accuracy performance, overshoot and number of oscillations, transient convergence speed and robustness to consecutive step changes in

R.-E. Tudoroiu
Mathematics and Informatics, University of Petrosani, Petrosani, Romania
e-mail: tudelena@mail.com

M. Zaheeruddin
Building Civil and Environmental Engineering, Concordia University, Montreal, Canada
e-mail: zaheer@encs.concordia.ca

S. M. Radu
Mechanical and Electrical Engineering, University of Petrosani, Petrosani, Romania
e-mail: sorin_mihai_radu@yahoo.com

D. D. Burdescu
Software Engineering Department, University of Craiova, Craiova, Romania
e-mail: dburdescu@yahoo.com

N. Tudoroiu (✉)
Engineering Technologies Department, John Abbott College, Sainte-Anne-de-Bellevue, Canada
e-mail: ntudoroiu@gmail.com

G. A. Tsihrintzis and L. C. Jain (eds.), *Machine Learning Paradigms*, Learning and Analytics in Intelligent Systems 18, https://doi.org/10.1007/978-3-030-49724-8_9

215

the input setpoints. The effectiveness of this new control approach is demonstrated by extensive simulations conducted in MATLAB software environment followed by a rigorous performance analysis by comparison to MATLAB simulation results obtained for a same case study by a standard two input and a single output digital PI controller.

Keywords Centrifugal chiller · System identification · ARMAX model · Neural models · ANFIS models · PI controller

9.1   Introduction

This paper continues further investigations to improve the results disseminated in the most recent international conference paper, FedCSIS' 2018, September 9–12, from Poznan, Poland. The novelty of the new modelling approach is a significant improvement of the single-input single-output (SISO) adaptive neural fuzzy inter-ference system (ANFIS) models developed in the conference paper applied for the same case study reconsidered in the new research paper, more precisely a centrifugal chiller, one of the most used in a variety of control systems applications from Heating Ventilation and Air Conditioning (HVAC) field [1].

Basically, the main contribution of this paper is a significant enhancement of the linear discrete-time Autoregressive Moving Average with exogenous input (ARMAX) models introduced in the mentioned conference paper, generated by using the well-known least squares estimation (LSE) method [1, 2].

Their improved versions, i.e. the ANFIS models developed in the previous conference paper for two SISO control loops, corresponding to an appropriate multi input multi-output (MIMO) centrifugal chiller plant decomposition, are extended in the new approach to multi-input single-output (MISO) control loops models. The MISO models are preferred due to their considerable simplicity and rapid real time implementation. The new proposed closed-loop control structures are built as hybrid combination of two core models, one of them is an artificial neural network (ANN) and the second one is a fuzzy logic (FL) system model as those well documented in [3–5]. The ANN system models perform several computations with a high processing capacity and adaptation on parallel, distributed or local information. Thus, the ANN system model is designed to mimic the human brain systems in building architectural structures, learning and operating techniques.

Nowadays, we are truly fascinated by the promising future of applying Artificial Intelligence (AI) technology almost everywhere by widespread implementation of practical speech, machines learning and translations, autonomous vehicles and robotics in households and industry. At the same time, as one of the AI's main areas, ANNs have also made significant advances in architectural design, learning and operating techniques, and can thus solve problems that are hard to solve or difficult to calculate traditional [3]. Furthermore, the ANNs have been widely accepted by scientists due to their accuracy and ability to develop complex nonlinear models and to solve a wide variety of tasks.

Some preliminary results obtained and related to ANN in our academic research activity over the years can be found by the reader in [4].

The FL modelling technique as a powerful tool for the formulation of expert knowledge is a combination of imprecise information from different sources. As is mentioned in [1] the FL is a suitable modelling approach for a "*complicated system without knowledge of its mathematical description*".

An interesting FL modelling approach we have developed in the research papers [5, 6] to improve the performance of a hybrid fuzzy sliding mode observer estimator that controls the speed of a dc servomotor in closed-loop.

In the present research paper, the proposed MISO hybrid closed-loop control structure is an integrated structure of a Proportional Integral (PI) digital controller and a MISO ANFIS model of a centrifugal chiller system. The centrifugal chiller is a multi-input, multi-output (MIMO) plant of a great complexity and high nonlinearity dynamics, as is developed in detail in the research paper [2]. As is stated in [1, 2], the centrifugal chillers are the most widely used devices since "*they have high capacity, reliability, and require low maintenance*". An extensive literature review made in [2] has shown a significant amount of work done in a classic way on transient and steady state modelling of centrifugal chillers, as in [7–16]. Additionally, it can be enlarged with the most recent investigations on a combined hybrid control ANFIS system model for single-input, single-output (SISO) or for MIMO models, integrated in different closed-loop control structures with centrifugal chillers plants.

An interesting integrated ANFIS modelling methodology is developed in [17] where the learning capability of ANN is integrated to the knowledge aspect of fuzzy inference system (FIS) to "*offer enhanced prediction capabilities rather than using a single methodology independently*". Similar, in [18] a combined neuro-fuzzy modelling technique is integrated to develop a fault detection, diagnosis and isolation (FDDI) strategy for a centrifugal chiller control system. Concluding, there is a great opportunity for us to use the research preliminary results obtained in this field, such that to explore, develop and implement in real-time the most suitable and intelligent modelling and control strategies applicable on centrifugal chillers control systems. The remainder of the paper is structured as follows. In Sect. 9.2 closed-loop simulations results for nonlinear chiller plant are presented. In Sect. 9.3 two MISO ARMAX and ANFIS models are generated and are shown the model validation simulations results compare to the input-output measurements data collected by an extensive number of MATLAB simulations in closed-loop for the nonlinear centrifugal chiller plant. In Sect. 9.4 is developed an improved PI ANFIS closed-loop control strategy of the MIMO centrifugal chiller plant control system as an intelligent combined hybrid structure of ANFIS model generated in Sect. 9.3 and a standard digital PID controller. The implementation of simulation results in real-time on the MATLAB R2018b software platform of both digital PI ANFIS MISO closed-loop control strategies of the centrifugal chiller plant are shown in the same section with a rigorous performance analysis by comparison to the first two PI closed-loop control strategies introduced at the beginning of Sect. 9.4. Finally, the Sect. 9.5 concludes the relevant contributions of this research paper.

9.2 Centrifugal Chiller System Decomposition—Closed-Loop Simulations

In a systemic description the centrifugal chiller plant is considered as an open loop-oriented object with a dynamic of high complexity. To generate accurate linear discrete time polynomials ARMAX models, an appropriate setup in open or closed loop of centrifugal chiller plant that collects an input-output measurements dataset is required. In this research paper is preferred a standard centralized MIMO PI closed-loop nonlinear centrifugal chiller control system setup represented by an interconnection of two main decoupled SISO standard PI closed loops control subsystems. As is shown in [1, 2] the first control subsystem is a chilled water temperature standard PI control loop inside an evaporator, and the second one is a refrigerant liquid level standard PI control loop inside a condenser. Details on full nonlinear centrifugal chiller dynamic model under investigation and some preliminary results on ARMAX modeling approach can be found in Annex 1 of our research work [2], pp. 299–305. The MIMO closed-loop PI centralized centrifugal chiller SIMULINK dynamic model implemented on MATLAB R2018b software and useful blocks details are shown in Figs. 9.1, 9.2 and 9.3, similar those developed in [2].

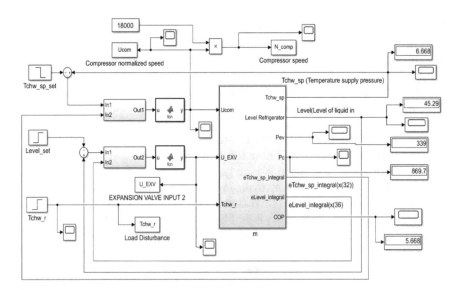

Fig. 9.1 SIMULINK model of centralized centrifugal chiller in closed-loop (see [2])

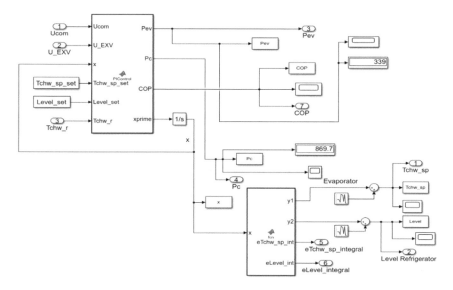

Fig. 9.2 SIMULINK models of the PI control blocks (MATLAB Function) and Evaporator and Condenser subsystems outputs block (MATLAB Function) (see [2])

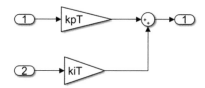

Fig. 9.3 SIMULINK model of the PI control blocks (see [2]). Legend: kpT is the proportional component of Temperature PI control law, and kiT is the integral component of the Temperature PI control law. For level control PI law, the index subscripts are replaced by kpL, kiL

Based on this SIMULINK model through extensive closed-loop simulations is collected the most appropriate input-output measurements dataset required to build two linear MISO polynomials ARMAX models for the both fully decoupled MISO control loops attached to the nonlinear centrifugal chiller plant. The benefit of this decomposition is to build two linear accurate ARMAX models, of low order and of a great implementation simplicity.

These models are capable to capture entire dynamics of the overall nonlinear MIMO chiller plant under various operating conditions, effectively to give more flexibility for designing suitable closed-loop control strategies.

In Figs. 9.4 and 9.5 are shown the two-measurement input-output datasets for both closed-loops control subsystems, Evaporator and Condenser. In the both figures the inputs and the outputs are denoted by:

$u_1 = $ Ucom, the relative speed of the compressor

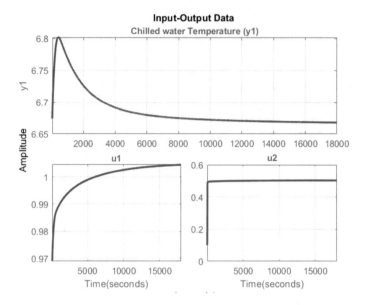

Fig. 9.4 The MISO Evaporator closed loop input—output measurements dataset

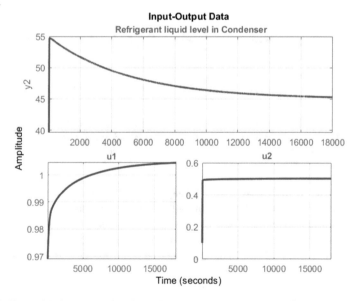

Fig. 9.5 The MISO Condenser closed loop input—output measurements dataset

u_2 = u_EXV, the opening of the expansion valve

y_1 = Tchw, the output chilled water temperature in Evaporator

y_2 = L, the output refrigerant liquid level in Condenser.

9.3 MISO ARMAX and ANFIS Models of MIMO Centrifugal Chiller Plant

In this section we build the both MISO (two-input and single output) models of the open-loop Evaporator and Condenser control subsystems based on the input-output measurements dataset of the centralized MIMO chiller plant closed-loop control system collected in the previous Sect. 9.2.

9.3.1 MISO ARMAX and ANFIS Evaporator Subsystem Models

Basically, *armax* is a useful function from MATLAB's System Identification Toolbox to estimate the parameters of the linear discrete time polynomials ARMAX models, based on the well-known least square errors (LSE) identification method, introduced also in [1, 2].

In addition, the System Identification Toolbox provides the option to simulate data obtained from a physical process and generates *IDDATA* objects into an appropriate MATLAB format that packs the input-output measurements dataset.

Also, the *armax* MATLAB function is based on a prediction error method and requires the polynomial ARMAX model orders. Additionally, a pure transport delay of the signal flow in the feedback path from measurement sensors to each closed-loop controller is specified as a new argument of *armax* MATLAB function. The ARMAX models are "*inherently linear*" and the advantage is to design the "*model structure and parameter identification rapidly*", as is stated also in [1, 2]. To generate a MISO ARMAX model is preferred to split the input-output dataset samples into two subsets, first one is required for model prediction, and the second one consisting of the remaining samples is used for model validation.

Fundamentally, a general description of a noise corrupted linear MISO control system can be described by:

$$y = H_y^u u + H_y^e e \tag{9.1}$$

where u, y are the control system input vector and the scalar output, while e is an unmeasured white noise disturbance source [1, 2]. H_y^u and H_y^e denote the input-output

transfer matrix functions and the description of the noise disturbance respectively
that will be estimated via different ways of parametrization.

The discrete-time ARMAX model corresponding to the general description (9.1)
is given by:

$$A(q)y(t) = B_1(q)u_1(t) + B_2(q)u_2(t) + C(q)e(t)$$
$$A(q) = 1 + a_1 q^{-1} + \cdots + a_{n_a} q^{-n_a}$$
$$B_1(q) = b_{10} q^{-n_{1k}} + b_{11} q^{-n_{1k}-1} + \cdots + b_{1n_{1b}} q^{-n_{1k}-n_{1b}}$$
$$B_2(q) = b_{20} q^{-n_{2k}} + b_{21} q^{-n_{2k}-1} + \cdots + b_{2n_{2b}} q^{-n_{2k}-n_{2b}}$$
$$C(q) = 1 + c_1 q^{-1} + \cdots + c_{n_c} q^{-n_c} \tag{9.2}$$

where n_a, n_{1b}, n_{2b} and n_c denote the orders of the polynomials $A(q), B_1(q), B_2(q)$
and $C(q)$ while n_{1k}, n_{2k} are the pure transport delays in each input-output channel.

Here q is a forward shift time operator, i.e. $qy(t) = y(t+1), qu(t) = u(t+1), qe(t) = e(t+1)$ while q^{-1} is the backward shift time operator, i.e.:
$q^{-1}y(t) = y(t-1), q^{-1}u(t) = u(t-1), q^{-1}e(t) = e(t-1), t = kT, k \in \mathbb{Z}^+$
is for discrete time description, and T is the sampling time. In the ARMAX model
structure (9.2) is easy to identify the three specific terms, namely AR (Autoregressive)
corresponding to the $A(q)$-polynomial, MA (Moving average) related to the noise
$C(q)$-polynomial, and X denoting the "eXtra" exogenous input vector $B(q)u(t)$. In
terms of the general transfer functions H_y^u and H_y^e, the ARMAX model corresponds
to a following parameterization:

$$H_y^u(q) = \frac{[B_1(q) \ B_2(q)]}{A(q)}, H_y^e(q) = \frac{C(q)}{A(q)} \tag{9.3}$$

with common denominators, that in complex domain (frequency domain) $z = e^{Ts}, z \in C, s = \sigma + j\omega, j = \sqrt{-1}, \omega = 2\pi f$, and f is the signal frequency
measured in hertz.

Equation (9.3) can be written now as is following:

$$H_y^u(z) = \frac{[B_1(z) \ B_2(z)]}{A(z)}, H_y^e(z) = \frac{C(z)}{A(z)} \tag{9.4}$$

The Command Window provides the following useful information about the
MISO ARMAX model of chilled water temperature inside the Evaporator subsystem
generated in MATLAB:

mARMAX_T = Discrete-time ARMAX model:

$$A(z)y(t) = B(z)u(t) + C(z)e(t)$$
$$A(z) = 1 - 1.71z^{-1} + 0.7104z^{-2}$$
$$B_1(z) = 2.899z^{-2} - 2.73z^{-3} - 0.1684z^{-4}$$

$$B_2(z) = -0.003326z^{-2} + 0.01051z^{-3} - 0.006553z^{-4}$$
$$C(z) = 1 - 0.7521z^{-1}$$

Sample time: 1 s

Parameterization:

Polynomial orders: $n_a = 2$, $n_b = [2 \ 2]$, $n_c = 1$, $n_k = [2 \ 2]$

Number of free coefficients: 9

Status:

Estimated using ARMAX on time domain data.

Fit to estimation data: 95.25% (prediction focus)

FPE: 2.895e-06, MSE: 2.888e-06

The MISO ARMAX model of Evaporator open-loop control subsystem in the both prediction and validation phases, is shown in Fig. 9.6.

Concluding, the dynamics of the MISO ARMAX model is of fifth order, the dynamics of "*colored noise*" is of first order, and the input signal is delayed by one single time sample. Thus, the state space dimension of the original nonlinear centrifugal chiller dynamics, approximatively 39, is drastically reduced to fifth order, but still higher.

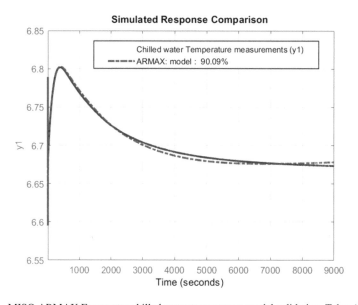

Fig. 9.6 MISO ARMAX Evaporator chilled water temperature model validation, Tchw (y1)

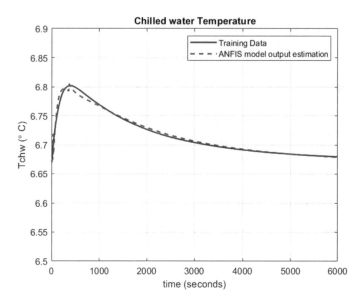

Fig. 9.7 MISO ANFIS Evaporator chilled water temperature model validation, Tchw

The validation of this model shown in Fig. 9.6 reveals a reasonable accuracy, namely 86.24%, where zvT denotes the subset of the last half of measurements dataset chosen for validation phase.

Invoking the MATLAB's *anfis* command, which can be opened directly to the MATLAB R2018b software package command line related to an Adaptive Neuro-Fuzzy training of Sugeno-type FIS, a SISO chilled water temperature model ANFIS is generated and the simulation results are shown in Fig. 9.7. The MATLAB's subroutine *anfis* uses a hybrid learning algorithm to identify the membership function parameters of MISO, Sugeno—type FIS model. A combination of LSE and back-propagation gradient descent methods are used for training FIS membership function parameters to model a given set of input-output (u, y) data.

For a set of data for input-output measurement data (u, y), for example, the following MATLAB R2018b code lines are the core of an ANFIS model:

options = genfisOptions('GridPartition');

options. NumMembershipFunctions = 5;

in_fis = genfis (u, options);

options = anfisOptions;

options. InitialFIS = in_fis;

options. Epoch Number = 30;

out_fis = anfis ([u y], options);

evalfis (u, out_fis)

The improved MISO ANFIS model corresponding to MISO ARMAX model of the Evaporator is shown in Fig. 9.7.

Comparing the MATLAB simulation results shown in Figs. 9.6 and 9.7 for both MISO models ARMAX and ANFIS, we can see a huge improvement in ANFIS model accuracy for chilled water temperature. This improvement will be a real advantage to build one of the most suitable PID ANFIS control strategy for the closed-loop Evaporator control subsystem.

9.3.2   MISO ARMAX and ANFIS Condenser Subsystem Models

Like the previous Sect. 9.3.1 the Command Window provides the following useful information about the SISO ARMAX model of refrigerant liquid level in Condenser subsystem generated in MATLAB:

mARMAX_L $=$ Discrete-time ARMAX model:
$$A(z)y(t) = B(z)u(t) + C(z)e(t)$$
$$A(z) = 1 - 1.49z^{-1} + 0.4894z^{-2}$$
$$B_1(z) = 1062z^{-1} - 2035z^{-2} + 973.9z^{-3}$$
$$B_2(z) = -7.481z^{-1} + 9.104z^{-2} - 3.678z^{-3}$$
$$C(z) = 1 - 0.9135z^{-1}$$

Sample time: 1 s

Parameterization:

Polynomial orders: $n_a = 2, n_b = [2\ \ 2], n_c = 1, n_k = [1\ \ 1]$

Number of free coefficients: 9

Status:

Estimated using ARMAX on time domain data.

Fit to estimation data: 99.28% (prediction focus)

FPE: 0.000279, MSE: 0.0002782

The MISO ARMAX model of Condenser open-loop control subsystem in the both prediction and validation phases, is shown in Fig. 9.8.

The model description reveals that the dynamics of the SISO ARMAX model is of second order, the dynamics of "*colored noise*" is of second order, and the input signal is delayed by one single time sample. Thus, the state space dimension of the original nonlinear centrifugal chiller dynamics, approximatively 39, is drastically reduced to the second lowest possible order.

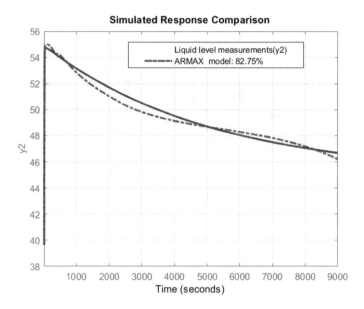

Fig. 9.8 MISO ARMAX Condenser refrigerant liquid level model validation, Level (y2)

The validation of this model shown in Fig. 9.8 reveals a great accuracy, at 92.19%, where zvL denotes the subset of the last half of measurements dataset chosen for validation phase.

Invoking also the *anfis* command, which can be opened directly to the MATLAB R2018b software package command line related to an Adaptive Neuro-Fuzzy training of Sugeno-type FIS, a MISO refrigerant liquid level model ANFIS is generated and the simulation results are shown in Fig. 9.9.

9.4 Centrifugal Chiller PID Closed-Loop Control Strategies—Performance Analysis

This section tries to build firstly a solid baseline of performance comparison and analysis for the proposed closed-loop control strategies developed in previous section by reviewing briefly the MATLAB SIMULINK simulation results obtained in [2] using a PI closed-loop controller for a nonlinear MIMO centrifugal chiller plant with a SIMULINK model represented in Fig. 9.1.

The MATLAB simulations results for chilled water Evaporator temperature and for refrigerant liquid level in the Condenser to a step tracking setpoint are presented in Figs. 9.10 and 9.11, and to a step changes in tracking setpoint are depictured in Figs. 9.12 and 9.13. It is worth noting for all four of the following figures, unlike the MATLAB simulation results obtained for the proposed PID closed-loop strategies, a

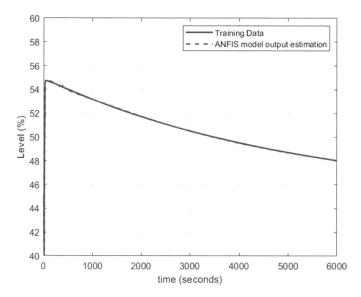

Fig. 9.9 MISO ANFIS Condenser refrigerant liquid level model validation, Level (y2)

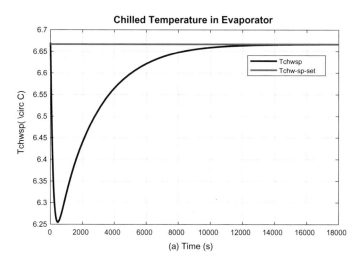

Fig. 9.10 PI MIMO Simulink model simulation results fora step tracking setpoint closed-loop Evaporator chilled water temperature (see [2])

very long transient response, especially for the chilled water temperature inside the Evaporator where the settling time reaches about 3 h.

Further, we analyse rigorously the closed-loop performance of the both proposed control strategies, the simulation results performed in real-time on a MATLAB R2018b attractive environment that are depictured in separate figures for different

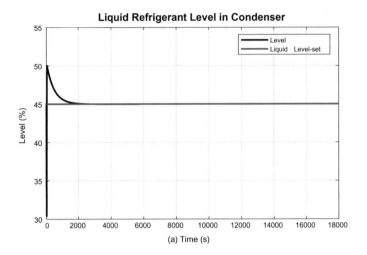

Fig. 9.11 PI MIMO Simulink model simulation results fora step tracking setpoint closed-loop Condenser refrigerant liquid level (see [2])

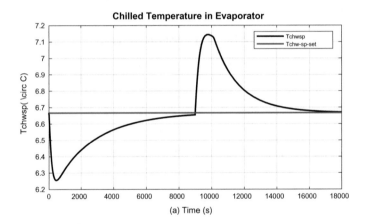

Fig. 9.12 PI MIMO Simulink model simulation results for step changes in tracking setpoint closed-loop Evaporator chilled water temperature (see [2])

time scales for the same measurements input-output data set. The MATLAB SIMULINK model in a matrix transfer function representation of the first digital PI closed-loop control strategy based on the two MISO ARMAX models developed in previous Sect. 9.3 is shown in Fig. 9.14, that correspond to the following equivalent state-space description of fifth order dynamics:

- For MISO ARMAX chilled water temperature inside the Evaporator control subsystem:

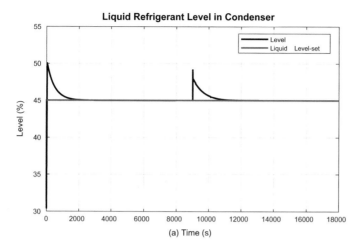

Fig. 9.13 PI MIMO Simulink model simulation results for step changes in tracking setpoint closed-loop Condenser refrigerant liquid level (see [2])

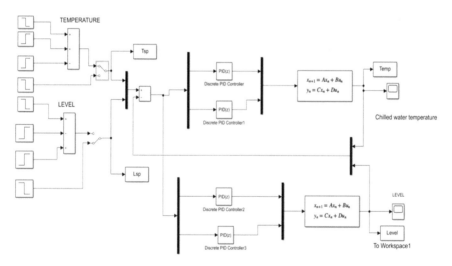

Fig. 9.14 PI MISO ARMAX Simulink models for Evaporator chilled water temperature and Condenser refrigerant liquid level closed-loop control subsystems

$$A = \begin{bmatrix} 0 & 0 & 0 \\ 1 & 0 & -0.7104 \\ 0 & 1 & 1.71 \end{bmatrix}, B = \begin{bmatrix} -0.0842 & -0.003276 \\ -1.365 & 0.005254 \\ 1.45 & -0.001663 \end{bmatrix}$$

$$C = \begin{bmatrix} 0 & 0 & 2 \end{bmatrix}, D = \begin{bmatrix} 0 & 0 \end{bmatrix} \tag{9.5}$$

- For refrigerant liquid level in Condenser control subsystem:

$$A = \begin{bmatrix} 0 & 0 & 0 \\ 1 & 0 & -0.4894 \\ 0 & 1 & 1.49 \end{bmatrix}, \ B = \begin{bmatrix} 30.43 & -0.1149 \\ -63.6 & 0.2845 \\ 33.2 & -0.2338 \end{bmatrix}$$

$$C = \begin{bmatrix} 0 & 0 & 32 \end{bmatrix}, \ D = \begin{bmatrix} 0 & 0 \end{bmatrix} \tag{9.6}$$

The state-space matrix representation for both systems with the dynamics given by the set of Eqs. (9.5) and (9.6) is written generally in the following form:

$$x(k + 1) = Ax(k) + Bu(k)$$
$$y(k) = Cx(k) + Du(k)$$

$$x(k) = \begin{bmatrix} x_1(k) \\ x_2(k) \\ x_3(k) \end{bmatrix}, u(k) = \begin{bmatrix} u_1(k) \\ u_2(k) \end{bmatrix}, y(k) = \begin{bmatrix} y_1(k) \\ y_2(k) \end{bmatrix} \tag{9.7}$$

where $x(k)$ denotes the state vector of third dimension, $u(k)$ is the input vector, and $y(k)$ represents the output vector. The sampling time for both MISO ARMAX linear discrete-time models is 1 s. In Figs. 9.15 and 9.16 is shown the tracking accuracy performance of step input response via extensive MATLAB simulations.

The proportional components of the tuning parameter of the digital PI controller block integrated in the control loop of chilled water temperature inside the Evaporator subsystem, situated in the top of the Fig. 9.14, are set to the following values: kp_{11} = 0.1, ki_{11} = 0.001, kp_{12} = 0.1, ki_{12} = 0.001. Similar, for the digital PI controller

Fig. 9.15 PI MISO ARMAX Evaporator chilled water temperature set point tracking performance, Tchw

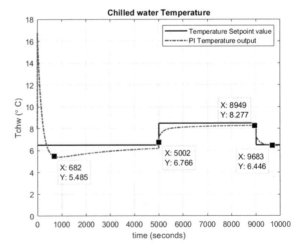

Fig. 9.16 PI MISO Condenser refrigerant liquid level set point tracking performance, Level

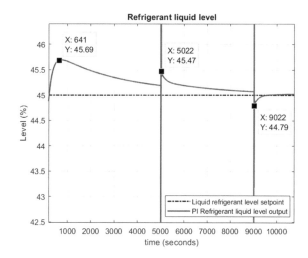

block from the bottom of the Fig. 9.14, corresponding to the refrigerant liquid level inside the Condenser, $kp_{11} = 0.1$, $ki_{12} = -1e-5$, $kp_{21} = -0.1$, and $kp_{22} = 1e-6$.

The MATLAB tracking accuracy performance simulation results for both corresponding digital PI MISO ANFIS improved models integrated into hybrid structures are shown in Figs. 9.17 and 9.18, for a step input setpoint response, and in Figs. 9.19 and 9.20, to prove the tracking accuracy performance robustness for two consecutive step changes in the input setpoints, when the refrigerant liquid is maintained to a constant level, i.e. 43%.

It is valuable to remark important improvements revealed by the MATLAB simulation results shown for both MISO digital PI ANFIS intelligent hybrid control structures.

Fig. 9.17 PIMISOANFIS Evaporator chilled water temperature set point tracking performance, Tchw

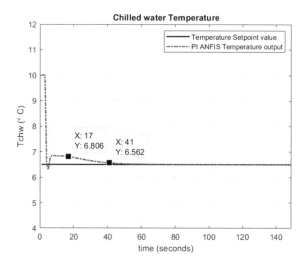

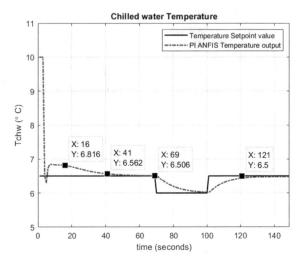

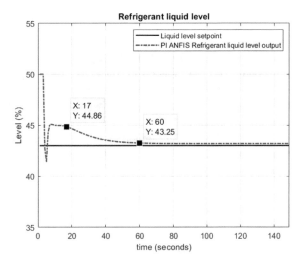

The last two figures, Figs. 9.19 and 9.20 represent a complete picture of the
great overall performance obtained by a PI ANFIS SISO control strategy applied on
Evaporator chilled water temperature in terms of high tracking accuracy, very fast
transient speed, and great robustness to the step changes in the tracking setpoints
compared to the MATLAB simulation results obtained in the previous first MISO
ARMAX digital PI control structure. It is also worthy to remark the more realistic
behavior during the transient of the MISO ANFIS digital PI control strategy for
which the refrigerant liquid level into Condenser follows tightly the variations of the
chilled water temperature inside the Evaporator, reaching both the steady state in
almost same time. The MATLAB simulations reveal a huge transient settling time
for first control strategy compared to a very fast convergence transient speed reached

Fig. 9.20 Robustness of PI MISO ANFIS Condenser refrigerant liquid level to changes in set point tracking performance, Level

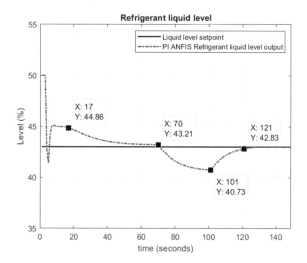

by second control strategy. Also, the overshoot and number of oscillations are much greater in the first control structure compared to the second one.

The proportional and integral tuning parameters of both ANFIS PI controllers are adjusted to the following values: kp1 = 0.001 and ki1 = 0.01, for Evaporator chilled water temperature closed-loop control subsystem, and kp2 = −0.0001, ki2 = 0.01 respectively, for refrigerant liquid level into Condenser control subsystem. For readers, it is also worth to mention that all MATLAB simulations are performed for a sampling time Ts = 1 s.

However, a single oscillation we notice in chilled water temperature inside the Evaporator and in the refrigerant liquid level into Condenser during the transient. The over shoot as an amplitude that is not significant for the overall performance of the controlled Condenser.

9.5 Conclusions

This research paper is an extension and in the same time a dissemination of the preliminary results obtained by authors during the years in this research field. The novelty of the paper consists of the improvement brought to the linear discrete-time polynomials ARMAX models of a MIMO centrifugal chiller plant of high nonlinearity and complexity, especially due to its high state-space dimensionality. The MISO ANFIS models attached to a MIMO centrifugal chiller plant are built based on same input-output measurements data as those used to generate the ARMAX models. AMISO digital PI ANFIS control strategy structure is based on ANFIS models and represents an intelligent hybrid combination of an adaptive neuro-fuzzy structure and a digital PI controller. A performance comparison of both control strategies developed in this research paper reveals that the second MISO ANFIS digital PI control strategy outperforms the first MISO ARMAX digital PI control strategy performance by a significant improvement in terms of tracking accuracy, overshoot, transient convergence speed and robustness to the step changes in tracking input setpoints.

Acknowledgments Research funding (discovery grant) for this project from the Natural Sciences and Engineering Research Council of Canada (NSERC) is gratefully acknowledged.

References

1. R-E. Tudoroiu, M. Zaheeruddin, N. Tudoroiu, D.D. Burdescu, MATLAB Implementation of an Adaptive Neuro-Fuzzy Modeling Approach applied on Nonlinear Dynamic Systems–a Case Study, in *Proceedings of the Federated Conference on Computer Science and Information Systems, Poznan, Poland*, vol. 15 (2018), pp. 577–583
2. N. Tudoroiu, M. Zaheeruddin, S. Li, R.-E. Tudoroiu, Design and implementation of closed-loop PI control strategies in real-time MATLAB simulation environment for nonlinear and linear ARMAX models of HVAC centrifugal chiller control systems. Adv. Sci. Technol. Eng. Syst. J. **3**(2), 283–308 (2018)
3. J.M. Zurada, *Introduction to Artificial Neural Networks*, 1st edn. (West Publishing company, St. Paul, USA, 1992)
4. M. Zaheeruddin, N. Tudoroiu, Neuro-PID tracking control of a discharge air temperature system, Elsevier. Energy Convers. Manag. **45**, 2405–2415 (2004)
5. S.M. Radu, R.-E. Tudoroiu, W. Kecs, N. Ilias, N. Tudoroiu, Real time implementation of an improved hybrid fuzzy sliding mode observer estimator. Adv. Sci. Technol. Eng. Syst. J. **2**(1), 214–226 (2017)
6. R-E. Tudoroiu, W. Kecs, M. Dobritoiu, N. Ilias, S-V. Casavela, N. Tudoroiu, Real-time implementation of DC servomotor actuator with unknown uncertainty using a sliding mode observer, ACSIS, Gdansk, Poland, vol. 8 (2016) pp. (2016)
7. S. Bendapudi, J.E. Braun et al., Dynamic model of a centrifugal chiller system-model development, numerical study, and validation. ASHRAE Trans. **111**, 132–148 (2005)
8. A. Beyene, H. Guven et al., Conventional chiller performances simulation and field data. Int. J. Energy Res. **18**, 391–399 (1994)
9. J.E. Braun, J.W. Mitchell et al., Models for variable-speed centrifugal chillers. ASHRAE Trans. (1987). (New York, USA)
10. M.W. Browne, P.K. Bansal, Steady-state model of centrifugal liquid chillers. Int. J. Refrig. **21**(5), 343–358 (1998)
11. J.M. Gordon, K.C. Ng, H.T. Chua, Centrifugal chillers, thermodynamic modeling and a diagnostic case study. Int. J. Refrig. **18**(4), 253–257 (1995)
12. P. Li, Y. Li, J.E. Yaoyu, Seem: Modelica Based Dynamic Modeling of Water-Cooled Centrifugal Chillers, in *International Refrigeration and Air Conditioning Conference*, Purdue University (2010), pp. 1–8
13. P. Popovic, H.N. Shapiro: Modeling study of a centrifugal compressor. ASHRAE Trans. (1998). (Toronto, Canada)
14. M.C. Svensson, Non-steady-state modeling of a water-to-water heat pump unit, in *Proceedings of 20th International Congress of Refrigeration*, Sydney, Australia (1999)
15. D.J. Swider, M.W. Browne, P. Bansal, V. Kecman, Modelling of vapour-compression liquid chillers with neural networks. Appl. Thermal Eng. J. **21**(3), 311–329 (2001)
16. H. Wang, S. Wang, A mechanistic model of a centrifugal chiller to study HVAC dynamics. Build. Serv. Eng. Res. Technol. **21**(2), 73–83 (2000)

17. M. Gholamrezaei, K. Ghorbanian, Application of integrated fuzzy logic and neural networks to the performance prediction of axial compressors. J. Power Energy **229**(8), 928–947 (2015)
18. Q. Zhou, S. Wang, F. Xiao, A novel strategy for the fault detection and diagnosis of centrifugal chiller systems, HVAC & R Res. J. **15**(1), 57–75 (2009)

Chapter 10
A Review of Deep Reinforcement Learning Algorithms and Comparative Results on Inverted Pendulum System

Recep Özalp, Nuri Köksal Varol, Burak Taşci, and Ayşegül Uçar

Abstract The control of inverted pendulum problem that is one of the classical control problems is important for many areas from autonomous vehicles to robotic. This chapter presents the usage of the deep reinforcement learning algorithms to control the cart-pole balancing problem. The first part of the chapter reviews the theories of deep reinforcement learning methods such as Deep Q Networks (DQN), DQN with Prioritized Experience Replay (DQN+PER), Double DQN (DDQN), Double Dueling Deep-Q Network (D3QN), Reinforce, Asynchronous Advanced Actor Critic Asynchronous (A3C) and Synchronous Advantage Actor-Critic (A2C). Then, the cart-pole balancing problem in OpenAI Gym environment is considered to implement the deep reinforcement learning methods. Finally, the performance of all methods are comparatively given on the cart-pole balancing problem. The results are presented by tables and figures.

Keywords Deep Q networks · Deep Q-network with prioritized experience replay · Double deep Q-network · Double dueling Deep-Q network · Reinforce · Asynchronous advanced actor critic asynchronous · Synchronous advantage actor-critic · Inverted pendulum system

R. Özalp · N. K. Varol · A. Uçar (✉)
Department of Mechatronics Engineering, Firat University, 23119 Elazig, Turkey
e-mail: agulucar@firat.edu.tr

R. Özalp
e-mail: rozalp@firat.edu.tr

N. K. Varol
e-mail: 171134109@firat.edu.tr

B. Taşci
Vocational School of Technical Sciences, Firat University, 23119 Elazig, Turkey
e-mail: btasci@firat.edu.tr

G. A. Tsihrintzis and L. C. Jain (eds.), *Machine Learning Paradigms*,
Learning and Analytics in Intelligent Systems 18,
https://doi.org/10.1007/978-3-030-49724-8_10

10.1   Introduction

In recent years, the modern control engineering has witnessed to significant developments [1]. PID control is known as the state of the art but it has been still commonly using in the industry. Model predictive control, the state feedback control, adaptive controller, and optimal controller have been developed [1, 2]. Most of these controllers require system dynamics and system modeling. Therefore, the control structure that do not require the system model and the inverse model are searched. The reinforcement learning methods have been shown to be a promising intelligence alternative ones to classical control methods [1, 2].

Reinforcement learning is the third stream of machine learning methods which allow to the agent to find the best action from its own experiences by trial and error [3–7]. Unlike supervised and unsupervised learning, the data set is labeled inexplicitly by reinforcement signal. Thanks to self-learning property of agent, the reinforcement learning methods have been increasing its importance day by day. At many areas such as production control, game theory, finance, communication, autonomous vehicles, robotic, the reinforcement learning methods have been successfully applied [8–15].

Literature includes many reinforcement methods such as tabular Q-learning, Sarsa, actor-critic methods, and policy gradient [4]. After DeepMind introduced AlphaGo, at the game of Go, the reinforcement learning algorithm beat the professional Go players Lee Sedol with the score of 4-1 in 2016 [16], many new reinforcement learning methods based on the Deep Neural Networks (DNNs) have appeared [17–21]. Convolutional Neural Networks (CNNs) and Long Short-Term Memory Network (LSTM) that are the variants of DNNs use large data amounts [22–28]. For example, object detection and recognition algorithms are trained using millions of images such as ImageNet [29]. On the other hand, the methods don't use the hand-crafted features that is difficult to determine at high dimensional state space or don't use any feature extraction method. The advantages increased the application range of the conventional reinforcement learning. The best successful applications consist of the control of mobile and humanoid robots including the sensors such as camera, LIDAR, IMU, and so on [30–36].

In this chapter, Deep Q Networks (DQN) [17, 18], DQN with Prioritized Experience Replay (DQN+PER) [37], Double DQN (DDQN) [20], Double Dueling DQN (D3QN) [22], Reinforce [38], Asynchronous Advanced Actor Critic Asynchronous (A3C) [39] and synchronous Advanced Actor Critic Asynchronous (A2C) [39, 41] are applied on the cart-pole balancing problem on OpenAI GYM environment [42]. The performance of the deep reinforcement learning algorithm in the problem are compared. Some control parameters such as rise time, setting time, max return, and mean reward are evaluated for comparing aim.

The chapter is organized as follows. In Sect. 10.2, the fundamentals of reinforcement learning are introduced and followed by the theory of the deep reinforcement learning algorithms. Section 10.3 describes the cart-pole balancing problem. The results of the simulation with the discussions are given in Sect. 10.4. Section 10.5 summarizes the conclusions.

10.2   Reinforcement Learning Background

Reinforcement learning is designed by using Markov Decision Processes (MDP) described in Sect. 10.2.1 [3, 4]. An agent learns its behavior by trial and error rule and bys interacting with the environment. Agent selects an action of a_t according to the s_t state at time t and then receives the r_t feedback reward signal corresponding to this action. It starts to receive the next state, s_{t+1}. An illustration of this general structure is given in Fig. 10.1.

10.2.1   Markov Decision Process

An MDP is a discrete time stochastic control process for optimizing decision-making under uncertainty [3]. In an MDP, the environment dynamics are defined as the transition probability distribution from one state s to the next state s' after taking a below as:

$$p\big(s_{t+1} = s', r_{t+1} = r | s_t, a_t, r_t, s_{t-1}, a_{t-1}, r_{t-1}, \ldots, s_1, a_1, r_1, s_0, a_0\big)$$
$$= p\big(s_{t+1} = s', r_{t+1} = r | s_t, a_t\big). \tag{10.1}$$

The expression which is depend on only the present state and action is known as Markov Property. Transition probabilities give the environment information as:

$$\sum_{s_{t+1} \in S} \sum_{r \in R} p(s_{t+1}, r_{t+1} | s_t, a_t) = 1, \forall s_t \in S, a_t \in A \tag{10.2}$$

$$p(s_{t+1}, | s_t, a_t) = \sum_{r \in R} p(s_{t+1}, r_{t+1} | s_t, a_t) \tag{10.3}$$

Fig. 10.1 A general reinforcement learning structure [4]

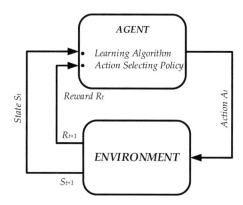

$$r(s_t, a_t) = \sum_{r \in R} r \sum_{s_{t+1} \in S} p(s_{t+1}, r_{t+1} | s_t, a_t). \tag{10.4}$$

An MDP is a five-tuple of S, A, P, R, where S is the space of possible states, A is the space of possible actions, P: $S \times A \times S \rightarrow [0,1]$ is the transition probability from one state s_t to the next state s_{t+1}, R: $S \times A \rightarrow \mathbb{R}$ is the reward function specifying the reward r as taking an action from one state s to the next state s_{t+1}, and $\gamma \in [0, 1)$ is an exponential discount factor.

The goal of agent is to maximize the total discounted accumulated return from each state s_t as below:

$$R_t = \sum_{k=0}^{\infty} \gamma^k r_{t+k+1} \tag{10.5}$$

Thanks to the discount factor γ, it is given more importance to short-term rather than long-term reward [4]. γ is chosen about zero to give the significance to the rewards received in the near future while γ is chosen about one for the future rewards.

A policy π is defined as a strategy to choose an action given a state s_t. A deterministic policy is a function converting a state to an action while a stochastic one is a distribution mapping to probabilities actions according to the state.

The expected return value relating to a state s_t under a policy π is given by a value function $V_\pi(s)$

$$V_\pi(s) = \mathbb{E}[R_t | s_t = s] = \mathbb{E}_\pi \left[\sum_{k=0}^{\infty} \gamma^k r_{t+k+1} | s_t = s \right]. \tag{10.6}$$

The value function in (10.5) is easily estimated by using dynamic programming [3] in terms of the form:

$$V_\pi(s) = \mathbb{E}_\pi \left[\sum_{k=0}^{\infty} \gamma^k r_{t+k+1} | s_t = s \right]$$

$$= \sum_{a \in A} \pi(s, a) \sum_{s' \in S} P_{ss'}^a \left[R_{ss'}^a + \gamma V_\pi(s') \right]. \tag{10.7}$$

Using the Bellman equation [5], the optimal state-action value function, Q-function, following the policy π over both the states and the actions are defined as similar to (10.7) as below:

$$Q_\pi(s, a) = \mathbb{E}_\pi [R_t | s_t = s, a_t = a]$$

$$= \mathbb{E}_\pi \left[\sum_{k=0}^{\infty} \gamma^k r_{t+k+1} | s_t = s, a_t = a \right]$$

$$= \sum_{s' \in S} P^a_{ss'} \left[R^a_{ss'} + \gamma V_\pi (s') \right]. \tag{10.8}$$

Given the transition probabilities and reward values, the optimal Q-function can be solved recursively by using dynamic programming algorithms [3]. One of the algorithms is the value iteration. In the value iteration, Q-values are first estimated in (10.8) and then the value function is updated by maximizing the Q-value by $V_\pi (s_t) = \max\limits_{a_t} Q_\pi (s_t, a_t)$. Until the variation between the former and present values of the value function is small for all states, the algorithm is run and the optimal policy is determined [4].

The optimal policy is determined by using dynamic programming methods thanks to the recursive value function. On the other hand, when the transition probabilities and reward values are not previously known, in order to determine to be taken which the action in a state the reinforcement learning is used. In the reinforcement learning, the agent interacts with the environment taking different actions in a state and approximates a mapping from states to the actions.

The reinforcement learning consists of model-based and model free methods. Model-based methods generate a model based on the reduced interactions with the environment whereas model-free methods directly learn from the experiences by via of the interaction with the environment. It does not need to the models of the transition probability and reward function. Q-learning is a kind of model-free reinforcement learning algorithm [6]. For each state and action pair, a Q-value is iteratively estimated as:

$$Q_{t+1}(s_t, a_t) = Q_t(s_t, a_t) + \eta \left[r_t + \gamma \left[\max_{a'} Q_t \left(s_{t+1}, a_{t+1} | s_t = s, a_t = a \right) \right] - Q_t(s_t, a_t) \right] \tag{10.9}$$

where η is learning rate.

10.2.2 Deep-Q Learning

Tabular Q-learning generates Q-matrix using state-action pairs. The algorithm provides fine results in not large environments and continuous state-action pairs. In fact, the Q-learning algorithm is considered as the problem of approximation of Q-function at the continuous states. Generally, the feed-forward neural networks (FNNs) [7, 43] are used for constructing Q-function and the gradient descent methods are used at the training stage of the FNNs for finding the optimum solution to this problem. However, FNNs are not sufficient at complex environments with large dimensional and discrete or continuous state-action. Hence, to get rid of this disadvantage, DQNs were proposed [17, 18]. DQNs are constructed by using DNN. CNNs and LSTMs that are kinds of DNNs, were successfully used at many application

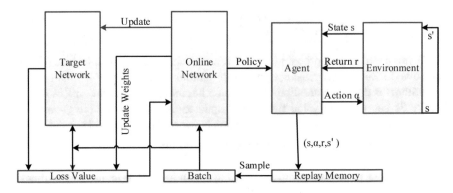

Fig. 10.2 The general framework of DQN

areas without time-consuming feature extraction processing for both large image and sensor inputs [25–36, 44].

Figure 10.2 shows the principles of DQN. In DQN, the network is trained by using the gradient descent method similar to supervised learning [7]. However, the targets are previously unknown. Hence, in order to generate the targets, the network parameters from (i-1) th iteration are first fixed and the target is then approximated by means of the discounted value of the estimated maximum Q-value using the fixed parameters for next state-action pair and the addition of the current reward.

The target is defined as

$$y_t = r_t + \gamma \max_{a_{t+1}} Q_t(s_{t+1}, a_{t+1}; w_{i-1}) \tag{10.10}$$

where w_i are the network parameters at ith iteration.

The loss function to be optimized is defined as

$$L_i(w_i) = \mathbb{E}_\pi \left[(y_t - Q(s_t, a_t; w_i))^2 \right] \tag{10.11}$$

where y_t and $Q_t(s_t, a_t; w_i)$ mean target network output and main network output, respectively. Similar to the supervised learning, the gradient of the loss function is calculated as

$$\nabla_{w_i} L_i(w_i) = \mathbb{E} \left[(y_t - Q(s_t, a_t; w_i)) \nabla_{w_i} Q(s_t, a_t; w_i) \right]. \tag{10.12}$$

In DQN, the exploration defines that agent tries out all actions in which previously unknown at the start of training stage. On the other hand, the exploitation express that agent uses the chosen action. If it is maximized discounted total accumulated return, it is called as the acting greedily. DQN is an off-policy method as it learns a greedy policy, i.e., it selects the highest valued action in each state. When an action is re-realized many times, it is expected that the reward for a state-action pair decreases and DQN prefers the action tried the least frequently. However, this exhibits the case of missing some better actions and there is no exploration. This is called exploration-exploitation dilemma. There are two exploration methods as direct and indirect. The direct exploration techniques use the past memory while the indirect ones don't use them. Direct methods have scale polynomials with the size of the state space while undirected method scales in general exponentially with the size of the state space at finite deterministic environments. In direct methods, Shannon information gain is maximized by exploration.

Direct exploration is not used generally in the model-free reinforcement with high dimensional state space [4]. Famous indirect approaches are ϵ-greedy and softmax exploration which is also called as Boltzmann exploration. The ϵ in ϵ-greedy is an adjusted to a constant determining the probability of taking a random action. At the beginning of the training stage, ϵ value is initialized with large value for exploration and then is generally annealed down to near to 0.1 for exploiting. The Boltzmann method uses an action with weighted probabilities instead of taking random action or optimal action. Boltzmann softmax equation is

$$P_t(a) = \frac{\exp(Q_t(a)/\tau)}{\sum_i^n \exp(Q_t(i)/\tau)} \tag{10.13}$$

where τ is the temperature parameter controlling spread of distribution of softmax function. At the beginning of the training stage, all actions are treated equally and then sparsely distributed as in (10.13).

In DQN, the neural networks cause to some instability problems due to the correlations of consecutive observations and small changes between Q-value. In [17, 18], two solutions were proposed: experience replay and frozen target network. In experience replay solution, the generated experience $E_{t=}(s_t, a_t, r_t, s_{t+1})$ by agent is stored in replay memory over many episodes. Mini batches samples are randomly drawn from the replay memory for training the DQN. This achieves uncorrelated samples and updates with small variance.

The updating of parameters at every time step causes lead to oscillations or obstructions in learning. The frozen target network solution [18] proposes that the weights of the target network has kept frozen at most of the time as the networks of the policy network are updated. Later, it is updated to approximate the policy network.

10.2.3 Double Deep-Q Learning

DQN uses same Q values for both selecting and evaluating an action [20]. This may exhibit overestimation of action values generating positive bias at the cases such as inaccuracies and noise. DDQN was proposed to alleviate the limitation of DQN [20]. DDQN uses two separate networks to select and evaluate the action. In Double

DQN, first the action is selected by maximization operation at the online network. Second, Q-value for the selected action are estimated at the target network. The target function of DDQN is written similar to target function in (10.7) of DQN

$$y_t = r_t + \gamma Q_t \left(s_{t+1}, \max_{a_{t+1}} Q_t \left(s_{t+1}, a_{t+1}; w_t^V \right); w_t^T \right) \tag{10.14}$$

where w_t^V and w_t^T are the parameters of value network and target network, respectively.

DQN or DDQN uses an experience replay uniformly sampled from the replay memory. There is not the order of the significance of experience. To give more importance to some experiences, the Prioritized Experience Replay (PER) are proposed in [37]. Some important transitions are replayed. Moore and Atkeson [45] used PER to get rid of the bias and to increase the performance. A solution for PER is given as

$$p = (error + e)^a \tag{10.15}$$

where an error is mapped to a priority. $e > 0$ is a constant and a is in the range $[0, 1]$ and assign with the importance to the error, 0 means that all samples is with equal priority.

10.2.4 Double Dueling Deep-Q Learning

D3QN was proposed to obtain faster convergence than DQN [21]. This architecture has one bottom casing structure and then are separated into two streams for estimating the state value function $V(s_t)$ and the associated advantage function $A(s_t, a_t)$ as in Fig. 10.3. The bottom casing structure is constructed like to DQN and DDQN except for the last fully connected layer. The Q-function is estimated combining two streams [21, 33, 46].

$$Q \left(s_t, a_t; w, w^V, w^B \right) = V \left(s_t; w, w^V \right) + A \left(s_t, a_t; w, w^A \right) \tag{10.16}$$

where w are the parameters of bottom casing structure and w^V and w^A are the parameters of the advantage and value streams with fully connected layers. The Q-function is estimated as

$$Q \left(s_t, a_t; w, w^V, w^B \right) = V \left(s_t; w, w^V \right) + \left(A \left(s_t, a_t; w, w^A \right) - \max_{a_{t+1}} A \left(s_t, a_t; w, w^A \right) \right) \tag{10.17}$$

In order to achieve better stability, the maximum is rescaled with average,

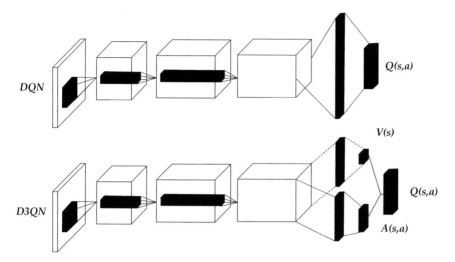

Fig. 10.3 DQN and D3QN network structure [21]

$$Q\left(s_t, a_t; w, w^V, w^B\right) = V\left(s_t; w, w^V\right) + \left(A\left(s_t, a_t; w, w^A\right) - \frac{1}{|\mathcal{A}|} A\left(s_t, a_t; w, w^A\right)\right)$$
$$(10.18)$$

where $|\mathcal{A}|$ is the number of actions.

10.2.5 Reinforce

In contrast to the value-function based methods represented by Q-learning, Policy search methods are based on gradient based approaches [38]. The methods use the gradient of the expected return J for updating the parameters w_i of stochastic policy $\pi^w\left(s_k, a_k; w\right)$.

$$w_i = w_i + \eta \nabla_w J \qquad (10.19)$$

Reinforce is one of the methods to estimate the gradient $\nabla_w J$ [38]. In this method, likelihood ratio is defined as:

$$\nabla_w B^w(l) = B^w(l) \nabla_w log B^w(l) \, J \qquad (10.20)$$

where $B^w(l)$ is the distribution of parameters w.

The expected return for the parameter set w is written as:

$$J^w = \sum_l B^w(l) R(l) \qquad (10.21)$$

where $R(l)$ is the return in episode l. $R(l) = \sum_{k=1}^{K} a_k r_k$ where a_k is a weighting factor at the step k. Generally, a_k is selected the equal to the discount factor in (10.5) or $1/K$.

$$\nabla_w J^w = \sum_l \nabla_w B^w(l) R(l) = \sum_l B^w(l) \nabla_w log B^w(l) R(l) = \mathbb{E}\{\nabla_w log B^w(l) R(l)\}$$
$$(10.22)$$

The distribution over a stochastic policy $\pi^w(s_k, a_k)$ is calculated:

$$B^w(l) = \sum_{k=1}^{K} \pi^w(s_k, a_k) \qquad (10.23)$$

$$\nabla_w log J^w = \mathbb{E}\left\{ \left(\sum_{k=1}^{K} \nabla_w log \pi^w(s_k, a_k) \right) R(l) \right\}. \qquad (10.24)$$

In the application, b_k called as baseline is subtracted from the return term and the result gradient is obtained as:

$$\nabla_w log J^w = \mathbb{E}\left\{ \left(\sum_{k=1}^{K} \nabla_w log \pi^w(s_k, a_k) \right) (R(l) - b_k) \right\}. \qquad (10.25)$$

10.2.6 Asynchronous Deep Reinforcement Learning Methods

The actor-critic reinforcement algorithms generally consist of two parts [39, 40]. The actor part presents the policy $\pi(a_t|s_t; w'_a)$ and the critic part presents the value function, $V(s_t; w_c)$ in which w'_a and w_c are the set of the parameters of the actor and the critic, respectively. The actor is responsible of providing the probalities for each action as the critic is responsible of the value for the state. As in Fig. 10.4, the agent chooses an action a_t in state s_t and the behaviour is then reinforceded if the error of the value function is positive.

A3C uses multiple agents. Each agent has its own parameters and a copy of the environment. Asynchronous in A3C emphasis for multiple agents to asynchronously execute in parallel. These agents interact with their own environments. Each of them is controlled by a global network. In A3C, the advantage is defined as

Fig. 10.4 Actor-Critic
reinforcement model

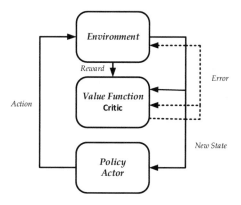

$$A(a_t, s_t) = R_t - V(s_t) \qquad (10.26)$$

where R_t is the total return at time step t and $V(s_t)$ is the value function called as the baseline. The critic is optimized with respect to the loss:

$$Loss = (R_t - V(s_t; w_c))^2. \qquad (10.27)$$

The actor is trained using the gradient in (10.28) including the addition of is the entropy of the policy π, $H\big(\pi\big(s_t; w_a'\big)\big)$ to improve exploration

$$\nabla_{w_a'} \log\pi\big(a_t|s_t; w_a'\big)(R_t - V(s_t; w_c)) + \beta \nabla_{w_a'} H\big(\pi\big(s_t; w_a'\big)\big) \qquad (10.28)$$

where $\log\pi\big(a_t|s_t; w_a'\big)$ means the log probability for taking an action a_t in the state s_t following the policy π and β is a hyperparameter controlling the strength of the entropy regularization.

Since A3C applies the bootstrapping, the variance is reduced and learning accelerates. A3C is applied at both continuous and discrete action spaces. The feedforward and recurrent network structures can be used in A3C. Generally, A3C uses a CNN with one softmax output and one linear output for the actor and for value function, respectively.

A2C is a synchronous, deterministic kind of A3C [40]. As in Fig. 10.5, in A2C, each agent previously complete its episode and then the policy and value function network are updated. A2C can work more effectively on GPUs and faster than A3C.

10.3   Inverted Pendulum Problem

In this chapter, the problem of the inverted pendulum in the classical control theory is considered. The system consists of a pole anchored on a cart moving in two dimension. The cart can move left or right on a flat surface. The aim of the control

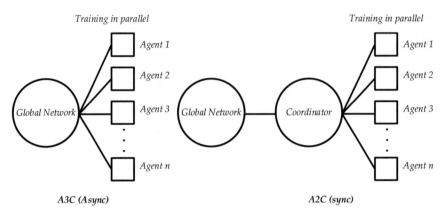

Fig. 10.5 The structure of A3C versus A2C [47]

problem is to balance on the upright position on its mass center of cart the pole. The system is a nonlinear due to the gravity and unstable. Hence, its control is difficult. In order to balance the pole on the cart, the cart has to move by applying a force to the of left and right side of equilibrium point. Figure 10.6 illustrates a setup of the inverted pendulum problem and its parameters.

The system in Fig. 10.6 has the state-space form in

$$
\begin{bmatrix} \dot{x}_1 \\ \dot{x}_2 \\ \dot{x}_3 \\ \dot{x}_4 \end{bmatrix} = \begin{bmatrix} x_3 \\ x_4 \\ \dfrac{m\left(g\cos(x_2)-lx_4^2\right)\sin(x_2)+u}{M+m\sin^2(x_2)} \\ \dfrac{g\sin(x_2)}{l} + \dfrac{\cos(x_2)}{l}\,\dfrac{m\left(g\cos(x_2)-lx_4^2\right)\sin(x_2)+u}{M+m\sin^2(x_2)} \end{bmatrix}, \tag{10.29}
$$

where $x = \begin{pmatrix} x_1 & x_2 & x_3 & x_4 \end{pmatrix}$ is the cart position, pendulum angle, cart velocity, and pendulum velocity, M and m mean the masses of the cart and pendulum, respectively, g is the gravitational acceleration, and u is the external force.

Fig. 10.6 The system of the inverted pendulum [48]

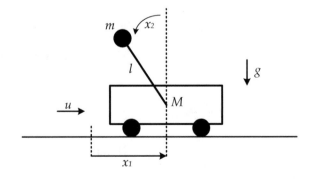

The reinforcement learning does not need to system model or inverse model for control.

10.4   Experimental Results

In this chapter, we used "Cartpole-v1" from OpenAI Gym environment [42]. In this environment, g, M, m, l was determined as 9.8, 1.0.1. 0.5. In the cart-pole balancing problem, it is applied two forces or actions with the magnitude of −1 and 1. The maximum duration of episode is 500.

Each episode is finalized when the pole passes more than 15 degrees from the pivot point, the pole falls down or when the cart moves more than 2.4 units from center. If the cart position and pole angle are not violated, the reward is 1 and the otherwise the reward is 0. We generated all deep reinforcement learning algorithms using Keras and Tensorflow in Python. Especially, the performance of the networks depends on an appropriate setting of hidden layer neuron number. There is no rule for the parameter setting. In this chapter, taking into consideration, we evaluated the network performances of hidden layer neuron number of 24, 32, and 64 since we observed that the training time is too long for bigger numbers and moreover it doesn't increase the performance also. All network structure and the other hyper parameters of (DQN, DQN with PER, DDQN), D3QN, and (Reinforce, A3C, and A2C) are given in Tables 10.1, 10.2, 10.3, 10.4, 10.5, respectively. In the networks, the optimization method was selected as Adam [49]. Mean Square Error (MSE) and CCE (CSE) were used as the loss function [49]. Rectifier Linear Units (Relu) and softmax activation functions were used [49].

In the experiments, all deep reinforcement learning network structures were evaluated on the cart-pole balancing problem. All experiments were repeated 10 times and their average learning curves were obtained. In order to further evaluate the obtained results, the following performance statistics such as mean return, maximum return, rise time, and settling time were used. In the control performance criteria, the rise time is defined as the episode number required to rise from 10 to 90% of the steady state value, whereas the settling time is defined as the episode number required for the steady state value to reach and remain within a given band.

The results of the networks are presented in Table 10.6. The biggest value of mean return was obtained as 207.12 by using DDQN. On the other hand, the smallest values of rise time and settling time were obtained as 40 and 80 by using DDQN. It was observed that DDQN was sufficient for cart-pole balancing problem in that training speed and the received reward. It is also seen from the figures that the second winner of the race is DQN in that mean reward value, the rise time and the settling time. Figures 10.7 and 10.8 show the learning curves of the networks. The figures show that the obtained maximum return is 500 and all the networks reached to 500. In addition, the learning curves of DQN and DDQN has not oscillation and reached to steady-state value in a fast way.

Table 10.1 The network
structure and
hyperparameters of DQN,
DQN with PER, and DDQN

Hyperparameter	Value
Discount factor	0.99
Learning rate	0.001
Epsilon	1.0
Epsilon decay	0.999
Epsilon min	0.01
Batch size	64
Train start	1000
Replay memory	2000
a	0.6
e	0.01
Network structure	
Hidden layer	2
Hidden activation	Rectifier Linear Units
Output activation	Linear
Output number	2
Optimizer method	Adam
Loss	MSE

Table 10.2 The networks
structure and
hyperparameters of Reinforce

Hyperparameter	Value
Discount factor	0.99
Learning rate	0.001
Network structure	
Hidden layer number	2
Hidden activation	Relu
Output activation	Softmax
Output number	2
Optimizer method	Adam
Loss	CCE

In the experiments, it was observed that the performance of A3C and A2C especially affected from the weight initialization and provided different results at each run although it was tried the different weight initialization methods such as Xavier and He Normal [49]. As can be seen from Table 10.6, the best hidden neuron number of DQN, DQN with PER, DDQN, D3QN, Reinforce, A3C, and A2C were obtained as 32, 24, 32, 32, 64, 32, and 64.

Table 10.3 The networks structure and hyperparameters of D3QN

Hyperparameter	Value
Discount factor	0.95
Learning rate	2.5e-4
Epsilon decay	0.99
Replay memory	2000
Tau	1
Network structure	
Hidden layer	2
Hidden activation	Relu
Output activation	Linear
Output number	2
Optimizer method	Adam
Loss	MSE

Table 10.4 The networks structure and hyperparameters of A3C

Hyper parameter	Value
Discount factor	0.99
Actor learning rate	0.001
Critic learning rate	0.001
Threads	8
Network structure of actor part	
Hidden layer number	2
Hidden activation	Relu
Output activation	Softmax
Output number	2
Optimizer method	Adam
Loss	CCE
Network structure of critic part	
Hidden layer/Node number	2
Hidden activation	Relu
Output activation	Linear
Output number	1
Optimizer method	Adam
Loss	MSE

10.5   Conclusions

In this chapter, the control problem of cart-pole balancing was taken into consideration. The main aim of the proposed chapter is to review deep reinforcement learning

Table 10.5 The networks structure and hyperparameters of A2C

Hyperparameter	Value
Discount factor	0.99
Actor learning rate	0.001
Critic learning rate	0.005
Network structure of actor part	
Hidden layer number	2/24
Hidden activation	Relu
Output activation	Softmax
Output number	2
Optimizer method	Adam
Loss	CCE
Network structure of critic part	
Hidden layer/node number	2
Hidden activation	Relu
Output activation	Linear
Output number	1
Optimizer method	MSE

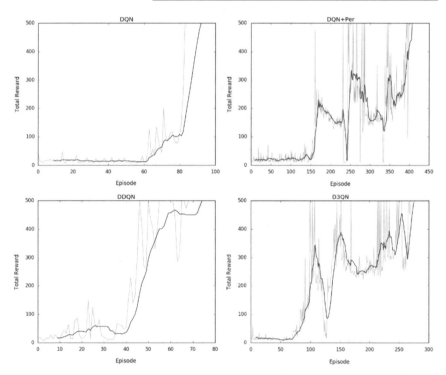

Fig. 10.7 Experimental results of DQN, DQN with Per, DDQN, D3QN

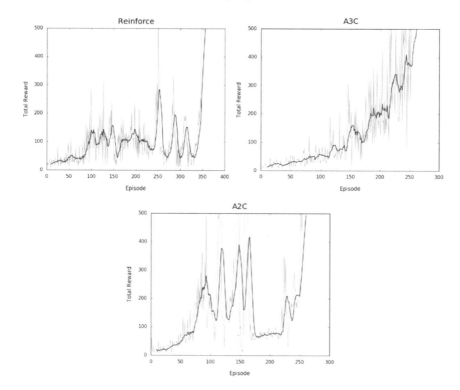

Fig. 10.8 Experimental results of Reinforce, A3C, A2C

methods in the literature and to illustrate the performances of the methods on the cart-pole balancing problem.

This chapter firstly describes the state of the art deep reinforcement learning methods such as DQN, DQN with PER, DDQN, D3QN, Reinforce, A3C, and A2C. Secondly the different network structures are constructed. The cart-pole balancing problem in the OpenAI Gym Environment is finally solved by using the generated methods in Keras.

The effectiveness of the deep reinforcement learning methods is shown by means of the accumulated reward, average reward, max reward, rise time, and settling time. Experimental results demonstrated that the DDQN performed significantly well in the control. It was observed that DDQN achieved the best values of the mean return, rise time and settling time for this cart-pole balancing problem, although it is known that the methods such as Reinforce, A3C, and A2C have fast and high accuracy. Meanwhile, the experimental result showed that the network structure is important for excellence performance.

The future work will focus on complex problems to test the performance of the deep reinforcement learning methods including both data and image inputs.

Table 10.6 Performance comparison of DQN, DQN with Per, DDQN, D3QN, Reinforce, A3C, A2C as to mean reward, max reward, rise time, and settling time

Method	Hidden neuron number	Mean reward	Max reward	Rise time	Settling time
DQN	24	112.64	500	114	140
	32	86.79	500	83	100
	64	179.02	500	141	180
DQN + Per	24	137.47	500	232	300
	32	147.57	500	161	430
	64	206.34	500	250	300
DDQN	24	140.30	500	108	180
	32	**207.12**	500	**46**	**80**
	64	191.64	500	99	180
D3QN	24	122.02	500	120	320
	32	201.04	500	99	280
	64	244.09	500	125	400
Reinforce	24	126.43	500	364	450
	32	120.02	500	293	450
	64	102.50	500	250	355
A3C	24	156.62	500	257	430
	32	191.50	500	244	255
	64	161.17	500	240	500
A2C	24	130.35	500	202	440
	32	101.45	500	191	379
	64	179.44	500	78	260

Acknowledgements This work was supported by the Scientific and Technological Research Council of Turkey (TUBITAK) grant numbers 117E589. In addition, GTX Titan X Pascal GPU in this research was donated by the NVIDIA Corporation.

References

1. K. Ogata, Y. Yang, *Modern Control Engineering*, vol. 4 (London, 2002)
2. A.E. Bryson, *Applied Optimal Control: Optimization, Estimation and Control* (Routledge, 2018)
3. M.L. Puterman, *Markov Decision Processes: Discrete Stochastic Dynamic Programming* (Wiley, 2014
4. R.S. Sutton, A.G. Barto, *Reinforcement Learning: An Introduction* (MIT Press, 2018)
5. K. Vinotha, Bellman equation in dynamic programming. Int. J. Comput. Algorithm **3**(3), 277–279 (2014)
6. C.J. Watkins, P. Dayan, Q-learning. Mach. Learn. **8**(3–4), 279–292 (1992)
7. S.S. Haykin, *Neural Networks and Learning Machines* (Prentice Hall, New York, 2009)

8. P. Abbeel, A. Coates, M. Quigley, A.Y. Ng, An application of reinforcement learning to aerobatic helicopter flight, in *Advances in Neural Information Processing Systems* (2007), pp. 1–8)

9. Y.C. Wang, J.M. Usher, Application of reinforcement learning for agent-based production scheduling. Eng. Appl. Artif. Intell. **18**(1), 73–82 (2005)

10. L. Peshkin, V. Savova, Reinforcement learning for adaptive routing, in *Proceedings of the 2002 International Joint Conference on Neural Networks. IJCNN'02 (Cat. No. 02CH37290)*, vol. 2 (IEEE, 2002), pp. 1825–1830. (2002, May)

11. X. Dai, C.K. Li, A.B. Rad, An approach to tune fuzzy controllers based on reinforcement learning for autonomous vehicle control. IEEE Trans. Intell. Transp. Syst. **6**(3), 285–293 (2005)

12. B.C. Csáji, L. Monostori, B. Kádár, Reinforcement learning in a distributed market-based production control system. Adv. Eng. Inform. **20**(3), 279–288 (2006)

13. W.D. Smart, L.P. Kaelbling, Effective reinforcement learning for mobile robots, in *Proceedings 2002 IEEE International Conference on Robotics and Automation (Cat. No. 02CH37292)*, vol. 4 (IEEE, 2002), pp. 3404–3410. (2002, May)

14. J.J. Choi, D. Laibson, B.C. Madrian, A. Metrick, Reinforcement learning and savings behavior. J. Financ. **64**(6), 2515–2534 (2009)

15. M. Bowling, M. Veloso, *An Analysis of Stochastic Game Theory for Multiagent Reinforcement Learning* (No. CMU-CS-00-165). (Carnegie-Mellon University Pittsburgh Pa School of Computer Science, 2000)

16. D. Silver, A. Huang, C.J. Maddison, A. Guez, L. Sifre, G. Van Den Driessche, J. Schrittwieser, I. Antonoglou, V. Panneershelvam, M. Lanctot, S. Dieleman, Mastering the game of Go with deep neural networks and tree search. Nature, **529**(7587), 484 (2016)

17. V. Mnih, K. Kavukcuoglu, D. Silver, A. Graves, I. Antonoglou, D. Wierstra, M. Riedmiller, Playing atari with deep reinforcement learning (2013). arXiv:1312.5602

18. V. Mnih, K. Kavukcuoglu, D. Silver, A.A. Rusu, J. Veness, M.G. Bellemare, A. Graves, M. Riedmiller, A.K. Fidjeland, G. Ostrovski, S. Petersen Human-level control through deep reinforcement learning. Nature **518**(7540), 529 (2015)

19. H.V. Hasselt, Double Q-learning, in *Advances in Neural Information Processing Systems* (2010), pp. 2613–2621

20. H. Van Hasselt, A. Guez, D. Silver, Deep reinforcement learning with double q-learning, in *Thirtieth AAAI Conference on Artificial Intelligence* (2016, March)

21. Z. Wang, T. Schaul, M. Hessel, H. Van Hasselt, M. Lanctot, N. De Freitas, Dueling network architectures for deep reinforcement learning (2015). arXiv:1511.06581

22. J. Sharma, P.A. Andersen, O.C. Granmo, M. Goodwin, Deep q-learning with q-matrix transfer learning for novel fire evacuation environment (2019). arXiv:1905.09673

23. A. Krizhevsky, I. Sutskever, G.E. Hinton, Imagenet classification with deep convolutional neural networks, in *Advances in Neural Information Processing Systems* (2012), pp. 1097–1105

24. A. Karpathy, G. Toderici, S. Shetty, T. Leung, R. Sukthankar, L. Fei-Fei, Large-scale video classification with convolutional neural networks, in *Proceedings of the IEEE Conference on Computer Vision and Pattern Recognition* (2014), pp. 1725–1732

25. T.N. Sainath, O. Vinyals, A. Senior, H. Sak, Convolutional, long short-term memory, fully connected deep neural networks, in *2015 IEEE International Conference on Acoustics, Speech and Signal Processing (ICASSP)* (IEEE, 2015), pp. 4580–4584. (2015, April)

26. O. Abdel-Hamid, A.R. Mohamed, H. Jiang, G. Penn, Applying convolutional neural networks concepts to hybrid NN-HMM model for speech recognition, in *2012 IEEE International Conference on Acoustics, Speech And Signal Processing (ICASSP)* (IEEE, 2012), pp. 4277–4280. (2012, March)

27. A. Uçar, Y. Demir, C. Güzeliş, Moving towards in object recognition with deep learning for autonomous driving applications, in *2016 International Symposium on INnovations in Intelligent SysTems and Applications (INISTA)* (IEEE, 2016), pp. 1–5. (2016, August)

28. H.A. Pierson, M.S. Gashler, Deep learning in robotics: a review of recent research. Adv. Robot. **31**(16), 821–835 (2017)

29. L. Fei-Fei, J. Deng, K. Li, ImageNet: constructing a large-scale image database. J. Vis. **9**(8), 1037–1037 (2009)
30. A.E. Sallab, M. Abdou, E. Perot, S. Yogamani, Deep reinforcement learning framework for autonomous driving. Electron. Imaging **2017**(19), 70–76 (2017)
31. Y. Zhu, R. Mottaghi, E. Kolve, J.J. Lim, A. Gupta, L. Fei-Fei, A. Farhadi, Target-driven visual navigation in indoor scenes using deep reinforcement learning, in *2017 IEEE International Conference on Robotics and Automation (ICRA)* (IEEE, 2017), pp. 3357–3364. (2017, May)
32. S. Levine, C. Finn, T. Darrell, P. Abbeel, End-to-end training of deep visuomotor policies. J. Mach. Learn. Res. **17**(1), 1334–1373 (2016)
33. L. Xie, S. Wang, A. Markham, N. Trigoni, Towards monocular vision based obstacle avoidance through deep reinforcement learning (2017). arXiv:1706.09829
34. X.B. Peng, M. van de Panne, Learning locomotion skills using deeprl: does the choice of action space matter?, in *Proceedings of the ACM SIGGRAPH/Eurographics Symposium on Computer Animation* (ACM, 2017), p. 12. (2017, July)
35. G. Kahn, A. Villaflor, B. Ding, P. Abbeel, S. Levine, Self-supervised deep reinforcement learning with generalized computation graphs for robot navigation, in *2018 IEEE International Conference on Robotics and Automation (ICRA)* (IEEE, 2018), pp. 1–8. (2018, May)
36. T. Hester, M. Vecerik, O. Pietquin, M. Lanctot, T. Schaul, B. Piot, D. Horgan, J. Quan, A. Sendonaris, I. Osband, G. Dulac-Arnold, Deep q-learning from demonstrations, in *Thirty-Second AAAI Conference on Artificial Intelligence* (2018, April)
37. T. Schaul, J. Quan, I. Antonoglou, D. Silver, Prioritized experience replay (2015). arXiv:1511.05952
38. R.J. Williams, Simple statistical gradient-following algorithms for connectionist reinforcement learning. Mach. Learn. **8**(3–4), 229–256 (1992)
39. V. Mnih, A.P. Badia, M. Mirza, A. Graves, T. Lillicrap, T. Harley, D. Silver, K. Kavukcuoglu, Asynchronous methods for deep reinforcement learning, in *International Conference on Machine Learning*, pp. 1928–1937. (2016, June)
40. A.V. Clemente, H.N. Castejón, A. Chandra, Efficient parallel methods for deep reinforcement learning (2017). arXiv:1705.04862
41. R.R. Torrado, P. Bontrager, J. Togelius, J. Liu, D. Perez-Liebana, Deep reinforcement learning for general video game AI, in *2018 IEEE Conference on Computational Intelligence and Games (CIG)* (IEEE, 2018), pp. 1–8. (2018, August)
42. G. Brockman, V. Cheung, L. Pettersson, J. Schneider, J. Schulman, J. Tang, W. Zaremba, Openai gym (2016). arXiv:1606.01540
43. Y. Demir, A. Uçar, Modelling and simulation with neural and fuzzy-neural networks of switched circuits. COMPEL Int. J. Comput. Math. Electr. Electron. Eng. **22**(2), 253–272 (2003)
44. Ç. Kaymak, A. Uçar, A Brief survey and an application of semantic image segmentation for autonomous driving, in *Handbook of Deep Learning Applications* (Springer, Cham, 2019), pp. 161–200
45. A.W. Moore, C.G. Atkeson, Memory-based reinforcement learning: efficient computation with prioritized sweeping, in *Advances in Neural Information Processing Systems* (1993), pp. 263–270
46. R. Özalp, C. Kaymak, Ö. Yildirum, A. Uçar, Y. Demir, C. Güzeliş, An implementation of vision based deep reinforcement learning for humanoid robot locomotion, in *2019 IEEE International Symposium on INnovations in Intelligent SysTems and Applications (INISTA)* (IEEE, 2019), pp. 1–5 (2019, July)
47. https://lilianweng.github.io/lil-log/2018/04/08/policy-gradient-algorithms.html. Accessed 9 Sept 2019
48. C. Knoll, K. Röbenack, Generation of stable limit cycles with prescribed frequency and amplitude via polynomial feedback, in *International Multi-Conference on Systems, Signals & Devices* (IEEE, 2012), pp. 1–6. (2012, March)
49. I. Goodfellow, Y. Bengio, A. Courville, *Deep Learning.* (MIT Press, 2016)

Part V
Deep Learning in Feature Vector Processing

Chapter 11
Stock Market Forecasting by Using Support Vector Machines

K. Liagkouras and K. Metaxiotis

Abstract Support Vector Machine (SVM) is a well established technique within machine learning. Over the last years, Support Vector Machines have been used across a wide range of applications. In this paper, we investigate stock prices forecasting by using a support vector machine. Forecasting of stock prices is one of the most challenging areas of research and practice in finance. As input parameters to the SVM we utilize some well-known stock technical indicators and macroeconomic variables. For evaluating the forecasting ability of SVM, we compare the results obtained by the proposed model with the actual stocks movements for a number of constituents of FTSE-100 in London.

Keywords Support vector machines · Stock price forecasting · Technical indicators · Macroeconomic variables

11.1 Introduction

Forecasting is the process of utilizing historical data for predicting future changes. Although managers and businesses around the world use forecasting to help them take better decisions for future, one particular business domain has been benefited the most from the development of different forecasting methods: the stock market forecasting. Forecasting of stock prices has always been an important and challenging topic in financial engineering, due to the dynamic, nonlinear, complicated and chaotic in nature movement of stock prices.

The emergence of machine learning [10] and artificial intelligence techniques has made it possible to tackle computationally demanding models for stock price

K. Liagkouras (✉) · K. Metaxiotis
Decision Support Systems Laboratory, Department of Informatics, University of Piraeus, 80, Karaoli & Dimitriou Str. 18534 Piraeus, Greece
e-mail: kliagk@unipi.gr

K. Metaxiotis
e-mail: kmetax@unipi.gr

© The Editor(s) (if applicable) and The Author(s), under exclusive license to Springer Nature Switzerland AG 2020
G. A. Tsihrintzis and L. C. Jain (eds.), *Machine Learning Paradigms*, Learning and Analytics in Intelligent Systems 18, https://doi.org/10.1007/978-3-030-49724-8_11

forecasting [7]. Among the various techniques that have been developed over the last years we distinguish approaches that stem from artificial neural networks (ANNs) [9, 27] and support vector machines (SVMs) [21, 29], because they have gained an increasing interest from academics and practitioners alike. ANN is a computing model whose layered structure resembles the structure of neurons in the human brain [2]. A number of studies examine the efficacy of ANN in stock price forecasting. Below, we provide a concise presentation of recent research finding in the field.

Adhikari and Agrawal [1] propose a combination methodology in which the linear part of a financial dataset is processed through the Random Walk (RW) model and the remaining nonlinear residuals are processed using an ensemble of feedforward ANN (FANN) and Elman ANN (EANN) models. The forecasting ability of the proposed scheme is examined on four real-world financial time series in terms of three popular error statistics. Pan et al. [25] presented a computational approach for predicting the Australian stock market index using multi-layer feed-forward neural networks. According to the authors their research is focused on discovering an optimal neural network or a set of adaptive neural networks for predicting stock market prices.

According to Ou and Wang [23] an important difficulty that is related with stock price forecasting is the inherent high volatility of stock market that results in large regression errors. According to the authors [23] compared to the price prediction, the stock direction prediction is less complex and more accurate. A drawback of ANNs is that the efficiency of predicting unexplored samples decreases rapidly when the neural network model is overfitted to the training data set. Especially this problem is encountered when we are dealing with noisy stock data that may lead ANNs to formulate complex models, which are more prone to the over-fitting problem.

Respectively, a considerable number of studies utilize approaches that are based on SVM for stock price forecasting [34]. Rosillo et al. [26] use support vector machines (SVMs) in order to forecast the weekly change in the S&P 500 index. The authors perform a trading simulation with the assistance of technical trading rules that are commonly used in the analysis of equity markets such as Relative Strength Index, Moving Average Convergence Divergence, and the daily return of the S&P 500. According to the authors the SVM identifies the best situations in which to buy or sell in the market.

Thenmozhi and Chand [29] investigated the forecasting of stock prices using support vector regression for six global markets, the Dow Jones and S&P500 from the USA, the FTSE-100 from the UK, the NSE from India, the SGX from Singapore, the Hang Seng from the Hong Kong and the Shanghai Stock Exchange from China over the period 1999–2011. The study provides evidence that stock markets across the globe are integrated and the information on price transmission across markets, including emerging markets, can induce better returns in day trading.

Gavrishchaka and Banerjee [6] investigate the limitations of the existing models for forecasting of stock market volatility. According to the authors [6] volatility models that are based on the support vector machines (SVMs) are capable to extract information from multiscale and high-dimensional market data. In particular, according to the authors the results for SP500 index suggest that SVM can efficiently work with high-dimensional inputs to account for volatility long-memory and multiscale effects and is often superior to the main-stream volatility models.

Özorhan et al. [24] examine the problem of predicting direction and magnitude of movement of currency pairs in the foreign exchange market. The authors make use of Support Vector Machine (SVM) with a novel approach for input data and trading strategy. In particular, the input data contain technical indicators generated from currency price data (i.e., open, high, low and close prices) and representation of these technical indicators as trend deterministic signals. Finally, the input data are also dynamically adapted to each trading day with genetic algorithm. The experimental results suggest that using trend deterministic technical indicator signals mixed with raw data improves overall performance and dynamically adapting the input data to each trading period results in increased profits.

Gupta et al. [8] presents an integrated approach for portfolio selection in a multi-criteria decision making framework. The authors use Support Vector Machines for classifying financial assets in three pre-defined classes, based on their performance on some key financial criteria. Next, they employ Real-Coded Genetic Algorithm to solve the multi-criteria portfolio selection problem.

According to Cortes and Vapnik [4] the SVMs often achieves better generalization performance and lower risk of overfitting than the ANNs. According to Kim [12] the SVMs outperform the ANNs in predicting the future direction of a stock market and yet reported that the best prediction performance that he could obtain with SVM was 57.8% in the experiment with the Korean composite stock price index 200 (KOSPI 200). Two other independent studies, the first by Huang et al. [11] and the second by Tay and Cao [28] also verify the superiority of SVMs over other approaches when it comes to the stock market direction prediction. Analytically, according to Huang et al. [11] a SVM-based model achieved 75% hit ratio in predicting Nihon Keizai Shimbun Index 225 (NIKKEI 225) movements.

A potential research limitation concerns the testing environment of the aforementioned studies. In particular, for the majority of the examined studies the testing was conducted within the in-sample datasets. Even in the cases that the testing was conducted in an out-of-sample testing environment, the testing was performed on small data sets which were unlikely to represent the full range of market volatility. Another difficulty in stock price forecasting with the SVMs lies in a high-dimensional space of the underlying problem. Indeed, the number of stock markets constituents can range from as few as 30–40 stocks for a small stock market, till several hundreds of stocks for a big stock market, which leads to a high dimensional space [20]. Furthermore, the bigger the examined test instance the bigger the requirements in terms of memory and computation time.

This paper is organized as follows. In Sect. 11.2, we provide an overview of the SVMs and describe how they are integrated in our model. In Sect. 11.3, we identify factors that influence the risk and volatility of stock prices. In Sect. 11.4, we present the proposed model for forecasting stock prices with SVMs. Finally, in Sect. 11.5, we discuss the experimental results and conclude the paper.

11.2 Support Vector Machines

Support Vector Machines were originally developed by Vapnik [30]. In general SVMs are specific learning algorithms characterized by the capacity control of the decision function and the use of kernel functions [31]. In its simplest form a Support Vector Machine is a supervised learning approach for discriminating between two separable groups $\{(\mathbf{x}; y)\}$, where the scalar target variable y is equal to either $+1$ or -1. The vector input variable $\mathbf{x}$ is arbitrary and it is commonly called "separating hyperplane" or otherwise plane in $\mathbf{x}$-space which separates positive and negative cases.

For the linearly separable case, a hyperplane separating the binary decision classes in the three-attribute case is given by the following relationship:

$$y = w_0 + w_1 x_1 + w_2 x_2 + w_3 x_3, \tag{11.1}$$

where y is the outcome, x_i are the attribute values, and there are four weights w_i. The weights w_i are determined by the learning algorithm. In Eq. (11.1), the weights w_i are parameters that determine the hyperplane. The maximum margin hyperplane can be represented by the following equation in terms of the support vectors:

$$y = b + \sum a_i y_i \mathbf{x}(i) \cdot \mathbf{x}, \tag{11.2}$$

where y_i is the class value of training example $\mathbf{x}(i)$. The vector $\mathbf{x}$ represents a test example and the vectors $\mathbf{x}(i)$ are the support vectors. In this equation, b and a_i are parameters that determine the hyperplane. Finally, for finding the support vectors and determining the parameters b and a_i a linearly constrained quadratic programming problem is solved.

For the nonlinearly separable case, a high-dimensional version of Eq. (11.2) is given by the following relationship:

$$y = b + \sum a_i y_i K(\mathbf{x}(i), \mathbf{x}), \tag{11.3}$$

The SVM uses a kernel function $K(\mathbf{x}(i), \mathbf{x})$ to transform the inputs into the high-dimensional feature space. There are some different kernels [32] for generating the inner products to construct machines with different types of nonlinear decision surfaces in the input space. Choosing among different kernels the model that minimizes the estimate, one chooses the best model. Figure 11.1 illustrates how a kernel function works. In particular, with the use of a kernel function K, it is possible to compute the separating hyperplane without explicitly carrying out the map into the feature space [33].

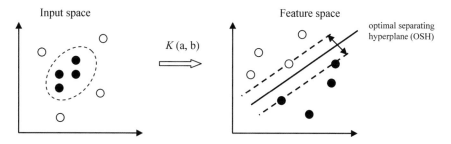

Fig. 11.1 Kernel functions in SVMs

11.3   Determinants of Risk and Volatility in Stock Prices

In stock prices there are two main sources of uncertainty. The first source of risk has to do with the general economic conditions, such as interest rates, exchange rates, inflation rate and the business cycle [14, 15]. None of the above stated macroeconomic factors can be predicted with accuracy and all affect the rate of return of stocks [13]. The second source of uncertainty is firm specific. Analytically, it has to do with the prospects of the firm, the management, the results of the research and development department of the firm, etc. In general, firm specific risk can be defined as the uncertainty that affects a specific firm without noticeable effects on other firms.

Suppose that a risky portfolio consists of only one stock (let say for example *stock* 1). If now we decide to add another stock to our portfolio (let say for example *stock* 2), what will be the effect to the portfolio risk? The answer to this question depends on the relation between *stock* 1 and *stock* 2. If the firm specific risk of the two stocks differs (statistically speaking *stock* 1 and *stock* 2 are independent) then the portfolio risk will be reduced [16]. Practically, the two opposite effects offset each other, which have as a result the stabilization of the portfolio return.

The relation between *stock* 1 and *stock* 2 in statistics is called correlation. Correlation describes how the returns of two assets move relative to each other through time [17]. The most well known way of measuring the correlation is the correlation coefficient (r). The correlation coefficient can range from -1 to 1. Figure 11.2, illustrates two extremes situations: Perfect Positive correlation ($r = 1$) and Perfect Negative correlation ($r = -1$).

Another well-known way to measure the relation between any two stocks is the covariance [18]. The covariance is calculated according to the following formula:

$$\sigma_{X,Y} = \frac{1}{N} \sum_{t=1}^{N} (X_t - \bar{X})(Y_t - \bar{Y}) \tag{11.4}$$

There is a relation between the correlation coefficient that we presented above, and the covariance. This relation is illustrated through the following formula:

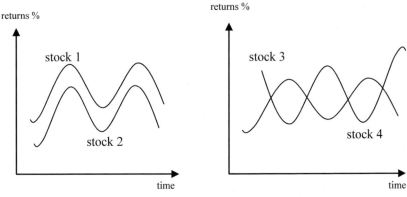

Fig. 11.2 Correlation between stocks

$$r_{X,\Upsilon} = \frac{\sigma_{X,\Upsilon}}{\sigma_X \sigma_Y} \tag{11.5}$$

The correlation coefficient is the same as the covariance, the only difference is that the correlation coefficient has been formulated in such way that it takes values from -1 to 1. Values of the correlation coefficient close to 1 mean that the returns of the two stocks move in the same direction, and values of the correlation coefficient close to -1 mean that the returns of the two stocks move in opposite directions. A correlation coefficient $r_{X,\Upsilon} = 0$ means that the returns of the two stocks are independent. We make the assumption that our portfolio consists 50% of *stock* 1 and 50% of *stock* 2. On the left part of the Fig. 11.2 because the returns of the two stocks are perfectly positively correlated the portfolio return is as volatile as if we owned either *stock* 1 or *stock* 2 alone. On the right part of the Fig. 11.2 the *stock* 3 and *stock* 4 are perfectly negatively correlated. This way the volatility of return of *stock* 3 is cancelled out by the volatility of the return of *stock* 4. In this case, through diversification we achieve risk reduction.

The importance of the correlation coefficient is indicated by the following formula:

$$\sigma_p^2 = w_1^2 \sigma_1^2 + w_2^2 \sigma_2^2 + 2w_1 w_2 r_{1,2}^2 \sigma_1 \sigma_2 \tag{11.6}$$

Equation 11.6 give us the portfolio variance for a portfolio of two stocks 1 and 2. Where w are the weights for each stock and $r_{1,2}$ is the correlation coefficient for the two stocks [19]. The standard deviation of a two—stocks portfolio is given by the formula:

$$\sigma_p = (w_1^2 \sigma_1^2 + w_2^2 \sigma_2^2 + 2w_1 w_2 r_{1,2}^2 \sigma_1 \sigma_2)^{1/2} \tag{11.7}$$

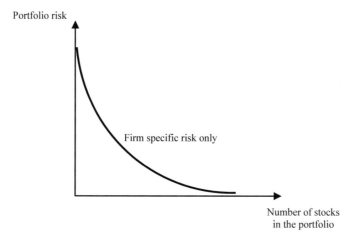

Fig. 11.3 Firm specific risk

From Eq. 11.7, it is obvious that the lower the correlation coefficient $r_{1,2}$ between the stocks, the lower the risk of the portfolio will be.

Obviously, if we continue to add stocks that are negatively correlated into the portfolio the firm-specific risk will continue to reduce. Eventually, however even with a large number of negatively correlated stocks in the portfolio it is not possible to eliminate risk [20]. This happens because all stocks are subject to macroeconomic factors such as inflation rate, interest rates, business cycle, exchange rates, etc. Consequently, no matter how well we manage to diversify the portfolio [22] it is still exposed to the general economic risk.

In Fig. 11.3 we can see that the firm specific risk can be eliminated if we add a large number of negatively correlated stocks into the portfolio. The risk that can be eliminated by diversification except from firm specific risk is called non systematic risk or diversifiable risk.

In Fig. 11.4 we can see that no matter how well diversified is the portfolio there is no way to get rid of the exposure of the portfolio to the macroeconomic factors. These factors related to the general economic risk are called market risk or systematic risk or non diversifiable risk.

11.4 Predictions of Stock Market Movements by Using SVM

11.4.1 Data Processing

The forecasting process requires the following steps: input of selected data, data pre-processing, training and solving support vectors, using test data to calculate

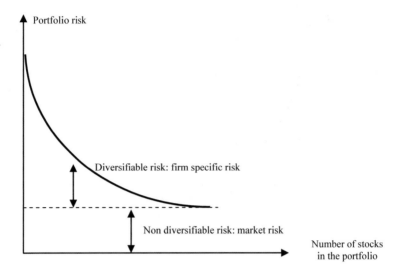

Fig. 11.4 Market risk or non diversifiable risk

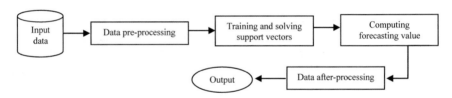

Fig. 11.5 The Forecasting process with SVM

forecasting values, data after-processing, and results analysis. Figure 11.5 illuminates the entire process.

For the purposes of this study, we used a dynamic training pool as proposed by Zhang [35]. Essentially, the training window will always be of the same constant size and 1, 5, 10, 15, 20, 25 and 30 days ahead predictions will be performed by using rolling windows to ensure that the predictions are made by using all the available information at that time, while not incorporating old data. Figure 11.6 illustrates how the dynamic training pool is implemented for the purposes of this study.

For the purposes of the present study we used the daily closing prices of 20 randomly selected constituents of FTSE-100 in London between, Jan. 2, 2018 and Dec. 31, 2018. There are totally 252 data points in this period of time. During the pre-processed phase the data are divided into 2 groups: training group and testing group. The 200 data points belong to training data and the remaining 52 data points are testing data. As shown in Fig. 11.6 we apply a dynamic training pool, which means that the training window will always be of the same constant size (i.e. 200 data points) and one-day-ahead predictions will be performed by using rolling windows.

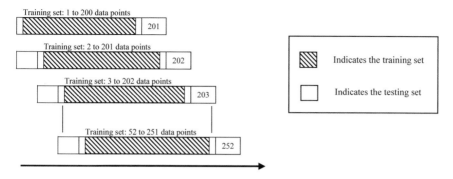

Fig. 11.6 The dynamic training pool for the case of 1-day ahead prediction

In this paper we treat the problem of stock price forecasting as a classification problem. The feature set of a stock's recent price volatility, index volatility, mean absolute error (MAE), along with some macroeconomic variables such as Gross National Product (GNP), interest rate, and inflation rate, are used to predict whether or not the stock's price 1, 5, 10, 15, 20, 25 and 30 days in the future will be higher (+1) or lower (−1) than the current day's price.

11.4.2 The Proposed SVM Model

For the purposes of this study we use the following radial kernel function:

$$K(x_i, x_k) = \exp\left(-\frac{1}{\delta^2} \sum_{j=1}^{n} (x_{ij} - x_{kj})^2 \right) \tag{11.8}$$

where δ is known as the bandwidth of the kernel function [12]. This function classifies test examples based on the example's Euclidean distance to the training points, and weights closer training points more heavily.

11.4.3 Feature Selection

In this study we use six features to predict stock price direction. Three of these features are coming from the field of macroeconomics and the other three are coming from the field of technical analysis. We opted to include three variables from the field of macroeconomics as it is well-known that macroeconomic variables have an influence on stock prices. For the purposes of this study we use the following macroeconomic variables: (a) Gross National Product (GNP), (b) interest rate, and (c) inflation rate.

Table 11.1 Features used in SVM

Feature name	Description	Formula
σ_s	The stock price volatility is calculated as an average over the past n days of percent change in a given stock's price per day	$\dfrac{\sum_{i=t-n+1}^{t} \frac{C_i - C_{i-1}}{C_{i-1}}}{n}$
σ_i	The index volatility is calculated as an average over the past n days of percent change in the index's price per day	$\dfrac{\sum_{i=t-n+1}^{t} \frac{I_i - I_{i-1}}{I_{i-1}}}{n}$
MAE	The mean absolute error (MAE) measures the average magnitude of the errors in a set of forecasts, without considering their direction	$\dfrac{1}{n}\sum_{i=1}^{n}\left| y_i - x_i \right|$
GNP%	Gross national Product (GNP). The formula for calculating the percent change in GNP rate looks like this	$\frac{GNP_i - GNP_{i-1}}{GNP_{i-1}} \times 100$
Interest rate%	Interest rate is the cost of borrowing money. The formula for calculating the percent change is interest rate is given by the following relationship	$\frac{IR_i - IR_{i-1}}{IR_{i-1}} \times 100$
Inflation rate%	Inflation rate is the percentage increase in general level of prices over a period. The formula for calculating the inflation rate is given by the following relationship	$\frac{CPI_i - CPI_{i-1}}{CPI_{i-1}} \times 100$

[a]Where C_i is the stock's closing price at time i. Respectively, I_i is the index's closing price at time i. In MAE y_i is the prediction and x_i the realized value. IR_i stands for interest rate at time i. Finally, CPI_i stands for Consumer Price Index at time i. We use these features to predict the direction of price change 1, 5, 10, 15, 20, 25 and 30 days ahead

According to a study by Al-Qenae et al. [3] it is found that an increase in inflation and interest rates have negative impact on stock prices, whereas an increase in GNP has positive effect on stock prices.

Respectively, we use the following three technical analysis indicators: (a) price volatility, (b) sector volatility and (c) mean absolute error (MAE). More details about the selected features are provided in Table 11.1.

11.5 Results and Conclusions

Figure 11.7 illustrates the mean forecasting accuracy of the proposed model in predicting stock price direction 1, 5, 10, 15, 20, 25 and 30 days ahead.

By observing Fig. 11.7, it is evident that the best mean forecasting accuracy of the proposed model is obtained for predicting stock price direction 1-day ahead. Furthermore, the forecasting accuracy falls drastically when the horizon increases. Indeed, the mean forecasting accuracy of the proposed model is slightly better than simple random guessing when it comes to predicting stock price direction 30-days ahead. This latest finding comes in support of the Efficient Markets Hypothesis [5], which posits that stock prices already reflect all available information and therefore technical analysis cannot be used successfully to forecast future prices. According

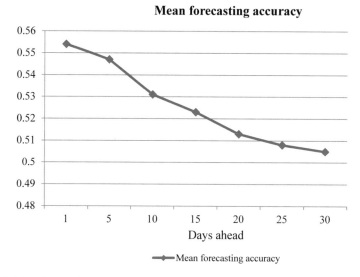

Fig. 11.7 Mean forecasting accuracy of the proposed model in predicting stock price direction

to the Efficient Markets Hypothesis, stock prices will only respond to new information and since new information cannot be predicted in advance, stock price direction cannot be reliably forecasted. Therefore, according to the Efficient Markets Hypothesis, stock prices behave like a random walk. To conclude the proposed model can be helpful in forecasting stock price direction 1–5-days ahead. For longer horizons, the forecasting accuracy of the proposed model falls drastically and it is slightly better than simple random guessing.

References

1. R. Adhikari, R.K. Agrawal, A combination of artificial neural network and random walk models for financial time series forecasting. Neural Comput. Appl. **24**(6), 1441–1449 (2014)
2. R. Aghababaeyan, N. TamannaSiddiqui, Forecasting the tehran stock market by artificial neural network. Int. J. Adv. Comput. Sci. Appl. Spec. Issue Artif. Intell. (2011)
3. R. Al-Qenae, C. Li, B. Wearing, The Information co earnings on stock prices: the Kuwait Stock Exchange. Multinatl. Financ. J. **6**(3 & 4), 197–221 (2002)
4. C. Cortes, V. Vapnik, Support-vector networks. Mach. Learn. **20**(3), 273–297 (1995)
5. E.F. Fama, Efficient capital markets: a review of theory and empirical work. J. Financ 25(2). in *Papers and Proceedings of the Twenty-Eighth Annual Meeting of the American Finance Association*, New York, N.Y, 28–30 December, 1969 (May, 1970), pp. 383–417
6. V.V. Gavrishchaka, S. Banerjee, Support vector machine as an efficient framework for stock market volatility forecasting, CMS **3**:147–160 (2006)
7. M. Gilli, E. Schumann, Heuristic optimisation in financial modelling. Ann. Oper. Res. **193**(1), 129–158 (2012)
8. P. Gupta, M.K. Mehlawat, G. Mittal, Asset portfolio optimization using support vector machines and real-coded genetic algorithm. J. Glob. Optim. **53**, 297–315

9. E. Guresen, G. Kayakutlu, T.U. Daim, Using artificial neural network models in stock market index prediction. Expert Syst. Appl. **38**(8), 10389–10397 (2011)

10. C.J. Huang, P.W. Chen, W.T. Pan, Using multi-stage data mining technique to build forecast model for Taiwan Stocks. Neural Comput. Appl. **21**(8), 2057–2063 (2011)

11. W. Huang, Y. Nakamori, S.-Y. Wang, Forecasting stock market movement direction with support vector machine. Comput. Oper. Res. **32**(10), 2513–2522 (2005)

12. K.-J. Kim, Financial time series forecasting using support vector machines. Neurocomputing **55**(1), 307–319 (2003)

13. K. Liagkouras, K. Metaxiotis, A new probe guided mutation operator and its application for solving the cardinality constrained portfolio optimization problem. Expert Syst. Appl. **41**(14), 6274–6290 (2014). Elsevier

14. K. Liagkouras, K. Metaxiotis, Efficient portfolio construction with the use of multiobjective evolutionary algorithms: Best practices and performance metrics. Int. J. Inf. Technol. Decis. Making **14**(03), 535–564 (2015). World Scientific

15. K. Liagkouras, K. Metaxiotis, Examining the effect of different configuration issues of the multiobjective evolutionary algorithms on the efficient frontier formulation for the constrained portfolio optimization problem. J. Oper. Res. Soc. **69**(3), 416–438 (2018)

16. K. Liagkouras, K. Metaxiotis, Multi-period mean–variance fuzzy portfolio optimization model with transaction costs. Eng. Appl. Artif. Intell. **67**(2018), 260–269 (2018)

17. K. Liagkouras, K. Metaxiotis, Handling the complexities of the multi-constrained portfolio optimization problem with the support of a novel MOEA. J. Oper. Res. Soc. **69**(10), 1609–1627 (2018)

18. K. Liagkouras, K. Metaxiotis, A new efficiently encoded multiobjective algorithm for the solution of the cardinality constrained portfolio optimization problem. Ann. Oper. Res. **267**(1–2), 281–319 (2018)

19. K. Liagkouras, K. Metaxiotis, Improving the performance of evolutionary algorithms: a new approach utilizing information from the evolutionary process and its application to the fuzzy portfolio optimization problem. Ann. Oper. Res. **272**(1–2), 119–137 (2019)

20. K. Liagkouras, A new three-dimensional encoding multiobjective evolutionary algorithm with application to the portfolio optimization problem. Knowl.-Based Syst. **163**(2019), 186–203 (2019)

21. C.J. Lu, T.S. Lee, C.C. Chiu et al., Financial time series forecasting using independent component analysis and support vector regression. Decis. Support Syst. **47**(2), 115–125 (2009)

22. K. Metaxiotis, K Liagkouras, Multiobjective evolutionary algorithms for portfolio management: a comprehensive literature review. Expert Syst. Appl. **39**(14), 11685–11698 (2012). Elsevier

23. P. Ou, H. Wang, Prediction of stock market index movement by ten data mining techniques. Mod. Appl. Sci. **3**(12), 28 (2009)

24. M.O. Özorhan, I.H. Toroslu, O.T. Sehitoglu, A strength-biased prediction model for forecasting exchange rates using support vector machines and genetic algorithms. Soft. Comput. **21**, 6653–6671 (2017)

25. H. Pan, C. Tilakaratne, J. Yearwood, Predicting Australian Stock market index using neural networks exploiting dynamical swings and intermarket influences. J. Res. Pract. Inf. Technol. **37**(1), 43–55 (2005)

26. R. Rosillo, J. Giner, D. Fuente, (2013) The effectiveness of the combined use of VIX and Support Vector Machines on the prediction of S&P 500. Neural Comput. Applic. **25**, 321–332 (2013)

27. E.W. Saad, D.V. Prokhorov, D.C. Wunsch et al., Comparative study of stock trend prediction using time delay, recurrent and probabilistic neural networks. IEEE Trans. Neural Netw. **9**(6), 1456–1470 (1998)

28. F.E.H. Tay, L. Cao, Application of support vector machines in financial time series forecasting. Omega **29**(4), 309–317 (2001)

29. M. Thenmozhi, G.S. Chand, Forecasting stock returns based on information transmission across global markets using support vector machines. Neural Comput. Applic. **27**, 805–824 (2016)

30. V.N. Vapnik, *Statistical Learning Theory* (Wiley, New York, 1998)
31. V.N. Vapnik, An overview of statistical learning theory. IEEE Trans. Neural Netw. **10**, 988–999 (1999)
32. L. Wang, J. Zhu, Financial market forecasting using a two-step kernel learning method for the support vector regression. Ann. Oper. Res. **174**(1), 103–120 (2010)
33. C. Wong, M. Versace, CARTMAP: a neural network method for automated feature selection in financial time series forecasting. Neural Comput. Appl. **21**(5), 969–977 (2012)
34. F.C. Yuan, Parameters optimization using genetic algorithms in support vector regression for sales volume forecasting. Appl. Math. **3**(1), 1480–1486 (2012)
35. Y. Zhang, Prediction of Financial Time Series with Hidden Markov Models, Simon Fraser University, 2004

Chapter 12
An Experimental Exploration of Machine Deep Learning for Drone Conflict Prediction

Brian Hilburn

Abstract Introducing drones into urban airspace will present several unique challenges. Among these is how to monitor and de-conflict (potentially high-density/low predictability) drone traffic. It is easy to imagine this task being beyond the capabilities of the current (human-based) air traffic control system. One potential solution lies in the use of Machine Learning (ML) to predict drone conflicts. The current effort, which took place under the umbrella of the European Commission's Horizon 2020 TERRA project, experimentally explored the use of ML in drone conflict prediction. Using a deep learning (multi-layer neural network) approach, this effort experimentally manipulated traffic load, traffic predictability, and look-ahead time. Testing the 'worst case scenario' relied on extremely limited data (aircraft instantaneous state only, without any intent information) and limited neural network architecture. Even so, results demonstrated (especially under structured traffic conditions) large potential ML benefits on drone conflict prediction. Binary classification accuracy ran generally above 90%, and error under the most demanding scenarios tended toward false positive (i.e. incorrectly predicting that a conflict would occur). Discussion considers how both (1) richer data (real-world vs stochastically generated) and (2) modelling architecture refinements (e.g. via recurrent neural networks, convolutional neural networks, and LSTM architectures) would likely bolster these observed benefits.

12.1 Introduction

The possible introduction of drone traffic into urban airspace has many in the air traffic management (ATM) community wondering how to accommodate such a novel new type of aircraft, whose potential numbers and unpredictability might overwhelm current human-based methods for managing air traffic [1, 2]. One possible solution

B. Hilburn (✉)
Center for Human Performance Research, Voorburg, The Netherlands
e-mail: brian@chpr.nl

© The Editor(s) (if applicable) and The Author(s), under exclusive license
to Springer Nature Switzerland AG 2020
G. A. Tsihrintzis and L. C. Jain (eds.), *Machine Learning Paradigms*,
Learning and Analytics in Intelligent Systems 18,
https://doi.org/10.1007/978-3-030-49724-8_12

273

lies in the use of Machine Learning (ML) techniques for predicting (and possibly resolving) drone conflicts in high density airspace.

The aim of this research was not to develop an optimised ML model per se, but to experimentally explore via low-fidelity offline simulations the potential benefits of ML for drone conflict prediction, specifically: how well can a (simple) ML model predict on the basis of instantaneous traffic pattern snapshot, whether that pattern will result in an eventual airspace conflict (defined as entry into a stationary prohibited zone)? Secondly, how is model performance impacted by such parameters as traffic level, traffic predictability, and 'look-ahead' time of the model?

12.1.1 Airspace and Traffic Assumptions

This effort started from several assumptions. First, we wanted to test the 'edge case,' or worst-case scenario. If ML was able to predict conflicts under the most challenging possible assumptions, we would have encouraging evidence that the real world results would only be better. For reasons of this analysis, our traffic assumptions therefore included the following [3]:

- Urban Air Mobility (UAM) scenario—envisions short flight times, and frequent trajectory changes;
- High traffic density—this is implicit in predicted future urban drone scenarios, but we intended to push the limits of traffic level;
- Lack of intent information—no flight plan information (regarding filed destination, speed, altitude, heading changes, etc.) would be available. Instead, only the minimal state information (instantaneous altitude, speed, and heading) would be provided;
- Random drone movements—ML conflict prediction would be trivial if all drone movements were completely predictable. In reality, VLL drone operations will have a fair amount of structure and determinism. However, we intentionally introduced a high level of randomness in drone movements, again to test the worst case scenario for ML;
- Prohibited airspace—was represented *as* static *no-go* regions (e.g., around security sensitive areas). This analysis included a single, static "no drone zone," and conflicts were defined as penetrations of this zone.

12.1.2 Methodological Assumptions

This effort set out to test ML conflict prediction using the most challenged methods. Specifically, this meant that whatever ML model we used must have no ability to look either forward or backward in time, nor make use of any other information beyond the simple instantaneous state of each drone. For research purposes, the conflict prediction problem was simplified to one of pattern recognition. We used a supervised learning approach, and in particular a fairly limited architecture: the standard deep learning (i.e. multi hidden layer) artificial neural net. Whereas enhancements to the neural net approach (including RNN, CNN, and LSTM enhancements) would be expected to show better time series processing and thus better classification performance, we employed a simpler neural net architecture, to establish baseline worst case model performance. Moreover, we set out to train different models (36 in all) so that model performance could be compared experimentally, to assess the impact of traffic level, traffic randomness, and look-ahead window range on ML conflict prediction performance.

12.2   A Brief Introduction to Artificial Neural Networks (ANNs)

Traditional computing architectures have several serious limitations. First, they rely on a central powerful processor, with all data being handled in a serial fashion. Serial processing requires that all knowledge be symbolically encoded at the time of programming, and in a discrete location. Certain types of problems, such as pattern recognition and nonlinear discrimination, are not handled well by serial methods.

In response to these limitations of serial computing, research began decades ago into "connectionist" models of computing that represent learning in a network of nodes ("neurons") and connections, meant to mimic the human brain [4]. ANNs consist of a group of neurons, each of which is a simple processor that can operate in parallel. Memory is created and strengthened by modifying connection weights (either positive or negative) between neurons, and connection weights are generally represented as a matrix. The principle underlying the behaviour of these neurons is that knowledge can be represented through the cooperation of relatively simple units, with each one comparing a threshold value to the sum of weighted inputs and producing in response a nonlinear output [5].

The concept of ANNs grew out of the development of the "Perceptron," a simple connectionist model that first combined adjustable inter-neuron weights, a feedforward mechanism for propagating these weights, and error feedback algorithm that compares output to the desired target value. With the system, weights are initially set to some random nonzero value, and the system then computes an error signal representing the difference between input and target. This error signal is then passed back through the system. The distributed pattern of (excitation and inhibition) activations, through error feedback and weight feedforward mechanisms, allows the network to either learn the relationship between predictor and target variables, or discover cluster patterns within the input factors.

At the most basic level, an ANN simply sums the weighted (excitation or inhibition) activations from inputs. Although Fig. 12.1 shows a single intermediate node, in most real-world examples there are multiple neurons at one or more "hidden layers."

Fig. 12.1 A basic "Perceptron" (left) and a feed forward neural network with one hidden layer (right)

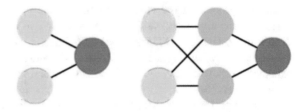

Fig. 12.2 Basic ANN operation

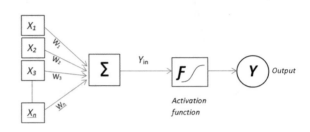

Table 12.1 Hypothetical input and target data for a neural network

Case	Obese?	Exercise?	Smoke?	High blood pressure?	Family history?	Heart attack	
1	1	0	0	1	0	0	*Training sub batch*
2	0	1	1	0	1	1	
3	0	1	0	0	1	0	
n-2	1	1	0	1	0	0	*Test sub batch*
n-1	0	0	0	1	0	1	
n	0	0	0	0	1	0	

This net summed activation is then passed through a nonlinear activation function over the net input (Fig. 12.2).

As used here, ANN modelling employs a supervised learning approach. Table 12.1 shows a hypothetical dataset linking five health factors (predictors) to incidence of heart attack (target). The batch, or complete dataset, is comprised of samples, or cases. In this example, each sample is an individual person (in our drone analysis, each sample was a traffic sample snapshot). Usually, ANN modelling involves randomly splitting the entire batch into training and test sub batches. After training the network with many repeated exposures to the training sub batch, performance of the trained network is validated by assessing prediction performance on the previously unseen testing sub batch.

Neural network modelling has advanced over the past few decades, for example through the introduction of "hidden layers" (intermediate neuron layers that combine weighted inputs to produce an output via an activation function, thereby allowing the network to solve the XOR problem), non-binary methods for weight adaptation (specifically, the least mean squares approach), and relaxation algorithms for pacing error correction steps. The most common and simple current-day ANN architecture incorporates feed-forward, back propagation, and hidden layers. As discussed later, recent architectural refinements to the neural network approach include *convolutional neural networks* (CNNs, for enhanced pattern recognition) [6], *recurrent neural networks* (RNNs, which incorporate enhanced memory) [7], and *long short term memory* models (LSTMs, which add a 'forget' function, and thereby allow a sliding window approach to time series processing) [8]. Certain (more complex) ANNs, such as those used in this analysis, are sometimes referred to as *deep learning* models, in which multiple hidden layers are used sequentially to refine feature extraction.

Notice that, although ANNS are now common enough that there are COTS software packages—and even Excel add-on modules—available to build ANNs, there is still a fair amount of 'black magic' involved in architecture design, including the specification of hidden layers, learning and error correction rates and algorithms, etc.

To be clear, there are two potential limitations of ANNs in the current context that are worth noting:

- **Opacity of results**—as noted earlier, some ML algorithms (ANNs included) can be robust, in that they will always find some solution. Interpreting the solution, however, can be difficult. The inner workings of the ANN are often opaque;
- **Limited ability to handle time series data**—traditional ANN approaches perform well at static pattern recognition tasks. They are not, however, so good at time series data (think of the step-by-step progression of aircraft trajectories in traffic). More sophisticated ML methods replace simple ANN feedforward mechanisms with internal neuron state ('memory' in RNNs), or even neuron internal state decay ('forgetting' in LSTMs).

A main strength of ANNs is that they can handle nonlinearity. This is important if targets and output do not have a linear relationship. It might be, for example, that: obese people have heart attacks; obese exercisers do not have heart attacks; obese smokers have heart attacks (regardless of exercise); yet skinny smokers have no heart attacks (regardless of smoking). Because of this nonlinearity, this is a case where traditional linear regression approaches would fail.

12.3 Drone Test Scenarios and Traffic Samples

The urban drone environment was represented by a 20×20 grid of 400 total cells. Each cell was either occupied or empty. Developmental testing established the number and size of restricted areas, so as to produce a reasonable number of Prohibited Zone (PZ) incursions. It was decided to use a single, stationary PZ, as shown in Fig. 12.3. One simplifying assumption was that altitude was disregarded, and drone movements were only considered in two dimensions (the PZ was assumed to be from the surface upward). Second, there were no speed differences between

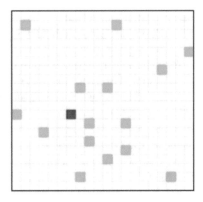

Fig. 12.3 Snapshot, traffic sample of 16 birthed drones (note PZ in red)

drones. Finally, conflicts were only defined as airspace incursions into the PZ, not as losses of separation between drones (drones were assumed to maintain vertical separation).

Traffic samples were built from three different kinds of drone routes, as shown in Fig. 12.4. Notice that drones could only fly on cardinal headings (North, South, East, or West). *Through-routes* transited the sector without any heading change. *TCP-routes* (i.e. Trajectory Change Point routes) added probabilistic heading changes to through-routes. After a random interval of 3–6 steps, TCP-route drones would either continue straight ahead, or make a 90° left/right heading change. The random TCP function was nominally weighted to 50% no heading change (i.e. continue straight ahead), 25% left turn, and 25% right turn. Finally, the ten possible *Bus-routes* (5 routes, flown in either direction) were pre-defined TCP trajectories. Bus-route drones all entered the sector after sample start time, except for bus-routes 9 and 10 (which flew a square pattern in the centre of the sector, and were birthed already on their route).

As discussed later, analysis compared "random" and "structured" route conditions, as an experimental manipulation. The random condition used TCP-routes

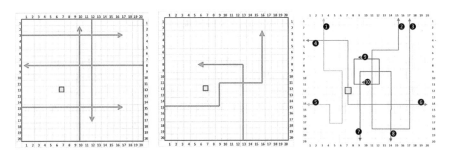

Fig. 12.4 Through-routes (left), TCP-routes (centre) and bus-routes (right)

exclusively. The structured condition used a random combination of through-routes and bus-routes.

Each traffic sample consisted of 40 time steps. First appearance of each drone was randomly timed to occur between steps 1 and 15. Each drone maintained current heading by advancing one cell per time step (no hovering). This meant that a through-route drone would transit the sector in 20 steps. Each traffic sample also consisted of 4, 8, or 16 birthed drones (this was also an experimental manipulation, as described later). Because birth time was randomized, the actual number of instantaneous in-sector drones could vary.

12.3.1 Experimental Design

Analysis used a $3 \times 2 \times 2$ experimental design and varied the following factors:

- **Aircraft count** (4 vs 8 vs 16)—the total number of birthed aircraft;
- **Look-ahead time** (Low vs High)—Snapshot time, in number of steps before conflict;
- **Traffic structure** (Low vs High)—Randomised vs semi-structured traffic flows.

Each of these factors is now described in more detail.

12.3.1.1 Aircraft Count

Aircraft count (4, 8, or 16) was manipulated as the total number of birthed aircraft. Note that instantaneous in-sector count might differ. Aircraft birth times were randomized within the first 15 of 40 total steps per traffic sample. Except for bus-routes number 9 and 10, which flew a square fixed pattern near the centre of the sector, all drones entered the sector at the perimeter. Sector exits were immediately extinguished, and no traffic from outside of the 20×20 grid was presented.

12.3.1.2 Look-Ahead Time

Because the aim was to train the neural network model using the weakest data (only current location), using a single snapshot of traffic, it was critical to examine how far in advance of the conflict the system should look. Should the ANN look two steps in advance? Four steps? Twenty steps? Notice that with a typical conflict prediction algorithm, we would expect the algorithm to perform better with a shorter time horizon (and less uncertainty in trajectories). However, the same logic does not necessarily apply to the ANN approach. It is possible, for example, that the neural network performs best by recognizing as early as possible when the pizza delivery drones first depart on their bus-route trajectories.

Fig. 12.5 Two look-ahead
intervals, for conflict (top)
and non-conflict (bottom)
traffic samples

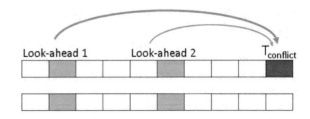

Timing of look-ahead snapshots could only be established post hoc, after traffic generation and conflict sample extraction. Figure 12.5 shows look-ahead snapshots for the conflict traffic sample at four and eight steps in advance, as an example. Notice that the look-ahead concept only applies to conflict samples. For non-conflict traffic samples, Fig. (12.5), a yoked procedure was used to randomly identify a snapshot time within the same general time range (to control somewhat for number and pattern of onscreen drones) used for the conflict sample.

Six separate look-ahead times were used (1–6 steps before $T_{conflict}$), for each combination of structure and aircraft number, and separate ANN models were developed for each of the 36 total combinations. For analysis, look-ahead time was generally collapsed into two levels of Low (1–3 steps ahead) and High (4–6 steps). However, fine grain analysis also assessed look-ahead as a continuous 1–6 steps, to see how prediction accuracy varied over time horizon.

12.3.1.3 Traffic Structure

The structure manipulation was also expected to impact traffic predictability. Again, the random condition used TCP-routes exclusively. The timing and direction of turns was randomized. The structured condition, however, used a combination of through-routes and bus-routes (nominally randomized at 50/50). Notice how much more predictable this structured condition makes the traffic. Of the 76 sector perimeter cells at which a drone might first appear in the structured condition, only 8 of these 76 cells can be associated with <u>either</u> bus-or through-routes. Any drone that enters at a perimeter cell other than these eight must be a through-flight, and its trajectory can be immediately determined (except for corner cells, which might head in one of two directions, and are therefore not immediately deterministic). This means that, given enough exposure, a neural network system could eventually determine in 64/76 or 84% of cases that a drone is a through-flight <u>on the very first step</u>. We therefore expected the structure manipulation to have a large impact on model performance.

12.3.2   ANN Design

Neural network modelling was done in NeuralDesigner v2.9, a machine learning toolbox for predictive analytics and data mining, built on the Open NN library. Modelling used a 400.3.1 architecture (i.e. 400 input nodes, a single hidden layer of 3 nodes, and a single binary output node), with standard feedforward and back propagation mechanisms, and a logistic activation function. Each of the 400 total cells was represented as an input node to the network. Each input node was simply coded on the basis of occupation, i.e. a given cell was either occupied (1) or empty (0). The output node of the ANN was simply whether the traffic pattern evolved into an eventual conflict (0/1). Maximum training iterations with each batch was set to 1000.

12.3.3   Procedures

The overall flow of the traffic generation, pre-processing, and ANN modelling process is shown in Fig. 12.6. Using a traffic generation tool, preliminary batches of 5000 traffic samples each were created. Separate batches were created for each combination of aircraft count and structure level. For each batch, samples were then automatically processed to identify conflict versus non-conflict outcomes, extract multiple look-ahead snapshots (for 1–6 steps) from conflict samples, and extract matching yoked snapshots from non-conflict samples. Target outputs were then labelled, and sample groups were fused into a final batch file. This batch file was then randomly split 60/40 into training and testing sub batch files. After training each of the 36 networks with its appropriate training sub batch file, each network was tested on its ability to classify the corresponding test sub batch file. The following section presents the results of this testing.

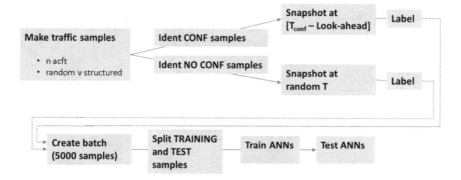

Fig. 12.6   Overview, traffic creation and model testing procedure

12.4   Results

12.4.1   Binary Classification Accuracy

The simplest performance measure is classification accuracy. That is, what percentage of samples was correctly classified as either conflict or no conflict? The ANN models each had a simple binary output: either an eventual conflict was predicted, or was not. This is a classic example of a binary classification task, which is characterized by two 'states of the world' and two possible predicted states. A binary classification table, as shown in Fig. 12.7, allows us to identify four outcomes: True Positive (TP), True Negative (TN), False Positive (FP), and False Negative (FN). In signal detection parlance, these outcomes are referred to, respectively, as: Hits, Correct Rejections, False Alarms, and Misses [9].

These four classification outcomes allow us to define the following rates:

- **Accuracy**—is the rate of proper classification, defined as: ACC = [TP + TN]/[TP + FN + TN + FP]
- **Error Rate** = 1 − ACC = [FN + FP]/[TP + FN + TN + FP]
- **True Positive Rate** (aka *sensitivity*) = TP/[TP + FN]
- **True Negative Rate** (aka *specificity*) = TN/[TN + FP]

For structured traffic, there seemed to be a ceiling effect on classification performance. Classification accuracy approached optimum (falling no lower than 0.948) regardless of traffic or look-ahead time. This means that, with structured traffic, the ANN model was able to predict almost perfectly which traffic samples would result in conflict. This was not surprising. As discussed earlier, under structured traffic the majority (84%) of drones would be predictable by the second step after sector entry. By step 3, the only uncertainty would be whether the other 16% were on through-routes or bus-routes.

Fig. 12.7 Binary classification outcomes

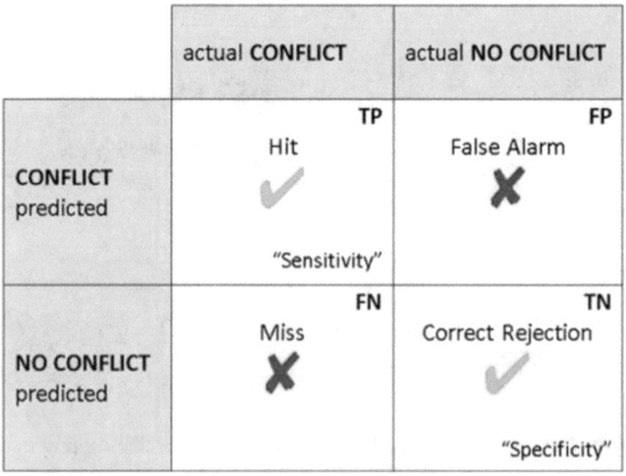

Random traffic, however, showed some variations in model performance. Classification performance with random traffic was still impressively high, ranging from 0.72 to 0.98, and generally well above chance levels (a Youdens Index value of 0.5, as shown in Table 12.2, would indicate a guessing level). However, under random traffic (Table 12.3) we began to see ML performance declines with both look-ahead time and traffic count (classification performance worsened with each), and a trend toward a three-way interaction between traffic, structure, and look-ahead.

Figure 12.8 shows the effect of both look-ahead and aircraft count, on overall classification accuracy. Data are somewhat collapsed in this view. Look-ahead (1–6) is binary split into Low (1–3) and High (4–6). Aircraft count includes only the extremes of 4 and 16. Besides a main effect of both look-ahead (longer look-ahead worsened performance) and aircraft count (higher count worsened performance), there is a slight trend toward a look-ahead x aircraft count interaction. Notice the interaction trend, whereby longer look-ahead had a greater cost under high traffic.

12.4.2   Classification Sensitivity and Specificity

For a finer-grained view, see the three panels of Fig. 12.9. These present classification performance under random traffic, for low, medium, and high aircraft count (from left to right panel). Each panel also shows the impact of look-ahead, from 1–6 steps. Notice that the pattern of Sensitivity (the TPR) and Specificity (TNR) decline vary by aircraft count. Basically, ML overall performance worsened with look-ahead time, but the underlying patterns (TPR, TNR) differed by aircraft count. For low traffic, Sensitivity fell disproportionately (i.e. the model tended toward FN rather than FP). For high traffic, Specificity fell (the system tended toward FP rather than FN). At the highest level, this interaction trend suggests that our ANN model tended to disproportionately false **positive** under the most demanding traffic samples.

12.4.3   The Extreme Scenario

To test one final, and even more challenging case, we generated traffic and trained/tested an ANN model, using random traffic, 24 aircraft, and a look-ahead of 6 (see Fig. 12.10).

Classification accuracy remained surprisingly high in this condition (in fact, slightly above the 16 drone sample). Also, this extreme scenario extended the trend toward Specificity decrement seen in the three panels of Fig. 12.9 (Table 12.4).

Table 12.2 Results summary table, binary classification performance STRUCTURED traffic

Structure	STRUCTURED																	
Traffic count	4	4	4	4	4	4	8	8	8	8	8	8	16	16	16	16	16	16
Lookahead	1	2	3	4	5	6	1	2	3	4	5	6	1	2	3	4	5	6
Classification accuracy	0.981	0.971	0.978	0.963	0.991	0.991	0.986	0.971	0.968	0.948	0.982	0.989	0.98	0.982	0.99	0.973	0.991	0.999
Error rate	0.019	0.028	0.021	0.037	0.009	0.009	0.014	0.029	0.032	0.052	0.018	0.01	0.024	0.018	0.01	0.027	0.008	0.003
Sensitivity (TP)	0.978	0.966	0.971	0.969	0.993	0.986	1	0.975	0.974	0.957	0.985	0.997	0.986	0.99	0.99	0.987	0.999	0.997
Specificity (TN)	0.983	0.976	0.984	0.959	0.99	0.994	0.96	0.962	0.957	0.931	0.977	0.977	0.904	0.928	0.936	0.872	0.944	0.992
False positive rate (FP)	0.017	0.024	0.015	0.041	0.01	0.005	0.04	0.038	0.043	0.069	0.023	0.023	0.096	0.072	0.064	0.128	0.056	0.008
False negative rate (FN)	0.024	0.034	0.029	0.031	0.007	0.014	0	0.026	0.026	0.042	0.015	0.003	0.014	0.01	0.003	0.013	0.001	0.002
Youdens index	0.961	0.942	0.956	0.928	0.982	0.98	0.96	0.938	0.931	0.888	0.962	0.974	0.9	0.92	0.93	0.859	0.943	0.99
ROCAUC	0.995	0.996	0.994	0.99	0.996	00.991	0.991	0.995	0.99	0.977	0.996	0.993	0.978	0.982	0.987	0.968	0.927	0.988

Table 12.3 Results summary table, binary classification performance, RANDOM traffic

Structure	RANDOM																	
Traffic count	4	4	4	4	4	4	8	8	8	8	8	8	16	16	16	16	16	16
Lookahead	1	2	3	4	5	6	1	2	3	4	5	6	1	2	3	4	5	6
Classification accuracy	0.984	0.959	0.947	0.913	0.89	0.825	0.976	0.93	0.88	0.854	0.814	0.726	0.98	0.924	0.866	0.8	0.77	0.724
Error rate	0.016	0.041	0.053	0.087	0.11	0.175	0.024	0.07	0.116	0.146	0.186	0.274	0.02	0.076	0.134	0.2	0.233	0.275
Sensitivity (TP)	0.974	0.913	0.847	0.821	0.796	0.699	0.98	0.911	0.83	0.846	0.751	0.707	0.98	0.951	0.903	0.85	0.856	0.654
Specificity (TN)	0.986	0.97	0.971	0.935	0.913	0.855	0.973	0.941	0.914	0.86	0.849	0.736	0.981	0.869	0.791	0.704	0.588	0.765
False positive rate (FP)	0.014	0.03	0.029	0.065	0.087	0.144	0.027	0.059	0.086	0.142	0.151	0.264	0.019	0.131	0.209	0.3	0.412	0.235
False negative rate (FN)	0.026	0.087	0.153	0.179	0.204	0.301	0.02	0.089	0.17	0.154	0.249	0.293	0.022	0.048	0.096	0.152	0.143	0.346
Youdens index	0.931	0.883	0.818	0.757	0.709	0.555	0.954	0.851	0.74	0.704	0.6	0.443	0.96	0.82	0.695	0.552	0.444	0.419
ROCAUC	0.994	0.867	0.975	0.957	0.931	0.853	0.988	0.972	0.94	0.914	0.89	0.789	0.993	0.958	0.913	0.862	0.802	0.777

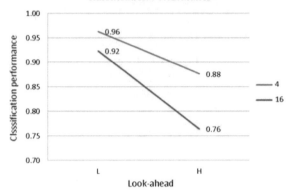

Fig. 12.8 Effect of aircraft count and look-ahead on overall classification performance, random traffic only

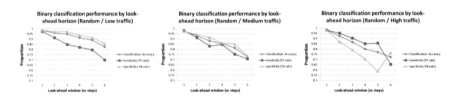

Fig. 12.9 Classification performance, sensitivity, and specificity, for random traffic

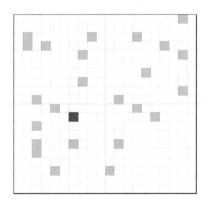

Fig. 12.10 The extreme scenario: 24 drones, random traffic, and a six-step look-ahead

Table 12.4 Results summary, binary classification performance, extreme scenario

Classification accuracy	0.739
Error rate	0.261
Sensitivity (TP)	0.705
Specificity (TN)	0.758
False positive rate (FP)	0.242
False negative rate (FN)	0.295
Youdens index	0.46
ROC AUC	0.8

12.4.4   ROC Analysis

The *Receiver Operating Characteristic* (ROC) curve [9] can be very useful for evaluating binary classification results because it integrates prediction accuracies across a few dimensions. Basically, ROC analysis plots TP rate (sensitivity) against FP rate (1-specificity) at various threshold settings (Fig. 12.11). The red 1:1 horizontal line of Fig. 12.11 indicates zero predictive capability (the model cannot distinguish between safe and conflict samples at all). As predictive value increases, the ROC curve (two curves are shown in blue) increases toward the upper left of the ROC graph. Predictive capability is generally expressed in ROC analysis as the *area under the curve* (AUC).

As shown in the last row of Table 12.2, ROC AUC showed a ceiling effect, with values generally close to 1 for structured traffic. AUC began to trail off slightly only in the most extreme scenarios: random traffic, high aircraft count, long look-ahead (Table 12.3). AUC drops off steadily with look-ahead, especially for 8 and 16 aircraft.

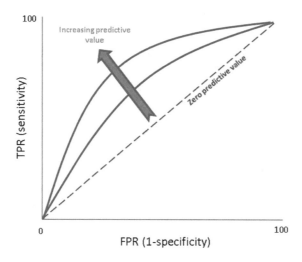

Fig. 12.11 The concept of the ROC curve

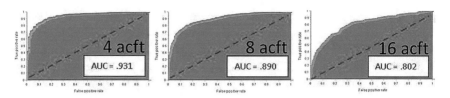

Fig. 12.12 AUC for random/5-step traffic, comparing 4 (left), 8(centre) and 16 (right) aircraft

As an example, see Fig. 12.12—AUC shows good discrimination performance for random 4/5-step samples (0.931, left), moderate performance for random 8/5-step samples (0.890, centre), and worst performance for random 16/5-step samples (0.802, right).

12.4.5 Summary of Results

In terms of conflict prediction performance, results from our ML model can be summarized as follows:

- With structured traffic, overall model performance was nearly perfect;
- With structured traffic, no effect of traffic count nor look-ahead could be found;
- For random traffic, the model still performed quite well;
- For random traffic, traffic count and (more so) look-ahead had a clear impact on overall classification accuracy;
- This look-ahead effect on accuracy revealed subtle differences in other parameters. For low traffic, Sensitivity (TP rate) declined more than did Specificity (TN rate). For high traffic, this was reversed, In other words, longer look-ahead worsened overall classification performance, but low traffic was biased toward false negatives, and high traffic was biased toward false positives;
- Even under the most challenging conditions (random, high traffic, long look-ahead), ML classification accuracy was still fairly high (76.5%);
- Even with random traffic, classification performance was generally well above guessing level, far better than chance;
- The ANN approach continued to show robust classification performance, even when presented an extreme case of 24 drones, random traffic, and long look-ahead.

12.5 Conclusions

Even with minimal data (i.e. nothing more than an instantaneous traffic snapshot), and a limited neural network architecture (without any explicit time series processing capability), neural network modelling demonstrated potential benefits in predicting and classifying drone trajectories in the urban environment. We would expect ML methods, once deployed, to show even better performance, for two main reasons.

First, and as noted elsewhere, advanced ML methods exist that can make better use of memory, time series processing, time delays, and memory erase functions. Static ANNs, which are limited in their ability to handle time series data, have clear limitations in predicting dynamic air traffic patterns. Related ML methods such as recurrent neural networks, convolutional neural networks, time delay neural networks, and long short term memory would all likely show better predictive performance.

Second, ML would presumably perform better with meaningful real world data, rather than stochastically generated flights. The neural network approach (as with all ML methods) assumes that there are some underlying patterns in the data which the network can uncover. However, the pattern and structure in our generated traffic were fairly low level. For example, our use of standard bus-route trajectories added some regularity to drone traffic movements. Nonetheless, this is a fairly shallow level of pattern and meaning. As an analogy, think of actual bus routes, that might run between residential and employment centres. The route itself is one level of pattern, but the direction and timing of movements also have some deeper meaning (to bring people to work at one time, and home at another). In our traffic sample, it is as if the buses run on the proper fixed routes, but are launched at random times and in random directions. Our paradox is therefore: how well can we assess machine learning for actual drone patterns, before there are actual drone patterns? Presumably, machine learning will do better once we have meaningful real-world data on drone traffic patterns, rather than randomly generated samples.

In summary, this analysis intentionally used limited data, and simple architectures, to enable experimental control over factors related to classification. Despite these limitations, neural network modelling provided encouraging first evidence that ML methods can be very useful in helping predict conflicts in the urban drone scenario.

Acknowledgements This research was conducted with support from the European Union's Horizon 2020 research and innovation programme, under the umbrella of the SESAR Joint Undertaking (SJU).

References

1. T. Duvall, A. Green, M. Langstaff, K. Miele, Air Mobility Solutions: What They'll Need to Take Off. McKinsey Capital Projects & Infrastructure, Feb (2019)
2. European Union, European Drones Outlook Study. SESAR Joint Undertaking. November (2016)
3. TERRA, UAS requirements for different business cases and VFR and other stakeholders derived requirements. SESAR Horizon 2020 TERRA Project, Deliverable 3.1 (2019)
4. J. Schmidhuber, Deep learning in neural networks: an overview. Neural Netw. **61**, 85–117 (2015)
5. I. Goodfellow, Y. Bengio, A. Courville, *Deep Learning* (MIT Press, Cambridge, Mass, USA, 2016)
6. Y. LeCun, Y. Bengio, Convolutional networks for images, speech, and time-series. In *The Handbook of Brain Theory and Neural Network*, ed. by M.A. Arbib (MIT Press, 1995)

7. G. Petnehází, Recurrent neural networks for time series forecasting (2019). arXiv:1901.00069
8. S. Hochreiter, J. Schmidhuber, *LSTM* can solve hard long time lag problems. In *Proceedings of the 10th Annual Conference on Neural Information Processing Systems, NIPS*, pp. 473–479, December 1996
9. D.M. Green, J.A. Swets, *Signal Detection Theory and Psychophysics* (Wiley, New York, 1966)

Chapter 13
Deep Dense Neural Network for Early Prediction of Failure-Prone Students

Georgios Kostopoulos, Maria Tsiakmaki, Sotiris Kotsiantis, and Omiros Ragos

Abstract In recent years, the constant technological advances in computing power as well as in the processing and analysis of large amounts of data have given a powerful impetus to the development of a new research area. Deep Learning is a burgeoning subfield of machine learning which is currently getting a lot of attention among scientists offering considerable advantages over traditional machine learning techniques. Deep neural networks have already been successfully applied for solving a wide variety of tasks, such as natural language processing, text translation, image classification, object detection and speech recognition. A great deal of notable studies has recently emerged concerning the use of Deep Learning methods in the area of Educational Data Mining and Learning Analytics. Their potential use in the educational field opens up new horizons for educators so as to enhance their understanding and analysis of data coming from varied educational settings, thus improving learning and teaching quality as well as the educational outcomes. In this context, the main purpose of the present study is to evaluate the efficiency of deep dense neural networks with a view to early predicting failure-prone students in distance higher education. A plethora of student features coming from different educational sources were employed in our study regarding students' characteristics, academic achievements and activity in the course learning management system. The experimental results reveal that Deep Learning methods may contribute to building more accurate predictive models, whilst identifying students in trouble soon enough to provide in-time and effective interventions.

G. Kostopoulos (✉) · M. Tsiakmaki · S. Kotsiantis · O. Ragos
Educational Software Development Laboratory (ESDLab), Department of Mathematics,
University of Patras, Patras, Greece
e-mail: kostg@sch.gr

M. Tsiakmaki
e-mail: m.tsiakmaki@gmail.com

S. Kotsiantis
e-mail: sotos@math.upatras.gr

O. Ragos
e-mail: ragos@math.upatras.gr

G. A. Tsihrintzis and L. C. Jain (eds.), *Machine Learning Paradigms*,
Learning and Analytics in Intelligent Systems 18,
https://doi.org/10.1007/978-3-030-49724-8_13

Keywords Deep learning · Educational data mining · Learning analytics · Student performance · Early prediction

13.1 Introduction

Educational Data Mining (EDM) is a constantly evolving interdisciplinary research field mainly centered on the development and application of data mining methods to datasets coming from a variety of learning environments [1]. EDM does not simply aim at extracting interesting and useful knowledge from students' raw data, but it is essentially focused on resolving important educational problems [2]. Depending on the problem to be resolved, EDM utilizes representative methods of different but interdependent fields, such as data mining, machine learning and statistics to name a few [3]. These methods include, but are not limited to, classification, regression, clustering, process mining, social network analysis and association rule analysis [1].

In recent years, the constant technological advances in computing power as well as in the processing and analysis of large amounts of data have given a powerful impetus to the development of a new research area. Deep Learning (DL) is a burgeoning subfield of machine learning which is currently getting a lot of attention among scientists offering considerable advantages over traditional techniques [4], such as transfer learning across relevant domains and automatic feature extraction [5]. Deep neural networks have already been successfully applied for solving a wide variety of tasks, such as natural language processing, text translation, image classification, object detection and speech recognition. A great deal of notable studies has recently emerged concerning the use of DL methods in the area of EDM and Learning Analytics (LA) [6]. Their potential use in the educational field constitutes a great challenge for researchers and opens up new horizons for processing and analyzing large volumes of data coming from varied educational settings [7].

In this context, the main purpose of the present study is to evaluate the efficiency of a deep dense neural network with a view to early predicting failure-prone students in distance higher education. A plethora of features coming from different educational sources are employed regarding students' characteristics, academic achievements and online activity in the course learning management system (LMS). The experimental results reveal that the proposed deep network prevails over familiar classification algorithms in terms of F_1-score and AUC (Area Under Curve). Moreover, it may serve as a tool for early estimation of students at risk of failure, thus enabling timely support and effective intervention strategies.

The remaining of the chapter is structured as follows: Sect. 13.2 reviews recent studies concerning the implementation of deep networks in the educational field. The proposed deep dense neural network is presented and described in Sect. 13.3. Section 13.4 includes the experiments carried out along with the relevant results, while Sect. 13.5 concludes the findings of the research.

13.2 Literature Review

Although deep neural networks have achieved notable results in a wide variety of tasks such as artificial grammar learning, automatic text generation, text translation, image classification and speech recognition [8], the number of studies exploring the effectiveness of DL architectures in the EDM area is really restricted [7]. A systematic literature review concerning the implementation of deep networks in the area of EDM and LA is presented in [6]. As a follow up to this review, a number of notable studies, which have emerged recently in the educational field, are discussed in the next few paragraphs. The main core of these studies concerns classification problems (binary or multiclass), such as the prediction of student performance, final grade or dropout from a distance learning course.

Guo et al. [9] developed a pre-warning system for students at risk of failing, called Students Performance Prediction Network (SPPN), on the base of a DL multiclass classification model. A six-layer feedforward network was built composed of an input layer, four hidden layers and an output layer. The input layer was fed with students' data and, consequently, an unsupervised learning algorithm was employed to pre-train the four hidden layers using the Rectified Linear Unit (ReLU) as activity function. After initialization of network weights and fine tuning of network through backpropagation, the softmax function was used in the five-neuron output layer to predict students' final score {O, A, B, C, D}. The results showed that SPPN acquired the highest accuracy values compared to three familiar classification algorithms such as Multi-Layer Perceptron (MLP), Naïve Bayes (NB) and Support Vector Machines (SVM).

Okubo et al. [10] built a Recurrent Neural Network (RNN) for predicting students' grades in a 15-week distance course. This network comprised an input layer, a hidden layer and an output one. The RNN was trained with backpropagation using time series students' log data per week, while a Long Short-Term Memory (LSTM) network was employed as a middle layer unit to preserve the backpropagated error through time. The accuracy of the model was higher than 90% in the sixth study week, thus indicating its efficiency for the early prediction of students at risk of failure.

Deep networks have been proven to be quite effective for predicting student dropout or final performance in Massive Open Online Courses (MOOCs), two typical and well-studied binary problems in the EDM field [11]. Wang et al. [12] proposed a DL model for predicting student dropout in MOOCs. Convolutional Neural Networks (CNNs) and RNNs were combined in a new architecture comprising an input layer, an output layer and six hidden layers (a recurrent, a fully connected, two convolutional and two pooling layers), while applying the ReLU activity function. The experimental results on data from the KDD Cup 2015 revealed that the proposed ConRec network showed a satisfactory performance compared to typical classification methods in terms of precision, recall, F_1-score and AUC.

In a similar work, Xing and Du [13] constructed a temporal DL model for estimating dropout probability of individual MOOC students on a weekly basis. More specifically, a personalized deep neural network was built for each student using forward propagation to train the weights and backward propagation with a batch gradient descent cost function to update the values of weights. The obtained results, considering accuracy and AUC as performance metrics, demonstrated the superiority of DL approach over three representative supervised methods: k-Nearest Neighbors (k-NN), SVM and C4.5 decision tree.

Quite recently, Wang et al. [14] formulated an LSTM-based RNN model for predicting student performance {pass, fail} when solving an open-ended programming exercise in a MOOC platform. The model was trained using as input several programs embedding sequences produced by students when attempting to solve an exercise, while the obtained results indicated an overall accuracy improvement of 5% compared to the baseline method. Moreover, the model was able to understand students' learning behavior and group them into different clusters.

In the same context, Kim et al. [15] implemented a deep learning-based algorithm, called GriNet, for early prediction of students' graduation performance {pass, fail} in two 8-week Udacity Nanodegree programs. The network consisted of an embedding layer, a Bidirectional LSTM (BLSTM) layer, a GMP layer, a fully connected layer and a softmax output layer. Student raw activity records along with their corresponding time stamp were transformed into a vector sequence and subsequently were passed on to BLSTM and GMP layers. Finally, the GMP output was fed sequentially into the output layer for computing students' log-likelihood and estimating their outcomes. The results indicated the predominant effectiveness of GriNet for the early prediction of at-risk students.

Whitehill et al. [16] proposed a deep and fully connected, feed-forward neural network architecture, utilizing the Net2Net training methodology, for predicting student certification in forty MOOC courses. Therefore, they varied the number of hidden layers, as well as the number of nodes per hidden layer, while measuring the prediction accuracy each time. The experimental outputs revealed that encompassing five hidden layers with five neurons in each hidden layer results in a highly accurate deep network.

In addition, there are a few studies related to recommendation tasks, sentiment analysis of students, Automatic Short Answer Grading (ASAG) and feature extraction. Tang et al. [8] applied RNN models in two datasets comprising: (a) essays written by students in a summative environment, and (b) MOOC clickstream data. The deep model consisted of two LSTM layers containing 128 hidden nodes each one and applied for either predicting the next word in a sequence of words, given a 'seed' context, or for automatically suggesting a learning resource in case of struggling while writing an essay.

Wei et al. [17] introduced a transfer learning framework, termed ConvL, for detecting confusion feeling, urgency priority and sentiment attitude expressed in MOOC forum posts. The ConvL architecture incorporated a CNN layer for extracting local contextual features and an LSTM layer for computing the representation of expressions by considering the long-term temporal semantic relationships of features. Three data sources from different domain areas were selected, while each problem was addressed as a binary classification problem with values {positive, negative}. In the same context, Zhou et al. [18] employed an extended LSTM deep network for building a full-path recommendation model. Zhang et al. [19] employed Deep Belief Networks (DBN) to facilitate the task of (ASAG). To this end, they utilized stacked Restricted Boltzmann Machine (RBM) layers for automatic extraction of the most representative features representing information about the answers, the questions and the students. Finally, Bosch et al. [7] proposed a DL network consisted of an input layer, two LSTM layers, a fully connected embedding layer, three hidden layers and an output layer. The network was initially trained on data regarding students' interaction within an e-learning platform and subsequently applied for automatic feature extraction and diagnosis of student boredom.

The above discussed studies employ mostly CNNs and LSTMs and are summarized in Table 13.1, whilst providing information about the structure and the source of the datasets used in each study, the data mining task, the methods used for comparing the performance of the constructed deep network and the evaluation metric employed.

13.3 The Deep Dense Neural Network

Deep learning has strong roots in Artificial Intelligence and has been developed as an independent and highly applicable field over the last few years [20]. DL architectures utilize models that consist of an input layer, an output layer and two or more hidden layers of nonlinear processing units that perform supervised or unsupervised feature extraction and transformation functions [7]. These layers are arranged in a hierarchical structure where the input of each layer is the output of the previous one, computed through an appropriate selected activation function. Each hidden layer corresponds to a different representation of the input space and provides a different abstraction level that is hopefully more useful and closer to the final goal [5].

An input layer, two dense layers and an output layer are the basis of the architecture of the proposed Deep Dense Neural Network, hereinafter referred to as DDNN (Fig. 13.1). The input layer consists of 14 input units corresponding to each one of the input features. The first hidden layer (L_2) has 12 hidden units and the second one (L_3) has 6 hidden units. Both layers L_2 and L_3 use the sigmoid activation function. The output layer (L_4) consists of a single neuron, while employing the softmax activation function and the Squared Hinge loss function [21] for predicting the output class. In addition, the network weights are initialized using the Xavier initialization method [22]. The configuration of the DDNN parameters is presented in Table 13.2.

Let $x_1, x_2, \ldots, x_{14}$ denote the inputs of the dataset, $w_{ij}^{(l)}$ the weights of the connection between the j node in l layer and the i node in the $l + 1$ layer and $a_i^{(l)}$ the output of the i node in the l layer. Therefore, the computation of the DDNN is given from the following equations:

$$a_i^{(2)} = f_1 \left(\sum_{j=1}^{14} w_{ij}^{(1)} x_j \right), i = 1, 2, \ldots, 12$$

Table 13.1 Summarizing the discussed DL studies in the EDM field

Paper	#Datasets	#Features	#Instances	Source data	Output feature	DM task	Compared methods	Metrics	Early prediction
Guo et al. [9]	1	N/S	120,000	High school	Final grade {O, A, B, C, D}	Classification	MLP, SVM, NB	Accuracy	No
Okubo et al. [10]	1	9	108	LMS	Final grade {A, B, C, D, F}	Classification	Multiple regression	Accuracy	Yes
Wang et al. [12]	1	8	120,542	MOOC	Dropout {yes, no}	Classification	SVM, LR, DT, RF, AdaBoost, Gaussian NB, Gradient Tree Boosting	Precision, Recall, F-score, AUC	Yes
Xing and Du [13]	1	13	3,617	MOOC	Dropout {yes, no}	Classification	SVM, k-NN, C4.5	Accuracy, AUC	Yes
Wang et al. [14]	1	N/S	79,553	MOOC	Performance {pass, fail}	Classification	Logistic Regression (LR)	Precision, Recall	No
	2	N/S	10,293	MOOC	Progress {A, B, C, D, E}	Clustering		F-score	–

(continued)

Table 13.1 (continued)

Paper	#Datasets	#Features	#Instances	Source data	Output feature	DM task	Compared methods	Metrics	Early prediction
Kim et al. [15]	1	N/S	1,853	MOOC	Performance	Classification	LR	AUC	Yes
	2	N/S	8,301	MOOC	{pass, fail}				
Whitehill et al. [16]	1	19	528,349	MOOC	Certification	Classification	LR	AUC	Yes
					{yes, no}				
Wei et al. [17]	1	N/S	9,879	MOOC	Confusion	Classification	Variants of CNN, LSTM	Accuracy	No
					{pos, neg}				
	2	N/S	5,184	MOOC	Urgency	Classification		Accuracy	No
					{pos, neg}				
	3	N/S	3,030	MOOC	Sentiment	classification		Accuracy	No
					{pos, neg}				
Zhou et al. [18]	1	N/S	100,000	E-learning	Learning path	Recommendation	Clustering	Accuracy	–
				Platform			Clustering and LSTM		
Zhang et al. [19]	1	N/S	16,228	E-learning	Automatic	modeling	LR, NB, SVB, DT,	Accuracy	–
				Platform	Short answer	Clustering	ANNs	Precision	
					Grading			Recall, AUC	

(continued)

Table 13.1 (continued)

Paper	#Datasets	#Features	#Instances	Source data	Output feature	DM task	Compared methods	Metrics	Early prediction
								F-score	
Bosch and Paquette [7]	1	3	94	E-learning Platform	Boredom	Classification	Gaussian NB, LR, DT	AUC	–
Tang et al. [8]	1	541	472	Kaggle	Next word	Time series	–	Accuracy	–
	2	25	8,094	MOOC	Resource	Recommendation	–	Accuracy	–

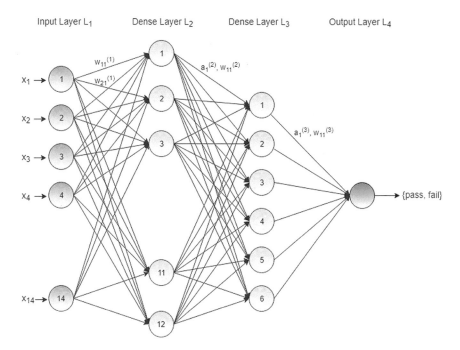

Fig. 13.1 DDNN architecture

Table 13.2 DDNN parameters configuration

Number of layers	4
Layer L_1 (input layer)	
Number of units	14
Layer L_2 (1st hidden layer)	
Number of units	12
Activation function	Sigmoid
Layer L_3 (2nd hidden layer)	
Number of units	6
Activation function	Sigmoid
Layer L_4 (output layer)	
Number of units	1
Activation function	Softmax
Loss function	Squared Hinge
Number of epochs	10
Weights initialization method	Xavier
Optimization algorithm	Stochastic gradient descent
Updater	Adam
Bias updater	Sgd

$$a_i^{(3)} = f_1 \left(\sum_{j=1}^{12} w_{ij}^{(2)} a_j^{(2)} \right), i = 1, 2, \ldots, 6$$

$$a_1^{(4)} = f_2 \left(\sum_{j=1}^{6} w_{1j}^{(3)} a_j^{(3)} \right)$$

where f_1, f_2 correspond to the sigmoid and softmax activation function respectively and are defined as follows:

$$f_1(x) = \frac{1}{1 + \exp(-x)}$$

$$f_2(x_i) = \frac{\exp(x_i)}{\sum_{j=1}^{2} \exp(x_j)}, i = 1, 2$$

13.4 Experimental Process and Results

The dataset used in this study involves 1,073 students having attended a full-year course of the Hellenic Open University. Data regarding students' characteristics, performance achievements and online activity in the LMS were gathered during the academic year 2013–2014. Therefore, the dataset comprises 14 input features in total (Table 13.3). The binary output feature 'Final' refers to students' performance {pass, fail} in the course examinations which take place at the end of the academic year. The distribution of the output classes {pass, fail} was 21.9% and 78.1% respectively.

For the purpose of our research the academic year was partitioned into four successive time-periods (hereinafter called steps) depending on the dates of the two first out of five optional contact sessions (Ocs) and the first two written assignments (Wri) submission deadline, as shown in Fig. 13.2. It should be mentioned that the only available information at the beginning of the course concerns gender and if an individual student was a first-year one or he/she failed to pass the final exams in the previous academic year and enrolled in the course again.

The experiments were executed in four steps according to the four time-periods described in Fig. 13.2. Each step included all available information about students gathered from the beginning of the academic year to the end of the specific time-period. Note that all the features were updated with new information over time, while the final 4th step took place before the middle of the academic year, thus allowing in-time support and effective interventions to failure-prone students.

The performance of the proposed deep network was tested at each one of the experiments steps against eight familiar classification algorithms, which are briefly presented below:

Table 13.3 Dataset features

Features	Description	Values
Gender	Student gender	{male, female}
NewStudent	A first-year student is characterized as a new one	{yes, no}
Wri_i, i = 1, 2	Student grade in the i-th written assignment	[0, 10]
Ocs_i, i = 1, 2	Student absence in the i-th optional contact session	{0, 1}
L_1	Total logins	Integer
L_2	Number of views in pseudo-code forum	Integer
L_3	Number of views in compiler forum	Integer
L_4	Number of views in module forum	Integer
L_5	Number of views in course forum	Integer
L_6	Number of views in course news	Integer
L_7	Number of posts in pseudo-code forum	Integer
L_8	Number of posts in compiler forum	Integer
L_9	Number of posts in module forum	Integer
L_{10}	Number of posts in course forum	Integer
Final	Student performance in the final course examinations	{pass, fail}

Fig. 13.2 Time-periods of the academic year

- The k-Nearest Neighbor (k-NN) instance-based learning algorithm [23], a nonparametric method which is widely used for both classification and regression problems.
- The C4.5 decision tree induction algorithm [24] introduced by Quinlan, which is based on the ID3 algorithm [25].
- The JRip rule-based algorithm, a java variant of the RIPPER (Repeated Incremental Pruning to Produce Error Reduction) algorithm [26], which is considered quite efficient, particularly for handling large-scale datasets with noisy data.
- Multilayer Perceptron (MLP) [27], a feedforward artificial neural network which employs backpropagation for training the model.
- The Naïve Bayes (NB) discriminative classifier, a very effective algorithm for discrete and continuous features, which is based on the Bayes theorem and works under the assumption that all features are independent given the class label [28].
- Random Forests (RFs) [29], a simple, powerful and robust ensemble method for both classification and regression problems.
- The Stochastic Gradient Descent (SGD) algorithm [30].

- The Sequential Minimal Optimization (SMO) classifier [31], a simple and fast algorithm for training Support Vector Machines (SVM) [32].

For evaluating the performance of classifiers, we considered two widely used metrics which are often preferred over accuracy in cases of imbalanced datasets [33] as in our case: F_1-score and AUC. F_1-score or balanced F-score is the harmonic mean of precision (p) and recall (r) and is defined as follows:

$$F_1\text{-score} = \frac{2pr}{p+r} = \frac{2TP}{FP + 2TP + FN}$$

where TP, FP, TN, FN correspond to true positive, false positive, true negative and false negative cases respectively and are computed from the following confusion matrix (Table 13.4).

AUC corresponds to the area under ROC (Receiver Operating Characteristics) curve and is considered to be a better metric than accuracy [34]. For binary classification problems, values of AUC near to 1 mean that the model separates the two output classes quite precisely.

At first, the dataset was partitioned randomly into 10 equally sized folds using the cross-validation resampling procedure. Therefore, 9 folds formed the training set for training the classification model and the other one was used as the testing set for evaluating the model. The procedure was repeated 10 times until all folds were utilized as the testing set and the results were averaged [35]. The experimental results regarding the F_1-score and AUC metrics are presented in Tables 13.5 and 13.6 respectively, while the highest scores in each one of the four steps are bold highlighted. Overall, it is observed that the proposed deep network achieves consistently better results than the other classification methods regardless of the performance metric applied, as well as the time-period.

F_1-score ranges from 0.782 in the first step to 0.819 in the fourth step (i.e. before the middle of the academic year). As regards the AUC metric, DDNN takes significant precedence over the rest classification methods, with values ranging from 0.804 to 0.882. Therefore, students at risk of failure may be early detected with a sufficient amount of accuracy in order to provide timely support actions and intervention strategies.

In addition, for detecting the specific differences of all classifiers, we applied the Friedman Aligned Ranks non-parametric test [36] and the Finner post hoc test [37]. According to the Friedman test, the algorithms are ordered from the best performer (lower ranking value) to the worst one. The DDNN method prevails in both cases,

Table 13.4 Confusion matrix

		Predicted	
		Fail	Pass
Actual	Fail	TP	FN
	Pass	FP	TN

Table 13.5 F$_1$-score results

Classifier	Step			
	1st	2nd	3rd	4th
DDNN	**0.782**	**0.802**	0.798	0.819
C4.5	0.766	0.790	0.791	0.799
RFs	0.746	0.772	0.784	0.782
SMO	0.736	0.762	0.774	**0.825**
3-NN	0.752	0.768	0.764	0.794
NB	0.768	0.786	0.785	0.794
MLPs	0.766	0.783	0.795	0.798
JRip	0.781	0.796	**0.800**	0.810
SGD	0.741	0.766	0.777	0.815

Table 13.6 AUC results

Classifier	Step			
	1st	2nd	3rd	4th
DDNN	**0.804**	**0.840**	**0.847**	**0.882**
C4.5	0.626	0.724	0.691	0.693
RFs	0.756	0.805	0.826	0.840
SMO	0.558	0.595	0.613	0.718
3-NN	0.740	0.747	0.770	0.796
NB	0.786	0.822	0.829	0.854
MLPs	0.800	0.808	0.826	0.841
JRip	0.648	0.655	0.691	0.726
SGD	0.584	0.601	0.619	0.703

as shown in Tables 13.7 and 13.8, while the null hypothesis (the means of the results of the algorithms are the same) is rejected (significance level $a = 0.1$). Finally, the results of the Finner post hoc test, using DDNN as control method, are presented in Tables 13.7 and 13.8. It is therefore evident that DDNN obtains the best results regarding both F$_1$-score and AUC, whilst all the other methods seem to be less competitive in both metrics.

13.5   Conclusions

In the present study, an effort was made to propose and evaluate the efficiency of a deep dense neural network for early prognosis of failure-prone students in distance higher education. Since the latest research on ML has placed a lot of emphasis on DL,

Table 13.7 Statistical results (F_1-score, $a = 0, 1$)

Algorithm	Friedman ranking	Finner post hoc test	
		Adjusted p-value	Null hypothesis
DDNN	4.25000	–	–
JRip	7.50000	0.66265	Accepted
C4.5	15.37500	0.15499	Accepted
MLPs	15.87500	0.15499	Accepted
NB	18.62500	0.08446	Rejected
SGD	24.25000	0.01447	Rejected
SMO	25.00000	0.01420	Rejected
RFs	27.50000	0.01076	Rejected
3-NN	28.12500	0.01076	Rejected

Table 13.8 Statistical results (AUC, $a = 0, 1$)

Algorithm	Friedman ranking	Finner post hoc test	
		Adjusted p-value	Null hypothesis
DDNN	2.75000	–	–
NB	8.50000	0.44022	Accepted
MLPs	9.62500	0.39534	Accepted
RFs	13.12500	0.21211	Accepted
3-NN	18.50000	0.05463	Rejected
JRip	24.62500	0.00663	Rejected
C4.5	25.37500	0.00636	Rejected
SGD	31.25000	0.00052	Rejected
SMO	32.75000	0.00045	Rejected

we addressed the problem of student performance prediction from a deep learning perspective. A plethora of 1,073 postgraduate student features were employed in our research regarding students' characteristics, academic achievements and activity in the course LMS. The experimental results revealed that the proposed deep network prevailed over familiar classification algorithms in terms of F_1-score and AUC. More specifically, F_1-score ranged from 0.782 to 0.819, while AUC values ranged from 0.804 to 0.882 before the middle of the academic year. Therefore, the proposed deep network may serve as an automatic warning tool for early estimation of students at risk of failure, enabling timely support and effective intervention strategies, with a view to improving their overall performance.

References

1. C. Romero, S. Ventura, Data mining in education. Wiley Interdiscip. Rev. Data Min. Knowl. Discov. **3**(1), 12–27 (2013)
2. C. Romero, S. Ventura, Educational data mining: a review of the state of the art. IEEE Trans. Syst. Man, Cybern. Part C (Appl. Rev.) **40**(6), 601–618 (2010)
3. N. Bousbia, I. Belamri, Which contribution does EDM provide to computer-based learning environments? in *Educational data mining* (Springer, 2014), pp. 3–28
4. Y. LeCun, Y. Bengio, G. Hinton, Deep learning. Nature **521**(7553), 436 (2015)
5. J. Patterson, A. Gibson, *Deep Learning: A practitioner's Approach* (O'Reilly Media, Inc., 2017)
6. O.B. Coelho, I. Silveira, Deep learning applied to learning analytics and educational data mining: a systematic literature review, in *Brazilian Symposium on Computers in Education (Simpósio Brasileiro de Informática na Educação-SBIE)*, vol. 28, no. 1, (2017) p. 143
7. N. Bosch, L. Paquette, Unsupervised deep autoencoders for feature extraction with educational data, in *Deep Learning with Educational Data Workshop at the 10th International Conference on Educational Data Mining* (2017)
8. S. Tang, J.C. Peterson, Z.A. Pardos, Deep neural networks and how they apply to sequential education data, in *Proceedings of the Third (2016) ACM Conference on Learning@ Scale* (2016), pp. 321–324
9. B. Guo, R. Zhang, G. Xu, C. Shi, L. Yang, Predicting students performance in educational data mining, in *2015 International Symposium on Educational Technology (ISET)* (2015), pp. 125–128
10. F. Okubo, T. Yamashita, A. Shimada, H. Ogata, A neural network approach for students' performance prediction, in *Proceedings of the Seventh International Learning Analytics & Knowledge Conference* (2017), pp. 598–599
11. P.M. Moreno-Marcos, C. Alario-Hoyos, P.J. Muñoz-Merino, C.D. Kloos, Prediction in MOOCs: a review and future research directions. IEEE Trans. Learn. Technol. (2018)
12. W. Wang, H. Yu, C. Miao, Deep model for dropout prediction in moocs, in *Proceedings of the 2nd International Conference on Crowd Science and Engineering* (2017), pp. 26–32
13. W. Xing, D. Du, Dropout prediction in MOOCs: using deep learning for personalized intervention, J. Educ. Comput. Res. (2018), p. 0735633118757015
14. L. Wang, A. Sy, L. Liu, C. Piech, Deep knowledge tracing on programming exercises, in *Proceedings of the Fourth (2017) ACM Conference on Learning Scale* (2017), pp. 201–204
15. B.-H. Kim, E. Vizitei, V. Ganapathi, GritNet: Student performance prediction with deep learning (2018). arXiv:1804.07405
16. J. Whitehill, K. Mohan, D. Seaton, Y. Rosen, D. Tingley, Delving deeper into MOOC student dropout prediction (2017). arXiv:1702.06404
17. X. Wei, H. Lin, L. Yang, Y. Yu, A convolution-LSTM-based deep neural network for cross-domain MOOC forum post classification. Information **8**(3), 92 (2017)
18. Y. Zhou, C. Huang, Q. Hu, J. Zhu, Y. Tang, Personalized learning full-path recommendation model based on LSTM neural networks. Inf. Sci. (Ny) **444**, 135–152 (2018)
19. Y. Zhang, R. Shah, M. Chi, Deep learning+student modeling+clustering: a recipe for effective automatic short answer grading, in *EDM* (2016), pp. 562–567
20. L. Deng, D. Yu, others, Deep learning: methods and applications. Found. Trends® in Signal Process. **7**(3–4), 197–387 (2014)
21. K. Janocha, W.M. Czarnecki, On loss functions for deep neural networks in classification (2017). arXiv:1702.05659
22. X. Glorot, Y. Bengio, Understanding the difficulty of training deep feedforward neural networks, in *Proceedings of the Thirteenth International Conference on Artificial Intelligence and Statistics* (2010), pp. 249–256
23. D.W. Aha, D. Kibler, M.K. Albert, Instance-based learning algorithms. Mach. Learn. **6**(1), 37–66 (1991)

24. J.R. Quinlan, *C4. 5: programs for machine learning* (Elsevier, 2014)
25. J.R. Quinlan, Induction of decision trees. Mach. Learn. **1**(1), 81–106 (1986)
26. W.W. Cohen, Fast effective rule induction, in *Machine Learning Proceedings 1995* (Elsevier, 1995), pp. 115–123
27. F. Rosenblatt, Principles of neurodynamics. Perceptrons and the theory of brain mechanisms (1961)
28. I. Rish, others, An empirical study of the naive Bayes classifier, in *IJCAI 2001 Workshop on Empirical Methods in Artificial Intelligence*, vol. 3, no. 22 (2001), pp. 41–46
29. L. Breiman, Random forests. Mach. Learn. **45**(1), 5–32 (2001)
30. J. Kiefer, J. Wolfowitz, others, Stochastic estimation of the maximum of a regression function. Ann. Math. Stat. **23**(3), 462–466 (1952)
31. J. Platt, Sequential minimal optimization: a fast algorithm for training support vector machines (1998)
32. O. Chapelle, V. Vapnik, O. Bousquet, S. Mukherjee, Choosing multiple parameters for support vector machines. Mach. Learn. **46**(1–3), 131–159 (2002)
33. M. Sokolova, N. Japkowicz, S. Szpakowicz, Beyond accuracy, F-score and ROC: a family of discriminant measures for performance evaluation, in *Australasian Joint Conference on Artificial Intelligence* (2006), pp. 1015–1021
34. J. Huang, C.X. Ling, Using AUC and accuracy in evaluating learning algorithms. IEEE Trans. Knowl. Data Eng. **17**(3), 299–310 (2005)
35. Y. Jung, J. Hu, AK-fold averaging cross-validation procedure. J. Nonparametr. Stat. **27**(2), 167–179 (2015)
36. J.L. Hodges, E.L. Lehmann, others, Rank methods for combination of independent experiments in analysis of variance. Ann. Math. Stat. **33**(2), 482–497 (1962)
37. H. Finner, On a monotonicity problem in step-down multiple test procedures. J. Am. Stat. Assoc. **88**(423), 920–923 (1993)

Part VI
Evaluation of Algorithm Performance

Chapter 14
Non-parametric Performance Measurement with Artificial Neural Networks

Gregory Koronakos and Dionisios N. Sotiropoulos

Abstract The large volume of information and data dictates the use of data analytics methods that provide the organizations with valuable insights to improve their performance. Data Envelopment Analysis (DEA) is the dominating non-parametric and "data driven" technique that is based on linear programming for performance assessment. While the computational effort in regular efficiency assessments is negligible, in large scale applications solving successively optimization problems increases dramatically the computational load. In this chapter, we develop a novel approach for large scale performance assessments, which integrates Data Envelopment Analysis (DEA) with Artificial Neural Networks (ANNs) to accelerate the evaluation process and reduce the computational effort. We examine the ability of the ANNs to estimate the efficiency scores of the fundamental DEA models as well as to capture the performance pattern of cross-efficiency that improves the discrimination power of DEA. We take into account the relative nature of DEA by securing that the DMUs used for training and testing the ANN are first evaluated against the efficient set. We carry out several experiments to explore the performance of the ANNs on estimating the DEA scores. Our experiments are of different size and contain data generated from different distributions. We illustrate that our approach, beyond the computational savings, estimates adequately the DEA efficiency scores that also satisfy on average the DEA properties, thus it can be a useful tool for the performance evaluation of large datasets.

G. Koronakos (✉) · D. N. Sotiropoulos
Department of Informatics, University of Piraeus, 80, Karaoli and Dimitriou, 18534 Piraeus, Greece
e-mail: gkoron@unipi.gr; gregkoron@gmail.com

D. N. Sotiropoulos
e-mail: dsotirop@unipi.gr

© The Editor(s) (if applicable) and The Author(s), under exclusive license
to Springer Nature Switzerland AG 2020
G. A. Tsihrintzis and L. C. Jain (eds.), *Machine Learning Paradigms*,
Learning and Analytics in Intelligent Systems 18,
https://doi.org/10.1007/978-3-030-49724-8_14

14.1 Introduction

The volatile environment and the presence of large volumes of information and data in most public and private sectors, render imperative the use of sophisticated data analytics methods that provide valuable insights and aid the decision making. In particular, firms and organizations resort to such methods to explore all aspects of their operations, understand the sources that may cause inefficiency and eventually improve their performance. The principal characteristic of the data analytics methods is that the obtained information derives exclusively from the data, i.e. the data speak for themselves.

A leading non-parametric and "*data driven*" technique for performance assessment is Data Envelopment Analysis (DEA). DEA does not require any pre-defined assumption about the functional relationship of the data and is employed for the evaluation of the relative efficiency of a set of homogeneous entities, called Decision Making Units (DMUs). Great attention has been paid to DEA from the research community as well as from operations analysts, because of its practical usefulness on providing performance measures and handling effectively the sources of inefficiency. Thus, the application field of DEA is wide range, for instance DEA has been applied for the performance evaluation of banks, airports, universities, industries, countries, stock market, etc.

Depending on the context of the performance assessment, the analysis may consist of either historical records of large scale or new data processed for real-time analytics, e.g. auditing, stock market, etc. While the computational demand in common efficiency assessment exercises is not an issue, the concern becomes sharp when the evaluated entities are numerous with high-cardinality data, the assessment needs to be performed quickly or the data stream in at high rates. Typically, applying DEA for classifying and scoring requires the successive solution of a linear program (LP) for each entity under evaluation. However, the size of the optimization problems depends on the number of the entities and the cardinality of the data that each of them involve. Dulá [15] described the assessment of fraud detection for credit cards as a dynamic data environment, where new entities arrive over time. As outputs are considered the transaction value, the special purchases and the distance from home, while the time between purchases is considered as input. Admittedly, this type of evaluation is computationally intensive and render impractical the traditional DEA methods, thus smart computational techniques should be developed to reduce the overall computational load and accelerate the analysis.

In the context of DEA, several pre-processors have been developed [10, 14] for the detection and classification of the entities to efficient and inefficient ones, as well as for the reduction of the optimization burden that follows. Although pre-processing the data before the optimization process is critical for huge-scale applications, it is only expected to classify the entities under evaluation and not to provide their efficiency scores. Thus, another research trend aims to deal with this issue in performance evaluation by means of machine learning techniques. Specifically by combining artificial neural networks (ANN) and DEA so as to estimate the efficiency of the entities in the presence of large scale data.

Athanassopoulos and Curram [6] generated an artificial production technology, based on a Cobb-Douglas production function, to compare the true efficiency of the evaluated units with DEA and ANNs. In the context of DEA, they employed the output oriented CCR and BCC models. They found that the BCC model was relatively superior comparing to CCR model and ANN. Also, they used empirical data of 250 commercial bank branches with 4 inputs and 3 outputs to compare DEA and ANN, but the ranking of DMUs obtained from the ANN is more similar with the corresponding one obtained from the CCR model. The annual performance of London underground for the years 1970–1994 was measured in [11]. The number of trains and workers were assumed as inputs, while the total train distance per year covered is assumed as output. It was concluded that the ANNs provide similar results to the CCR and the BCC models. Santin et al. [23] based on a non-linear production function with one input and one output carried out a comparison of ANNs and parametric and non-parametric performance evaluation methods. They noticed that ANNs are a promising alternative to traditional approaches under different observations number and noise scenarios. In [26] DEA and ANNs are integrated for the assessment of the branch performance of a big Canadian bank. They noticed that the efficiency scores obtained from the CCR model with the corresponding ones estimated by the ANN are comparable. In [17] a random set of DMUs was selected for training an ANN to estimate the CCR efficiency scores. Five large datasets were used to derive that the predicted efficiency scores from the ANN appear to be good estimates. In the same line of though, Hanafizadeh et al. [20] employed ANN to estimate the efficiency of a large set of mutual funds in USA so as to reduce the response time to the investors. DEA and ANNs are also used in conjunction in [2–5] for the estimation of the efficiencies of the DMUs. Notice that in these studies, ANNs are initially used to discover the relation between the inputs and outputs of the DMUs, i.e. the production function.

In this chapter, we develop a novel approach for large scale assessments, which integrates DEA with ANNs to estimate the efficiency scores of DMUs. In this way, the evaluation process is accelerated and the computational effort is reduced. A major drawback of the studies that estimate efficiency scores through ANNs is that they neglect the relative nature of the DEA estimator, i.e. the DEA efficient frontier (efficient DMUs) act as benchmark for scoring the non-frontier DMUs. Notice that the notion of efficiency is different between DEA and ANN. An ANN learns the central tendency of the data while DEA builds an empirical best practice (efficient) frontier that envelops them. In our approach, we assure that the DMUs used for training the ANN are first evaluated against the efficient set. For this purpose, we first identify the efficient DMUs of a given dataset by employing the *BuildHull* pre-processing procedure developed in [13]. Then, we investigate the ability of the ANNs to estimate the efficiency scores provided by the two milestone DEA models, namely the CCR [9] and the BCC [7]. We show that our approach estimates accurately the DEA efficiency scores of the CCR and the BCC models. We also verify that the estimated efficiency scores satisfy the properties of the aforementioned models, i.e. the estimated BCC scores are greater on average than or equal to the corresponding CCR ones. Notice that the fundamental DEA properties may not be satisfied at the individual unit level due to the approximation nature of the ANNs. In particular, the efficiency patterns learned by the ANNs are ultimately generated by a global non–linear approximation function that maps the DEA inputs and outputs to the corresponding DEA efficiency score. This is a major differentiation between the two methods, since within the DEA context the efficiency score for each DMU is derived from the optimal solution of a DMU–specific LP. In addition, we introduce the concept of DEA cross-efficiency into the ANNs to examine their ability to capture this performance pattern and improve their discriminating power. Indeed, the ANNs trained over the cross-efficiency pattern provide more accurate estimation of efficiency scores and better discrimination of DMUs than the corresponding ones obtained from ANNs for the CCR case. Notably, the cross-efficiency scores estimated by the ANNs satisfy the DEA properties on average and for each DMU individually. We illustrate our approach by generating synthetic datasets based on Normal and Uniform distributions and designing various experiments with different size (number of DMUs and inputs-outputs).

14.2   Data Envelopment Analysis

Data Envelopment Analysis was developed as an alternative to the econometric (parametric) approach for the efficiency measurement of production units by employing linear programming. The origins and the historical evolution of DEA are succinctly outlined in [18, 24]. In the econometric approach an explicit production function is assumed (e.g. Cobb-Douglas) and the parameters of this function are estimated to fit the observations. On the other hand, in DEA no assumptions are made for the underlying functional form that converts inputs to outputs. A schematic representation of a DMU is given in Fig. 14.1. Without the need of a priori knowledge of the production function but based only on the observed data of the units, DEA builds an empirical best practice production frontier that envelops them. This is the fundamental concept of DEA, i.e. to identify the best practice units to form an efficient frontier and act as benchmark for the non-frontier units. Using this information, the non-frontier (inefficient) units can be compared with the benchmarks and their level of efficiency can be measured. Therefore, the efficiency scores derived by DEA are relative rather than absolute.

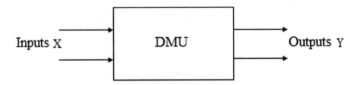

Fig. 14.1 A typical representation of a DMU

As noticed, the primary purpose of DEA is to identify the best production practice among the DMUs to derive an empirical production frontier. The best production practice is derived from any DMU of the production possibility set (PPS) that produces the highest possible levels of output given its level of inputs, or vice versa. The PPS, also known as technology, is related to the production process operated by the DMUs. It is a convex set that contains all the feasible levels of outputs that can be produced from the available levels of inputs, no matter if they are not observed in practice. These feasible input-output correspondences are obtained using interpolations between the observed input-output bundles of the DMUs.

The CCR model was introduced in [9] to measure the relative efficiency of DMUs under the constant returns to scale (CRS) assumption. The efficiency of a DMU is defined as the ratio of the weighted sum of outputs (total virtual output) to the weighted sum of inputs (total virtual input), with the weights being obtained in favor of each evaluated unit by the optimization process. The technical efficiency of a DMU can be measured by adopting either an input or an output orientation. If the conservation of inputs is of greater importance an input orientation is selected. Alternatively, if the expansion of outputs is considered more important according to the analyst's perspective an output orientation is selected.

Assume n DMUs, each using m inputs to produce s outputs. We denote by y_{rj} the level of the output $r \in [s]$ produced by unit $j \in [n]$ and x_{ij} the level of the input $i \in [m]$ used by unit j. The vector of inputs for DMU j is $X_j = \left[x_{1j}, \ldots, x_{mj} \right]^T$ and the vector of outputs is $Y_j = \left[y_{1j}, \ldots, y_{sj} \right]^T$. The input oriented CCR model that provides the efficiency for the DMU_{j0} is given below:

$$
\begin{aligned}
max \ e_{j_0} &= \frac{\omega Y_{j_0}}{\eta X_{j_0}} \\
s.t. & \\
\frac{\omega Y_j}{\eta X_j} &\leq 1, \quad j = 1, \ldots, n \\
\eta &\geq 0, \omega \geq 0
\end{aligned}
\tag{14.1}
$$

The model (14.1) is formulated and solved for each DMU in order to obtain its efficiency score. The variables $\eta = [\eta_1, \ldots, \eta_m]$ and $\omega = [\omega_1, \ldots, \omega_s]$ are the weights associated with the inputs and the outputs respectively. These weights are calculated in a manner that they provide the highest possible efficiency score for the evaluated DMU_{j0}. The constraints of model (14.1) limit the efficiency scores of the DMUs in the interval (0, 1]. The input oriented CCR model (14.1) can be transformed to a linear program by applying the Charnes and Cooper transformation [8]. The transformation is carried out by considering a scalar $t \in \Re^+$ such as $t\eta X_{j_0} = 1$ and multiplying all terms of model (14.1) with $t > 0$ so that $v = t\eta$ and $u = t\omega$. Model (14.2), presented in Table 14.1, is the linear equivalent of model (1).

Table 14.1 exhibits also the envelopment form (14.3) of the CCR model, which derives as the dual of the multiplier form (14.2). The output oriented variants of the CCR model in multiplier and envelopment forms, models (14.4) and (14.5) respectively, are also presented in Table 14.1. Notice that the efficiency scores provided

Table 14.1 Input and output oriented BCC models

	Multiplier form	Envelopment form
Input oriented	$\max e_{j_0} = uY_{j_0}$ $s.t.$ $vX_{j_0} = 1$ (14.2) $uY_j - vX_j \le 0, \ j = 1,\dots,n$ $v \ge 0, u \ge 0$	$\min \theta$ $s.t.$ $\theta X_{j_0} - X\lambda - s^- = 0$ (14.3) $Y_{j_0} - Y\lambda + s^+ = 0$ $\lambda \ge 0, s^- \ge 0, s^+ \ge 0$
Output oriented	$\min \dfrac{1}{e_{j_0}} = vX_{j_0}$ $s.t.$ $uY_{j_0} = 1$ (14.4) $uY_j - vX_j \le 0, \ j = 1,\dots,n$ $v \ge 0, u \ge 0$	$\max \varphi$ $s.t.$ $X_{j_0} - X\mu - \tau^- = 0$ (14.5) $\varphi Y_{j_0} - Y\mu + \tau^+ = 0$ $\mu \ge 0, \tau^- \ge 0, \tau^+ \ge 0$

from the input oriented model (14.2) and the output oriented model (14.4) are the same, i.e. there is a reciprocal relation between the two models.

Definition 14.1 (*CCR-Efficiency*) DMU_{j_0} is *CCR-efficient* if and only if $e_{j_0}^* = 1$ and there exists at least one optimal solution (v^*, u^*), with $u^* > 0$ and $v^* > 0$. Otherwise, DMU_{j_0} is *CCR-inefficient*.

The CCR model was extended to the BCC model by Banker et al. [7] to accommodate variable returns to scale (VRS). The incorporation of variable returns to scale in DEA allows for decomposing the overall efficiency to technical and the scale efficiency in contrast to the CCR model which aggregates them into a single measure. A structural difference of the CCR and the BCC models is the additional free of sign variable u_o in the multiplier form of the latter, which is the dual variable associated with the additional convexity constraint $(e\lambda = 1)$ in the envelopment form. Table 14.2 exhibits the input oriented and the output oriented BCC models.

The definition of efficiency is analogous to the Definition 14.1. The efficiency scores obtained from the BCC models are greater than or equal to the corresponding ones obtained from the CCR models. However, notice that the reciprocal relation between the input oriented model (14.6) and the output oriented model (14.8) is not available.

We provide in Fig. 14.2 a schematic representation of the frontiers under CRS and VRS assumptions for a case of seven DMUs, labelled A to G, which use one input to produce one output, as presented in Table 14.3.

In Fig. 14.2, the efficient frontier of the CCR model, which envelops all the data points (DMUs), is the line defined by the horizontal axis and the ray from the origin through DMU B. Under VRS assumption, the efficient frontier is determined by the DMUs A, B and C. The piecewise efficient frontier ABC, the horizontal line passing

Table 14.2 Input and output oriented BCC models

	Multiplier form	Envelopment form
Input oriented	$\max uY_{j_0} - u_0$ s.t. $vX_{j_0} = 1$ (14.6) $uY_j - u_0 - vX_j \leq 0, \ j = 1, \ldots, n$ $v \geq 0, u \geq 0$	$\min \theta$ s.t. $\theta X_{j_0} - X\lambda - s^- = 0$ $Y_{j_0} - Y\lambda + s^+ = 0$ (14.7) $e\lambda = 1$ $\lambda \geq 0, s^- \geq 0, s^+ \geq 0$
Output oriented	$\min vX_{j_0} - u_0$ s.t. $uY_{j_0} = 1$ (14.8) $uY_j - vX_j + u_0 \leq 0, \ j = 1, \ldots, n$ $v \geq 0, u \geq 0$	$\max \varphi$ s.t. $X_{j_0} - X\mu - \tau^- = 0$ $\varphi Y_{j_0} - Y\mu + \tau^+ = 0$ (14.9) $e\lambda = 1$ $\lambda \geq 0, \tau^- \geq 0, \tau^+ \geq 0$

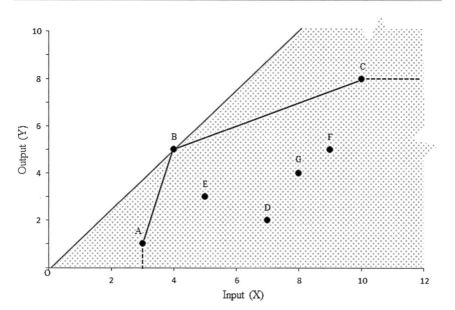

Fig. 14.2 Production possibility sets and efficient frontiers under CRS and VRS assumption

Table 14.3 Single input–output case

DMU	A	B	C	D	E	F	G
Input (X)	3	4	10	7	5	9	8
Output (Y)	1	5	8	2	3	5	4

through C and the vertical line passing through A envelop all the DMUs. As can be deduced, the assumption of the returns to scale affects the shape of the frontier and therefore the performance of the DMUs, since it is measured against the best practice units (frontier units). The VRS frontier is closer to the observations (DMUs) since the convexity constraint ($e\lambda = 1$) imposed in the BCC envelopment models leads to tighter envelopment of the observations (DMUs). In other words, the convexity constraint ($e\lambda = 1$) spans a subset of the CRS production possibility set.

The CCR and the BCC models provide efficiency scores in favor of each evaluated unit, i.e. this modeling approach grants the flexibility to each DMU to determine its own weights of the inputs and outputs in a manner that maximize its performance. However, this self-appraisal has received criticism because does not establish an objective reality, i.e. a common ground for assessment. To eliminate the effects of the variability of DEA optimal weights, the existence of unrealistic weight schemes as well as to establish a peer-appraisal, Sexton et al. [25] and Doyle and Green [12] introduced the concept of cross-efficiency. The cross-efficiency is employed in [25] for ranking the DMUs in the frame of a peer evaluation. In addition, the cross-efficiency can be utilized to improve the discrimination among the efficient DMUs. The basic notion of cross-efficiency is to expose every unit in the light of the others. The cross-efficiency is a two-phase process, where in phase I the efficiency of each DMU is measured by the CCR model or the BCC one. In phase II, the optimal values of the weights of all DMUs derived from phase I are applied for the calculation of the cross-efficiency scores for each unit. Finally, the cross-efficiency of each unit is calculated as the arithmetic mean of its cross-efficiency scores. Notice that the cross-efficiency scores are lower than or equal to the efficiency scores obtained from the conventional CCR or BCC model.

We provide below a tabular representation of the cross-efficiency concept based on the cross-efficiency matrix given in Doyle and Green [12]. The cross-efficiency matrix contains $n * n$ efficiency scores due to the n DMUs under evaluation. The efficiency of DMU j based upon the optimal weights calculated for DMU d is denoted as e_{dj}. These weights provide the highest level of efficiency e_{dd} for DMU d, the leading diagonal of the matrix contains the highest efficiency scores that the DMUs can attain. The cross-efficiency for a given DMU j is obtained as the arithmetic or the geometric mean down column j, denoted by $\overline{e_j}$ (Table 14.4).

Table 14.4 Cross–efficiency matrix

Rating DMU	Rated DMU			Averaged appraisal of peers
	1	…	N	
1	e_{11}	…	e_{1n}	A_1
⋮	⋮	⋱	⋮	⋮
N	e_{n1}	…	e_{nn}	A_n
	$\overline{e_1}$	…	$\overline{e_n}$	
	Averaged appraisal by peers (peers' appraisal)			

14.3 Artificial Neural Networks

ANNs and Multi-layer Perceptrons (MLPs), in particular, constitute quintessential pattern recognition models that are capable of addressing a wide spectrum of supervised, unsupervised and reinforcement learning tasks. Specifically, a vast volume [1] of the relative research has demonstrated that they are extremely efficient in confronting an unlimited range of pattern recognition tasks that fall within the general frameworks of classification, clustering and regression. Indeed, when regression problems are concerned, MLPs exhibit the universal approximation property [22] which suggests that any continuous function defined in a compact subspace with finitely many points of discontinuity can be approximated arbitrarily closely by a two-layer perceptron. This property of ANNs is of substantial importance in the context of this work since our primary objective is to investigate whether a global functional relationship may be sufficiently learned between the combined multi–dimensional space of DEA inputs–outputs and the (0, 1] interval of possible efficiency. The particularity, however, of this learning task lies upon the fact that DEA scores are actually acquired by a unit—specific optimization process.

14.4 Proposed Approach

We integrate DEA with ANNs to estimate the efficiency scores of DMUs instead of solving for each evaluated DMU a linear program to derive its efficiency score. In this way, we reduce the computational effort and accelerate the performance assessment. Initially, we employ the two milestone DEA models, namely the CCR and the BCC models under CRS and VRS assumption respectively, to examine the ability of ANNs to estimate the efficiency scores obtained from these models. In particular, we utilize the BCC input oriented model (14.6) and the BCC output oriented model (14.8). Since the CCR input oriented model (14.2) and the CCR output oriented model (14.4) provide the same efficiency scores due to the reciprocal relation that exists between them, in the analysis we do not distinguish between them and we refer to them as CCR scores. We also introduce the concept of DEA cross-efficiency into the ANNs to examine their ability to capture this performance pattern and improve their discriminating power. We confine the DEA cross-efficiency only under CRS assumption due to the reported problems under VRS assumption [21].

14.4.1 Data Generation—Training and Testing Samples

The data generation process realized in this work involves sampling from two probability density functions, namely the Uniform distribution and the Normal distribution. In particular, each sample corresponds to a $(m+s)$-dimensional random vector which is composed by the m inputs and s outputs associated with each DMU in the context of DEA. The constituent values of each random vector are jointly sampled, forming a sequence of independent datasets that are identically drawn from the aforementioned distributions.

We first identify the efficient DMUs under CRS and VRS assumption for every dataset by utilizing the *BuildHull* pre-processing procedure introduced in [13]. As noticed, applying DEA for classifying and scoring requires the consecutive solution of a linear program for each evaluated DMU. However, the *BuildHull* procedure provides fast the efficient set that is crucial for the performance evaluation, since this set is only required to compare and measure the efficiency scores of the rest DMUs. Then, based on this efficient set we score the DMUs that will be used for training in each ANN. In this way, we secure that the true efficiency scores are obtained for the DMUs in the training dataset. We also score the DMUs that will be used for testing in each ANN only for validation purposes. Notice that in the case of a demanding environment with streaming data, a new arrival may change the whole picture of the assessment, i.e. may affect the efficiency scores of all DMUs. Therefore, we propose that the newcomer DMU should be firstly compared with the existing efficient set. This can be accomplished by solving a small linear program for the newcomer DMU or by examining if the newcomer DMU dominates any of the efficient DMUs. If the newcomer DMU will enter the efficient set, then to consider the relative nature of DEA, we propose that the DMUs used for training and testing should be evaluated again and the ANN should be retrained.

Every dataset is randomly partitioned into two disjoint subsets of training and testing DMUs. The percentage of DMUs used for training the ANN in each run was set to 65% of the whole dataset. Thereby the remaining 35% was used for testing the approximation power of each ANN to estimate the DEA efficiency scores in each run. In other words, the efficiency scores of the 35% of the DMUs in each dataset are not derived through optimization process but directly as estimates of the corresponding trained ANN. This, in conjunction with the quick identification of the efficient set, entails the acceleration of the performance assessment and a significant reduction of the computational burden. Conclusively, our methodology facilitates the performance evaluation of high volume data.

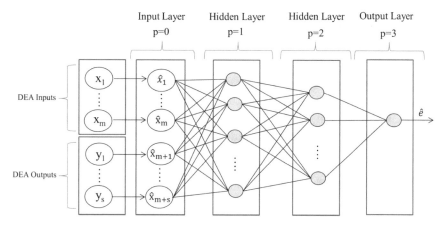

Fig. 14.3 Artificial neural network architecture

14.4.2 ANN Architecture and Training Algorithm

We employ a feed forward neural network consisting of an input layer, two non-linear hidden layers and a linear output layer as depicted in Fig. 14.3. Notice that for each DEA model we employ a specific ANN. Letting $0 \leq p \leq L$ with $L = 3$ be an indicator of each particular network layer, parameter k_p identifies the number of neurons at the p-th layer. Specifically, the zeroth layer of the utilized neural network architecture is fed with a k_0-dimensional input feature vector ($k_0 = m + s$) of the form $\hat{X} = \left[\hat{x}_1, \ldots, \hat{x}_m, \hat{x}_{m+1}, \ldots \hat{x}_{m+s}\right]^T$ such that the first m feature dimensions correspond to the inputs consumed by each DMU ($\hat{x}_q = x_q, 1 \leq q \leq m$) and the latter s feature dimensions are associated with the outputs produced by each DMU ($\hat{x}_q = y_{q-m}, m + 1 \leq q \leq k_0$). Both hidden layers are composed by a variable number of nodes depending on the dimensionality of the input feature space. Our primary concern was to generate a non-linear mapping of the combined inputs and outputs of each DMU to a lower-dimensional space where learning the DEA-based performance patterns (efficiency scores) may be conducted more efficiently [16]. Therefore, by taking into consideration the fact that throughout the experimentation process the dimensionality of the input space ranges within the [4, 20] interval, we decided that each hidden layer should contain half the number of neurons pertaining to the previous layer, such that $k_1 = \left[\frac{k_0}{2}\right]$ and $k_2 = \left[\frac{k_1}{2}\right]$. The third layer of the neural network contains a single node such that $k_L = 1$ producing the efficiency score for any given configuration of DEA inputs and outputs.

Each neuron q at the pth layer of the neural network for $1 \leq p \leq L$ with $1 \leq q \leq k_p$, is associated with a $(k_{p-1} + 1)$-dimensional weight vector of the form $W_q^{(p)} = \left[w_{q0}^{(p)}, w_{q1}^{(p)}, \ldots, w_{qk_{p-1}}^{(p)}\right]^T$. In particular, the synaptic weights $w_{qq'}^{(p)}$ with $1 \leq q' \leq k_{p-1}$ quantify the connection strength between the qth node at layer p and every other node q' at the $(p - 1)$th layer. The fist component of the weight vector $w_{q0}^{(p)}$ corresponds to the bias term which is associated with every node in the neural network accepting a constant input value equal to $+1$. Moreover, let $\hat{y}_q^{(p)}$ to denote the output value generated by the qth neuron at the pth layer with $1 \leq p \leq L - 1$ and $1 \leq q \leq k_p$. Thus, in order to maintain the same mathematical notation for the output of the nodes at the zeroth layer of the neural network as well as for the constant input value associated with each bias term, we may write that $\hat{y}_0^{(p)} = +1$ for $0 \leq p \leq L - 1$ and $\hat{y}_q^{(0)} = \hat{x}_q$. Therefore, the extended output of the pth network layer can be expressed as a $(k_p + 1)$-dimensional vector of the following form $\hat{Y}^{(p)} = \left[+1, \hat{y}_1^{(p)}, \ldots, \hat{y}_{k_p}^{(p)}\right]^T$. In this context, each node pertaining to a hidden layer of the neural network produces an intermediate output value $\hat{z}_q^{(p)}$ which is given by the inner product between the weight vector at the pth layer and the extended output vector at the $(p - 1)$th layer as $\hat{z}_q^{(p)} = W_q^{(p)T} \hat{Y}^{(p-1)}$. This intermediate value is subsequently passed through a non-linear activation function which provides the overall output for each hidden node as $\hat{y}_q^{(p)} = f\left(\hat{z}_q^{(p)}\right)$. The activation function utilized in this work for every hidden neuron is the hyperbolic tangent sigmoid transfer function which has the following analytical form:

$$f(x) = \frac{2}{1 + e^{-2x}} - 1 \tag{14.10}$$

Finally, the overall output of the neural network, providing the actual approximation of the efficiency score for a given pair of DEA inputs and outputs, may be formulated as:

$$\hat{e}(X_j, Y_j) = \sum_{q'=0}^{k_{L-1}} w_{q'}^{(L)} \hat{y}_{q'}^{(L-1)} \tag{14.11}$$

The optimal weight vector $\hat{W}_q^{(p)}$ for each node of the neural network, with $1 \leq p \leq L$ and $1 \leq q \leq k_p$, may be determined through the application of the backpropagation iterative scheme [19] on the subset of DMUs used for training given the corresponding DEA-based efficiency scores. In particular, by letting $\mathcal{D} = \left\{\left(\hat{X}_1, e_1\right), \ldots, \left(\hat{X}_M, e_M\right)\right\}$ to denote the subset of training DMUs along with the associated efficiency scores, where $\hat{X}_j = [X_j, Y_j]^T$, we can form the following error function:

$$E = \sum_{j=1}^{M} E_j = \sum_{j=1}^{M} \left(e_j - \hat{e}\left(\hat{X}_j\right)\right)^2 \tag{14.12}$$

In this setting, the optimal weight assignment for each node of the neural network will be obtained by the minimization of the previously defined error function with respect to that particular weight vector as:

$$\widehat{W}_q^{(p)} = \underset{W_q^{(p)}}{\text{argmin}} \; E\left(W_q^{(p)}\right) \tag{14.13}$$

The backpropagation training process provides the optimal solutions to the afore-mentioned optimization problem by implementing an iterative weight—updating scheme which is formulated according to:

$$W_q^{(p)}(t+1) = W_q^{(p)}(t) + \Delta W_q^{(p)} \tag{14.14}$$

The actual weight—updating vector $\Delta W_q^{(p)}$ for each neuron can be acquired by taking into consideration the dependence of the jth error term on any given weight vector captured by $\frac{\partial E_j}{\partial W_q^{(p)}} = \frac{\partial E_j}{\partial \widehat{Z}_{qj}^{(p)}} \frac{\partial \widehat{Z}_{qj}^{(p)}}{\partial W_q^{(p)}}$. Quantity $\widehat{Z}_{qj}^{(p)}$ is the intermediate output of the qth node at the pth layer when presented with jth training instance $\left(\widehat{X}_j, e_j\right)$. Likewise, $\widehat{Y}_j^{(p)}$ may be interpreted as the vector output of the pth network layer for the jth training pattern. Finally, setting $\delta_{qj}^{(p)} = \frac{\partial E_j}{\partial \widehat{Z}_{qj}^{(p)}}$ and evaluating the partial derivative of the intermediate node output with respect to the corresponding weight vector as $\frac{\partial \widehat{Z}_{qj}^{(p)}}{\partial W_q^{(p)}} = \widehat{Y}_j^{(p-1)}$ leads to:

$$\Delta W_q^{(p)} - \mu \sum_{j=1}^{M} \delta_{qj}^{(p)} \widehat{Y}_j^{(p-1)} \tag{14.15}$$

where μ is the adopted learning rate parameter.

14.5 Results

We illustrate our approach by conducting twelve experiments of different size, i.e. among the experiments we vary the number of DMUs and the number of inputs–outputs. Also, each experiment consists of fifty different datasets, where each dataset is obtained from Uniform or Normal distribution. In each experiment we carry out fifty runs as the number of datasets. The percentage of DMUs used for training the ANN in each run was set to 65% of the whole dataset. Thereby the remaining 35% was used for testing the approximation power of each ANN to estimate the DEA efficiency scores in each run. Notice that in this way, the efficiency scores of the 35% of DMUs (training dataset) in each run are not obtained from the optimization process, resulting in significant computational savings. The general setting of the experimentation process is presented in Table 14.5.

Table 14.5 DEA—Neural network experimentation framework

Number of DMUs	Data distribution						Training percentage	Number of datasets
	Uniform			Normal				
	(Inputs, Outputs)			(Inputs, Outputs)			65%	50
2000	(2, 2)	(3, 3)	(10, 10)	(2, 2)	(3, 3)	(10, 10)		
4000	(2, 2)	(3, 3)	(10, 10)	(2, 2)	(3, 3)	(10, 10)		

We applied the CCR and the BCC models as well as we calculated the cross-efficiency to each dataset for each experiment. The efficiency scores derived by the ANNs approximate accurately the DEA efficiency scores obtained from the aforementioned DEA models. Also, the efficiency scores estimated by the ANNs satisfy on average the fundamental DEA properties, as the DEA efficiency scores derived by the conventional DEA models. The ANNs capture sufficiently the performance pattern of cross-efficiency by providing efficiency scores that improve the discrimination among the DMUs, as the original DEA cross-efficiency scores.

The approximation ability of ANNs is quantified by three performance measures, namely the Mean Absolute Error (MAE), the Mean Squared Error (MSE) and the coefficient of determination denoted as R^2. Specifically, Tables 14.6, 14.7, 14.8, 14.9, 14.10, 14.11, 14.12, 14.13, 14.14, 14.15, 14.16, and 14.17 exhibit significantly low

Table 14.6 Experiment 1: Uniform distribution (2000 units, 2 inputs, 2 outputs)

DEA model	Average training performance			Average testing performance		
	MSE	MAE	R^2	MSE	MAE	R^2
CCR	0.0005177	0.0135022	0.9759971	0.0019624	0.0147677	0.9571647
Cross efficiency	0.0018474	**0.0000694**	**0.9981910**	**0.0002269**	**0.0019189**	**0.9966411**
BCC input oriented	0.0014292	0.0323119	0.9028719	0.0047293	0.0345475	0.8766674
BCC output oriented	**0.0004822**	0.0085362	0.9918879	0.0015638	0.0090114	0.9887132

Table 14.7 Experiment 2: Uniform distribution (2000 units, 3 inputs, 3 outputs)

DEA model	Average training performance			Average testing performance		
	MSE	MAE	R^2	MSE	MAE	R^2
CCR	0.0008054	0.0229716	0.9711431	0.0030821	0.0262057	0.95061097
Cross efficiency	0.0011957	**0.0000426**	**0.9995053**	**0.0001559**	**0.0012946**	**0.99898231**
BCC input oriented	0.0016829	0.0450134	0.9217769	0.0058899	0.0500299	0.89206884
BCC output oriented	**0.0006501**	0.0154462	0.9767874	0.0022674	0.0168690	0.96682448

Table 14.8 Experiment 3: Uniform distribution (2000 units, 10 inputs, 10 outputs)

DEA model	Average training performance			Average testing performance		
	MSE	MAE	R^2	MSE	MAE	R^2
CCR	0.0002365	0.0052746	0.9952290	0.0093718	0.0763349	0.1661345
Cross efficiency	0.0003431	**0.0000103**	**0.9999753**	**0.0001064**	**0.0009248**	**0.99966738**
BCC input oriented	0.0000350	0.0002519	0.9997417	0.0123801	0.0772409	−1.5144433
BCC output oriented	**0.0000154**	0.0001209	0.9997605	0.0067510	0.0382678	−4.1821253

Table 14.9 Experiment 4: Uniform distribution (4000 units, 2 inputs, 2 outputs)

DEA model	Average training performance			Average testing performance		
	MSE	MAE	R^2	MSE	MAE	R^2
CCR	**0.0003432**	0.0120852	0.9713836	0.0011649	0.0125429	0.9623906
Cross efficiency	0.0016750	**0.0000478**	**0.9975823**	**0.0001400**	**0.0016366**	**0.9965639**
BCC input oriented	0.0008870	0.0272546	0.9035072	0.00280971	0.0280332	0.8903747
BCC output oriented	0.0003943	0.0110337	0.9732129	0.0013837	0.0114903	0.95934719

Table 14.10 Experiment 5: Uniform distribution (4000 units, 3 inputs, 3 outputs)

DEA model	Average training performance			Average testing performance		
	MSE	MAE	R^2	MSE	MAE	R^2
CCR	0.0005491	0.0218733	0.9692263	0.0018673	0.0234194	0.9600186
Cross efficiency	0.0011238	**0.0000291**	**0.9994185**	**0.0001018**	**0.0011914**	**0.9990590**
BCC input oriented	0.0011497	0.0415489	0.9153117	0.0028097	0.0435698	0.9032587
BCC output oriented	**0.0004467**	0.0134169	0.9801585	0.0016214	0.0146866	0.9688336

Table 14.11 Experiment 6: Uniform distribution (4000 units, 10 inputs, 10 outputs)

DEA model	Average training performance			Average testing performance		
	MSE	MAE	R^2	MSE	MAE	R^2
CCR	0.0005093	0.0199125	0.9659402	0.0045122	0.0533186	0.6988247
Cross efficiency	0.0003291	**0.0000071**	**0.9999733**	**0.0000392**	**0.0005109**	**0.99990733**
BCC input oriented	0.0004001	0.0112380	0.9693166	0.0069929	0.0672786	−0.0612705
BCC output oriented	**0.0001676**	0.0051694	0.9693319	0.0041362	0.03559479	−1.3247827

Table 14.12 Experiment 7: Normal distribution (2000 units, 2 inputs, 2 outputs)

DEA model	Average training performance			Average testing performance		
	MSE	MAE	R^2	MSE	MAE	R^2
CCR	0.0001095	0.0032507	0.9967115	0.0004011	0.0034229	0.9943959
Cross efficiency	**0.0000981**	**0.0000036**	**0.9999903**	**0.0000131**	**0.0001061**	**0.9999836**
BCC input oriented	0.0002217	0.0062241	0.9482878	0.0007450	0.0064625	0.9443861
BCC output oriented	0.0001807	0.0054861	0.9323345	0.0006274	0.0056927	0.9276385

Table 14.13 Experiment 8: Normal distribution (2000 units, 3 inputs, 3 outputs)

DEA model	Average training performance			Average testing performance		
	MSE	MAE	R^2	MSE	MAE	R^2
CCR	0.0001610	0.0049786	0.9934146	0.0006103	0.0055370	0.9880753
Cross efficiency	**0.0000483**	**0.0000017**	**0.9999963**	**0.0000059**	**0.0000521**	**0.9999956**
BCC input oriented	0.0001783	0.0052404	0.9837195	0.0006452	0.0058177	0.9723390
BCC output oriented	0.0001545	0.0046729	0.9680146	0.0005875	0.0052000	0.9582981

Table 14.14 Experiment 9: Normal distribution (2000 units, 10 inputs, 10 outputs)

DEA model	Average training performance			Average testing performance		
	MSE	MAE	R^2	MSE	MAE	R^2
CCR	0.0000443	0.0011700	0.9952901	0.0013693	0.0113335	0.4260734
Cross efficiency	**0.0000184**	**0.0000005**	**0.9999994**	**0.0000037**	**0.0000329**	**0.9999976**
BCC input oriented	0.0000326	0.0007493	0.9318946	0.0013357	0.0104425	0.0694910
BCC output oriented	0.0000452	0.0012542	0.9540211	0.0010372	0.0082831	0.0986643

Table 14.15 Experiment 10: Normal distribution (4000 units, 2 inputs, 2 outputs)

DEA model	Average training performance			Average testing performance		
	MSE	MAE	R^2	MSE	MAE	R^2
CCR	**0.0000779**	0.0031663	**0.9970348**	0.0002588	0.0032685	**0.9956498**
Cross efficiency	0.0014794	**0.0000332**	0.9799813	**0.0000955**	**0.0014675**	0.9799726
BCC input oriented	0.0001616	0.0061362	0.9673725	0.0005156	0.0062954	0.9639029
BCC output oriented	0.0001028	0.0037516	0.9901382	0.0003240	0.0039070	0.9891171

Table 14.16 Experiment 11: Normal distribution (4000 units, 3 inputs, 3 outputs)

DEA model	Average training performance			Average testing performance		
	MSE	MAE	R^2	MSE	MAE	R^2
CCR	0.0001032	0.0044059	0.9936087	0.0003531	0.0046390	0.9914043
Cross efficiency	**0.0000475**	**0.0000012**	**0.9999955**	**0.0000039**	**0.0000492**	**0.9999951**
BCC input oriented	0.0001197	0.0047470	0.9833310	0.0004108	0.0050146	0.9639029
BCC output oriented	0.0000990	0.0040510	0.9885028	0.0003344	0.0043146	0.9850008

Table 14.17 Experiment 12: Normal distribution (4000 units, 10 inputs, 10 outputs)

DEA model	Average training performance			Average testing performance		
	MSE	MAE	R^2	MSE	MAE	R^2
CCR	0.0000603	0.0025160	0.9814983	0.0003774	0.0047189	0.9044561
Cross efficiency	**0.0000293**	**0.0000066**	**0.9999960**	**0.0000027**	**0.0000354**	**0.9999935**
BCC input oriented	0.0000468	0.0018402	0.9642113	0.0004332	0.0049653	0.7062816
BCC output oriented	0.0000514	0.0020357	0.9631799	0.0004345	0.0050021	0.7395385

error measurements for both MAE and MSE averaged over a sequence of fifty independent Runs for each experiment. The extremely high R^2 values obtained throughout the testing process demonstrate the ability of the ANNs to mimic the DMU-specific DEA optimization process in a global manner. Moreover, we observe that when the dimensionality of the ANN input space increases (sum of DEA inputs and outputs), the acquired R^2 values decrease, i.e. the approximation power of the ANNs declines. However, this effect is reduced when the number of DMUs is increased. For instance, the testing R^2 values derived from experiments with 2000 DMUs (Tables 14.8 and 14.14) are lower than the corresponding experiments with 4000 DMUs (Tables 14.11 and 14.17).

Figures 14.4, 14.5, 14.6, 14.7, 14.8, 14.9, 14.10, 14.11, 14.12, 14.13, 14.14, and 14.15 depict, for each experiment, the average efficiency scores over all units per run. In particular, we demonstrate that the fundamental properties of the DEA models are effectively transferred to the ANN framework. Indeed, the BCC efficiency scores estimated by the ANNs are on average greater than or equal to the corresponding CCR ones. Also, the estimated cross-efficiency scores are lower than or equal to the corresponding CCR efficiency scores when averaged over the units.

As noticed, the BCC and CCR efficiency scores estimated from ANNs may not satisfy the fundamental DEA properties at the DMU level. This is attributed to the

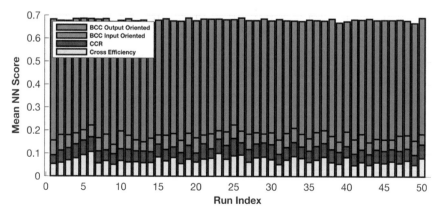

Fig. 14.4 Experiment 1: Mean ANN–based scores per run

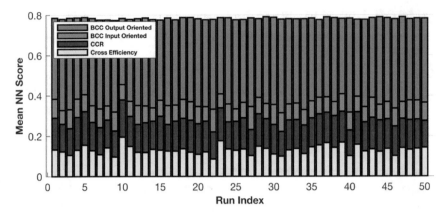

Fig. 14.5 Experiment 2: Mean ANN–based scores per run

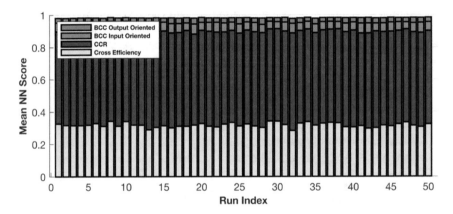

Fig. 14.6 Experiment 3: Mean ANN–based scores per run

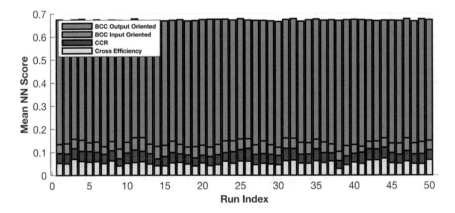

Fig. 14.7 Experiment 4: Mean ANN–based scores per run

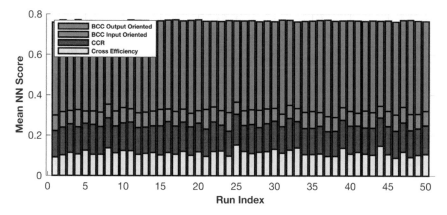

Fig. 14.8 Experiment 5: Mean ANN–based scores per run

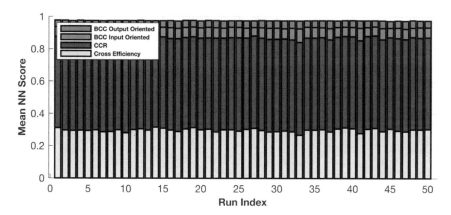

Fig. 14.9 Experiment 6: Mean ANN–based scores per run

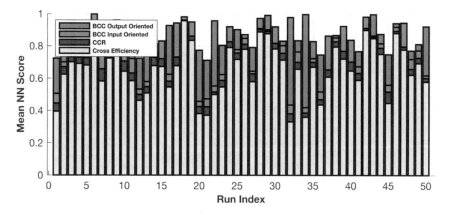

Fig. 14.10 Experiment 7: Mean ANN–based scores per run

328   G. Koronakos and D. N. Sotiropoulos

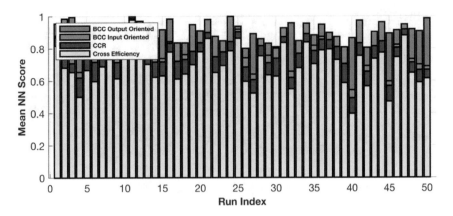

Fig. 14.11 Experiment 8: Mean ANN–based scores per run

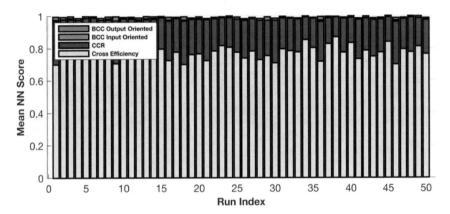

Fig. 14.12 Experiment 9: Mean ANN–based scores per run

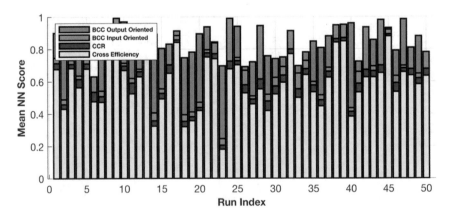

Fig. 14.13 Experiment 10: Mean ANN-based scores per run

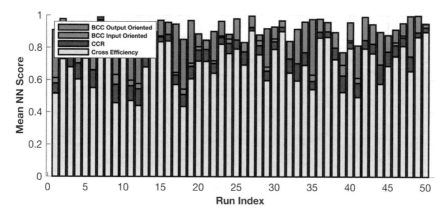

Fig. 14.14 Experiment 11: Mean ANN–based scores per run

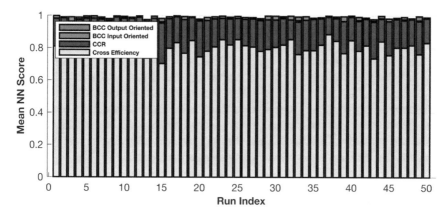

Fig. 14.15 Experiment 12: Mean ANN–based scores per run

conceptual differentiation between the DEA and ANNs. In particular, in DEA the efficiency score of each DMU is derived from the optimal solution of a DMU–specific LP while in ANNs a global non–linear approximation function maps the DEA inputs and outputs of each DMU to the corresponding efficiency score. To illustrate the above, we provide a detailed overview of two different Runs from two Experiments (1 and 7), which contain data obtained from the Uniform and the Normal distribution respectively. Figure 14.16 portray the estimated efficiency scores of the 2000 DMUs included in Run 5 of Experiment 1, it is evident that the fundamental DEA properties are not satisfied by all DMUs. The same holds for the estimated efficiency scores of the 2000 DMUs included in Run 5 of Experiment 7 (see Fig. 14.18), with data obtained from Normal distribution. On the contrary, the cross-efficiency scores estimated by the ANNs satisfy the DEA properties on

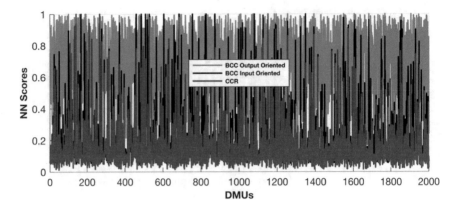

Fig. 14.16 Experiment 1-Run 5: ANN–based scores per unit for the CCR, BCC input oriented and BCC output oriented models

average and for each DMU individually, as depicted in Figs. 14.17 and 14.19 for the two aforementioned runs.

The experimentation reveals that the DEA cross-efficiency performance pattern can be effectively transferred to the ANNs. In Figs. 14.20, 14.21, 14.22, 14.23, 14.24, and 14.25, we provide the DMU rankings based on the CCR and cross-efficiency estimated scores, concerning six indicative experiments. It can be deduced that the approximated cross-efficiency scores and rankings are lower than or equal to the corresponding CCR ones. Thus, the improvement in the discrimination power conveyed by the DEA cross-efficiency concept is transferred to ANNs.

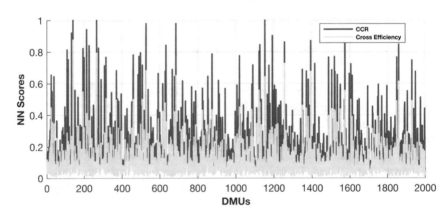

Fig. 14.17 Experiment 1-Run 5: ANN–based scores per unit for the CCR model and the cross–efficiency

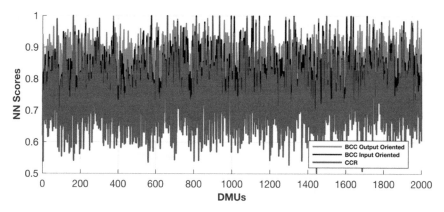

Fig. 14.18 Experiment 7-Run 5: ANN–based scores per unit for the CCR, BCC input oriented and BCC output oriented models

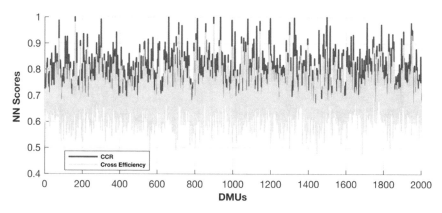

Fig. 14.19 Experiment 7-Run 5: ANN–based scores per unit for the CCR model and the cross–efficiency

14.6 Conclusion

In this chapter, we pursue via the proposed approach to provide theoretical and practical tools that accelerate the large scale performance assessments and offer significant computational savings. For this purpose, we integrated Data Envelopment Analysis with Artificial Neural Networks to estimate the efficiency scores of the units under evaluation. We considered the relative nature of the DEA by securing that the DMUs used for training and testing the ANN are first evaluated against the efficient set, which is obtained fast by employing a pre-processing procedure. We examined the ability of the ANNs to estimate the efficiency scores of the DEA models (CCR and BCC) as well as to capture the performance pattern of cross-efficiency that improves the discrimination power of DEA. We designed several

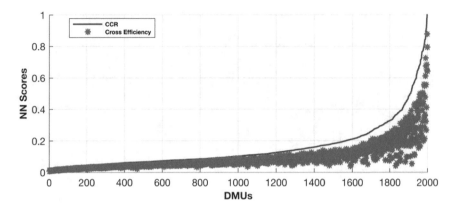

Fig. 14.20 Experiment 1-Run 5: ANN–based scores per unit for the CCR model and the cross–efficiency

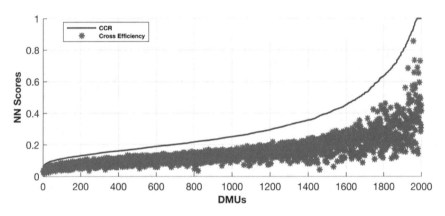

Fig. 14.21 Experiment 2-Run 5: ANN–based sorted scores per unit for the CCR model and the cross-efficiency

experiments of different size that contain data generated from different distributions. The results of experimentation illustrated that our approach effectively estimates the DEA efficiency scores. Also, the estimated scores satisfy on average the fundamental properties of the DEA milestone models, i.e. the estimated BCC scores are greater than or equal to the corresponding CCR ones. In addition, the ANNs trained over the cross-efficiency pattern provide more accurate approximation of the DEA cross-efficiency scores and better discrimination of DMUs than the corresponding ones trained over the CCR model. Finally, our approach is general and can be universally applied for the performance assessment of entities of various type and area of origin (e.g. health, privacy, stock market, etc.).

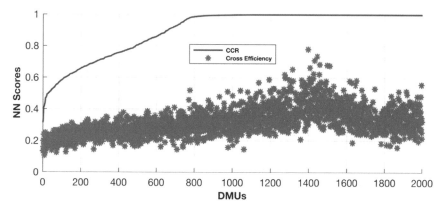

Fig. 14.22 Experiment 3-Run 5: ANN–based sorted scores per unit for the CCR model and the cross-efficiency

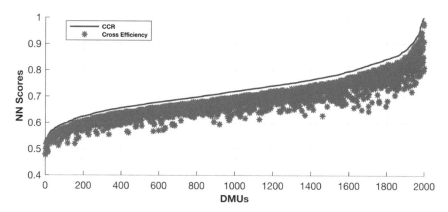

Fig. 14.23 Experiment 7-Run 5: ANN–based sorted scores per unit for the CCR model and the cross-efficiency

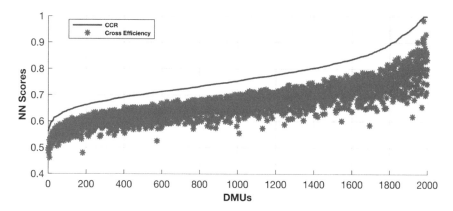

Fig. 14.24 Experiment 8-Run 5: ANN–based scores per unit for the CCR model and the cross–efficiency

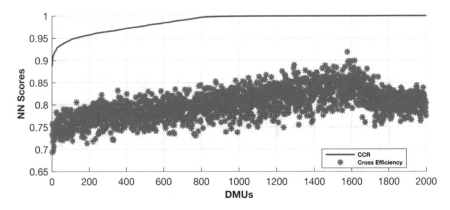

Fig. 14.25 Experiment 9-Run 5: ANN–based scores per unit for the CCR model and the cross–efficiency

Acknowledgment This research is co-financed by Greece and the European Union (European Social Fund—ESF) through the Operational Programme «Human Resources Development, Education and Lifelong Learning» in the context of the project "Reinforcement of Postdoctoral Researchers—2nd Cycle" (MIS-5033021), implemented by the State Scholarships Foundation (ΙΚΥ).

References

1. O.I. Abiodun, A. Jantan, A.E. Omolara, K.V. Dada, N.A. Mohamed, H. Arshad, State-of-the-art in artificial neural network applications: a survey. Heliyon **4**(11), e00938 (2018)
2. A. Azadeh, S.F. Ghaderi, M. Anvari, M. Saberi, H. Izadbakhsh, An integrated artificial neural network and fuzzy clustering algorithm for performance assessment of decision making units. Appl. Math. Comput. **187**(2), 584–599 (2007)
3. A. Azadeh, M. Saberi, M. Anvari, An integrated artificial neural network algorithm for performance assessment and optimization of decision making units. Expert Syst. Appl. **37**, 5688–5697 (2010)
4. A. Azadeh, M. Saberi, M. Anvari, An integrated artificial neural network fuzzy C-Means-Normalization Algorithm for performance assessment of decision-making units: the cases of auto industry and power plant. Comput. Indus. Eng. **60**, 328–340 (2011)
5. A. Azadeh, M. Saberi, R.T. Moghaddam, L. Javanmardi, An integrated data envelopment analysis-artificial neural network-rough set algorithm for assessment of personnel efficiency. Expert Syst. Appl. **38**, 1364–1373 (2011)
6. A.D. Athanassopoulos, S.P. Curram, A comparison of data envelopment analysis and artificial neural networks as tools for assessing the efficiency of decision making units. J. Oper. Res. Soc. **47**(8), 1000–1016 (1996)
7. R.D. Banker, A. Charnes, W.W. Cooper, Some models for estimating technical and scale inefficiencies in DEA. Manag. Sci. **32**, 1078–1092 (1984)
8. A. Charnes, W.W. Cooper, Programming with linear fractional functional. Naval Res. Logist. **9**, 181–185 (1962)
9. A. Charnes, W.W. Cooper, E. Rhodes, Measuring the efficiency of decision making units. Eur. J. Oper. Res. **2**, 429–444 (1978)
10. W.C. Chen, W.J. Cho, A procedure for large-scale DEA computations. Comput. Oper. Res. **36**, 1813–1824 (2009)

11. A. Costa, R.N. Markellos, Evaluating public transport efficiency with neural network models. Transp. Res. C **5**(5), 301–312 (1997)
12. J. Doyle, R. Green, Efficiency and cross-efficiency in DEA: derivations, meanings and uses. J. Oper. Res. Soc. **45**(5), 567–578 (1994)
13. J.H. Dulá, An algorithm for data envelopment analysis. INFORMS J. Comput. **23**(2), 284–296 (2011)
14. J.H. Dulá, F.J. Lopez, Algorithms for the frame of a finitely generated unbounded polyhedron. INFORMS J. Comput. **18**(1), 97–110 (2006)
15. J.H. Dulá, F.J. Lopez, DEA with streaming data. Omega **4**, 41–47 (2013)
16. S. Edelman, N. Intrator, Learning as extraction of low-dimensional representations. Psychol. Learn. Motiv. **36**, 353–380 (1997)
17. A. Emrouznejad, E. Shale, A combined neural network and DEA for measuring efficiency of large scale datasets. Comput. Indus. Eng. **56**, 249–254 (2009)
18. F.R. Forsund, N. Sarafoglou, On the origins of data envelopment analysis. J. Product. Anal. **17**, 23–40 (2002)
19. Goodfellow, I., Bengio, Y., Courville, A., Deep learning (MIT Press, 2016)
20. P. Hanafizadeh, H. Reza Khedmatgozar, A. Emrouznejad, M. Derakhshan, Neural network DEA for measuring the efficiency of mutual funds. Int. J. Appl. Decis. Sci. **7**(3), 255–269 (2014)
21. Lim, S., Zhu, J., DEA cross-efficiency evaluation under variable returns to scale. J. Oper. Res. Soc. **66**, 476–487
22. A. Pinkus, Approximation theory of the MLP model in neural networks. Acta numerica **8**, 143–195 (1999)
23. D. Santin, F.J. Delgado, A. Valiño, The measurement of technical efficiency: a neural network approach. Appl. Econ. **36**(6), 627–635 (2004)
24. L.M. Seiford, Data envelopment analysis: the evolution of the state of the art (197–1995). J. Product. Anal. **7**, 99–137 (1996)
25. T.R. Sexton, R.H. Silkman, A.J. Hogan, Data envelopment analysis: critique and extensions. New Direct. Program Eval. **1986**, 73–105 (1986). https://doi.org/10.1002/ev.1441
26. D. Wu, Z. Yang, L. Liang, Using DEA-neural network approach to evaluate branch efficiency of a large Canadian bank. Expert Syst. Appl. **31**(1), 108–115 (2006)

Chapter 15
A Comprehensive Survey on the Applications of Swarm Intelligence and Bio-Inspired Evolutionary Strategies

Alexandros Tzanetos and Georgios Dounias

Abstract For many decades, Machine Learning made it possible for humans to map the patterns that govern interpolating problems and also, provided methods to cluster and classify big amount of uncharted data. In recent years, optimization problems which can be mathematically formulated and are hard to be solved with simple or naïve heuristic methods brought up the need for new methods, namely Evolutionary Strategies. These methods are inspired by strategies that are met in flora and fauna in nature. However, a lot of these methods are called nature-inspired when there is no such inspiration in their algorithmic model. Furthermore, even more evolutionary schemes are presented each year, but the lack of applications makes them of no actual utility. In this chapter, all Swarm Intelligence methods as far as the methods that are not inspired by swarms, flocks or groups, but still derive their inspiration by animal behaviors are collected. The applications of these two sub-categories are investigated and some preliminary findings are presented to highlight some main points for Nature Inspired Intelligence utility.

Keywords Nature Inspired Algorithms · Swarm Intelligence · Bio-inspired algorithms

15.1 Introduction

Machine Learning (ML) derives its roots from the time that Alan Turing succeeded to decipher the encrypted messages of Enigma Machine in the World War II, in early 1940s. Eighty years later, many methods have been presented to identify the information that is included in raw data or to solve hard optimization problems.

A. Tzanetos (✉) · G. Dounias
Management and Decision Engineering Laboratory, Department of Financial and Management Engineering, School of Engineering, University of the Aegean, Chios, Greece
e-mail: atzanetos@aegean.gr

G. Dounias
e-mail: g.dounias@aegean.gr

© The Editor(s) (if applicable) and The Author(s), under exclusive license to Springer Nature Switzerland AG 2020
G. A. Tsihrintzis and L. C. Jain (eds.), *Machine Learning Paradigms*,
Learning and Analytics in Intelligent Systems 18,
https://doi.org/10.1007/978-3-030-49724-8_15

Through these decades, a lot of research has been done on the aim to model the way a human brain is taking into consideration aspects of the problem to conclude on the final decision. The field of Artificial Intelligence (AI) that aims to provide decision-making abilities to machines, such as computers, was named Computational Intelligence (CI) and consists of Fuzzy rule-based Systems [1, 2], Machine Learning [3–6], Artificial Neural Networks [7–9] and Evolutionary Algorithms.

Rule-based Systems use logic expressions to represent and action or decision as a result of others. In addition with fuzzy rules and fuzzy sets, these systems can take into consideration the relation between decision variables to model higher level decisions [10] and are called Fuzzy rule-based Systems. This kind of Computational Intelligent method has been used in Control Systems [11], in Knowledge Representation [12], in Computer Vision [13], in Image Detection [14], in Natural Language Processing [15] etc.

The first attempt to imprint the way of thinking of the human brain brought up a class of Computational Intelligence, named Artificial Neural Networks (ANN). A wide network of neurons comprises the human brain. Each movement of the body is a result of a pulse that is transferred to the corresponding body part through a sequence of neurons. Different sequences result to different body actions. This unique phenomenon is modeled in ANN. The input data of the problem form the first layer of the ANN, while the possible combinations of values of decision variables usually construct a number of hidden layers. The output layer provides the solution of the problem.

Accordingly, research was done in order to implement the adaptive capability of a human brain, which is a result from previous knowledge [16]. This new approach was called Machine Learning (ML). The most common type of ML is Supervised Learning, where the decision maker has to provide a wide set of data, named training set, so that the algorithm can develop a model that fits better on the given data. This method gave the opportunity to train a machine to recognize similarity between cases, resulting in classification of cases with multiple decision variables. However, when the clusters were not known, another way of clustering cases had to be found. Unsupervised Learning solved this problem.

At that time, forecasting models became valuable for time-depended data problems, such as stock market forecasting. The plethora of features (decision variables) that had to be taken into consideration made the models insufficient. Feature Selection came up to cope with this problem, as another ML method. The next years, many more ML methods made their appearance, such as Support Vector Machines (SVM) and Reinforcement Learning. Also, many more applications were introduced for ML methods, such as Recommender Systems, Anomaly Detection etc.

While Machine Learning and Artificial Neural Networks have been widely used for classification, clustering and forecasting, there are some problems that cannot be tackled with these methods. Optimization problems are usually described by mathematical equations and can be solved by the maximization or minimization of the objective function. Objective function can be either an equation that calculates one of the magnitudes of the problem or a man-made function that describes the relation between the important decision goals that should be optimized.

In 1975, Holland [17] presented the Genetic Algorithm (GA), which was inspired by the mutation and crossover mechanism that are met in the evolution of species in nature. The existence of ANN in conjunction with GA made a lot of scientists to look for inspiration in nature. The newly developed algorithms were initially entitled Evolutionary Algorithms (EAs) because in the algorithmic process they produce a solution and then keep evolving it for a number of steps. However, after some years, even more algorithms that were inspired by nature were presented, so they form a category of EAs, named Nature Inspired Algorithms (NIAs) or Nature Inspired Intelligence (NII).

In previous work [18], authors presented one of the categories of Nature Inspired Algorithms, which consists of schemes that are inspired by Physical Phenomena and Laws of Science. This chapter is focusing on the most known category of NII, namely Swarm Intelligence, while also points out the reasons that some methods should not be considered swarm-based. These methods, including others that are inspired by flora and fauna organisms, form the Organisms-inspired category of NII algorithms. The applications of both categories are presented and discussion is being done on the absence of applications in many schemes. Some preliminary findings on the nature-inspired metaphors are presented and discussed.

Similar works can be found in literature, but the interested reader cannot find a comprehensive collection of all published algorithms in the existing surveys. Surveys mentioning the most known Swarm Intelligence oriented schemes have been conducted by [19, 20]. In the work of [21], some applications of SI algorithms are mentioned and also 38 SI algorithms are collected. Another list with such schemes including a brief discussion on them can be found in [22]. A more detailed work has been done by Del Ser et al. [23]. A lot of Swarm Intelligent and Organisms-based algorithms are included and a discussion on their applications is done in Roy et al. [24].

All the aforementioned works focus on specific algorithms, usually the established schemes (Ant Colony Optimization, Particle Swarm Optimization etc.). Most of these works do not investigate the specific application area that the algorithms are more useful. Instead, some applications are mentioned and briefly discussed. In this chapter, all methods that are Swarm Intelligence oriented or Organisms-based can be found, along with some discussion on the scope of each algorithm. What is more, thorough research has been done to present the areas of application that such nature-inspired techniques can contribute to.

The following section explains what Nature Inspired Intelligence is and provides a comprehensive review on the methods that have been presented. In Sect. 15.3, each category's applications are presented per algorithm and the statistical results are discussed. The conclusion of this work is summarized in Sect. 15.4 and the future research goals are set.

15.2 Nature Inspired Intelligence

Nature Inspired Intelligence (NII) describes such algorithmic schemes that model phenomena met in nature, foraging and social behaviors between members of swarms, flocks and packs of animals, or strategies that animals, bugs and plants have developed to cope with daily problems. A more comprehensive definition of what Nature Inspired Intelligence is, can be found in Steer et al. [25], where it is stated that with the term nature refers "to any part of the physical universe which is not a product of intentional human design".

As a result, based on the above definition, Nature Inspired Algorithms (NIA) can be inspired by Physical Phenomena and Laws of Science [18], by Swarm Intelligence (SI) or by other strategies that are not embraced by big groups of animals, such as swarms, but a single unit of some species is using them for foraging or survival reasons. A very common mistake made by authors is to state that their proposed algorithm is a swarm intelligence method, while there is no such swarm behavior modeled. In this section, the nature inspired methods that can be categorized as Swarm Intelligence are separated out by these which are simple bio-inspired methods, which in this study are referred as Organisms-based.

As it can be seen in Fig. 15.1, both categories were rapidly grown from mid-2000s. Based on the popularity of Swarm Intelligence and the numerous Journals and Conferences that are publishing works on it, we would expect this category to be way larger than the other one. However, organisms have been inspiration for many more algorithms, since the number of published techniques is growing from 2012 and then. The 15 published Organisms-inspired algorithms in 2018 is remarkable,

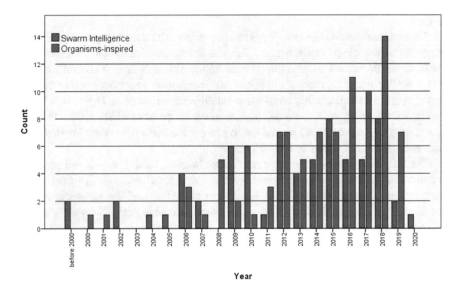

Fig. 15.1 Number of presented methods by year

while also brings up some concern. Are all these algorithms actually needed? This work is investigating this issue by looking into the applications that these algorithms solve.

In the analysis below, three main areas of problems are selected to divide the applications found; Optimization, Classification or Clustering and Other applications, such as forecasting. All problems that are coped with CI methods can be classified in one of these three categories. In a second level analysis, we focus on the problem areas of optimization. The majority of NII algorithms are developed to face optimization problems, whereas only a few have been presented as clustering methods, such as Big Bang-Big Crunch [26] and Black Hole algorithm [27].

Many engineering optimization problems, such as truss structure design and machine compound design, comprise the first category of optimization problems that NII algorithms found to be an efficient tool. Another wide area of applications is based on Finance, where portfolio optimization problems can be found. Then, a large area of applications is Operational Research problems. Classic known problems, such as the Traveling Salesman Problem (TSP) and Assignment Problem are included here. Energy problems, like the well-known Economic Load Dispatch (ELD), comprise the fourth area of applications. At last, all other problems that cannot be classified in one of the aforementioned categories are included in "Other" applications.

Papers that contain hybrid schemes were not taken into consideration. However, the best practice to enhance the performance of a nature-inspired algorithm is to find good hybrid schemes containing it. On the other hand, in the total number of publications with application were counted also papers with modifications of the initial algorithms that do not alter the algorithmic concept inspired by the physical analogue.

15.2.1 Swarm Intelligence

The term Swarm Intelligence (SI) was introduced by Beni and Wang [28] to express the collective behavior between robotic systems. However, this term is used to describe the algorithms, in which the agents collaborate to fulfil the objective of the problem solved. The first SI algorithm was proposed under the name of Stochastic Diffusion Search [29] and was inspired by the one-by-one communication strategy of a species of ants, named Leptothorax acervorum. Many applications including feature selection [30], object recognition [31], mobile robot self-localization [32], medical image recognition [33] and site selection for wireless networks [34] have been done through the years.

In 1992, Marco Dorigo proposed Ant Colony Optimization as Ant System in his PhD Thesis [35]. This algorithm was proposed mainly to solve discrete optimization problems, like Traveling Salesman Problem [36], Vehicle Routing Problems [37], Scheduling Problems [38, 39, 40], the Assignment Problem [41], the Set Problem [42], but has been also used in other applications, such as robotics [43], classification [44], protein folding [45] etc.

After three years, [46] proposed Particle Swarm Optimization (PSO) to tackle continuous space optimization problems. PSO has been used in a wide variety of problems, such as antennas design (H. [47], biomedical [48], communication networks [49], finance [50], power systems [51], robotics [52], scheduling [53], but also classification/clustering problems [54] and prediction [55] and forecasting applications [56].

All the above schemes are not included in the research presented below, because they correspond to established algorithms that have been implemented in numerous works to solve many different problems. More details about the application areas that they are more useful and their modifications can be seen in the surveys found in literature. What is more, Artificial Bee Colony and Bacterial Foraging Optimization are considered also established schemes and attempting to find all the papers that include an application of them will be impossible.

The rest of the schemes included in SI category are divided in three subcategories based on their characteristics; Foraging, Social Behavior and Other Swarm Behavior. Foraging describes the strategies followed by groups of animals to find their food source. In Social Behavior algorithms that model social characteristics like hierarchy between swarms and packs are included. Any other swarm behavior which cannot be encompassed in the aforementioned subcategories is referred as "Other Swarm Behavior".

It should be mentioned that in the work of Fister et al. [57], both Virtual Ant Algorithm and Virtual Bees are considered SI methods. However, in this work, schemes that do not fall into the aforementioned definition of Nature Inspired Algorithms have been excluded.

15.2.1.1 Foraging Methods

A lot of methods of Swarm Intelligence (SI) are inspired by foraging strategies of animals, birds, bugs or fishes. Initially, Ant Colony Optimization was the first SI algorithm which was based on the collaboration between organisms (ants) to find food sources. The next foraging strategy algorithm has been proposed in 2002 and was based on foraging of bacteria [58].

In 2004, the Beehive algorithm was proposed by Wedde et al. [59]. One year later, [60] presented Bee Colony Optimization trying to solve TSP routing problems. One more algorithm based on collective behavior of bees was proposed by Pham et al. [61]. The next year, [62] presented one of the well-known optimization algorithms, namely Artificial Bee Colony.

Another well-known and very useful algorithm was proposed by Yang [63], namely Bat algorithm. The same year, one more algorithm based on bees' behavior was presented [64]. The third method presented in 2010 that was inspired by foraging strategies was given the name Hunting Search [65].

Although Cockroach Swarm Optimization was presented for the first time in the 2nd International Conference on Computer Engineering and Technology in 2010, the paper was retracted and its initial work can be considered to be the one of 2011 [66].

In 2012, Blind, naked mole-rat algorithm was proposed by Shirzadi and Bagheri [67]. Inspired by eurygaster's foraging method, [68] proposed an algorithm by the same name, which they applied on a graph partitioning problem. Another scheme based on bees was presented by Maia et al. [69]. The same year, Wolf Search has been proposed by Tang et al. [70].

In 2013, [71] proposed African Wild Dog Algorithm as an engineering optimization method. In the same year, [72] presented Penguins Search Optimization Algorithm. At last, [73] came up with Seven-spot Ladybird Optimization algorithm at the same year.

Mirjalili et al. [74] presented another challenging optimization method, named Grey Wolf Optimizer. Also this year, [75] proposed Chicken Swarm Optimization. Both of these methods were applied on engineering optimization problems. In 2015, [76] proposed Cockroach Swarm Evolution which has previously been presented as Cockroach Genetic Algorithm by the same authors [77]. Also in 2015, [78] proposed Butterfly Optimization Algorithm as Butterfly Algorithm, giving its final name later on [79].

In 2016, [80] presented Dolphin Swarm Optimization Algorithm. Inspired by ravens, [81] proposed Raven Roosting Optimization Algorithm. Another algorithm proposed by Mirjalili and Lewis [82] was given the name Whale Optimization Algorithm. Also in 2016, Bird Swarm Optimization [83] was proposed.

In 2017, [84] proposed Spotted Hyena Optimizer applying it to engineering design problems. The same year, Whale Swarm Algorithm was presented by Zeng et al. [85]. With the inspiration coming from another fish, [86] proposed Yellow Saddle Goatfish algorithm the next year. Meerkats' foraging strategy inspired [87] to present Meerkat Clan Algorithm as a method to cope with TSP problems.

The most recently proposed scheme is Sailfish Optimizer [88], which has been applied on engineering design problems.

15.2.1.2 Social Behavior Methods

Another source of inspiration for researchers is the social structure and behavior among members of flocks, packs and groups. Inspired by the social behavior of fish, [89] proposed Fish-swarm algorithm. After four years, [90] proposed Glowworm Swarm-based Optimization. The social behavior of roaches was modeled by [91].

While the majority of bee-inspired algorithms are met in the previous subcategory, in 2009, [92] came up with the Bumblebees algorithm. Also in 2009, two of the most-known nature-inspired algorithms were presented, namely Cuckoo Search [93] and Firefly Algorithm [94]. The same year, Dolphin Partner Optimization was proposed by Shiqin et al. [95]. Inspired by locusts, [96] proposed a swarm intelligent algorithm.

In 2010, [97] proposed Termite Colony Optimization. The herding behavior of Krill inspired [98] to present an algorithm.

Inspired by the spiders, [99] proposed Social Spider Optimization. The same year, Swallow Swarm Optimization Algorithm was proposed [100]. Another species of spider inspired [101] to come up with Spider Monkey Optimization Algorithm.

In 2015, African Buffalo Optimization was proposed by [102] as a TSP optimization method. What is noteworthy in that year is the fact that two different algorithms based on Elephant's behavior were proposed [103, 104] and in both works Professor S. Deb was one of the authors. Also in 2015, [105] proposed See-See Partridge Chicks Optimization. The same year, [106] proposed Earthworm Optimization Algorithm, which was implemented to solve the TSC problem, where the goal is to find a combination of questions to satisfy constraints from item bank.

The Lion Optimization Algorithm was proposed by Neshat et al. [107]. The communication and social behavior of dolphins inspired [108] to present Dolphin Swarm Algorithm. Coyotes inspired [109] to come up with the homonym algorithm.

Finally, Emperor Penguins inspired [110] to propose Emperor Penguin Optimizer, but also [111] to present Emperor Penguins Colony.

15.2.1.3 Other Swarm Behaviors

In this subcategory, Particle Swarm Optimization (PSO) can be included, which is based on the way that large flocks of birds are moving. Such phenomena that model simultaneous movement of groups compose this subcategory. What is more, Stochastic Diffusion Search can be included here based on its physical analogue.

Although there are only a few implementations, Sheep Flocks Heredity Model algorithm [112] showed up almost twenty years ago. In 2007, another algorithm based on bacteria was presented [113], named Bacterial Swarming Algorithm.

Garcia and Pérez [114] came up with Jumping Frogs Optimization, which was applied on p-Median problem. One year later, [115] proposed Artificial Searching Swarm Algorithm. Filho et al. [116] presented Fish School Search being inspired by the swarming behavior of fishes, where the flock would turn as a unity trying to avoid enemies and find food.

Hierarchical Swarm Model has been presented in 2010 by Chen et al. [117]. Based on migration behavior of birds [118] proposed Migrating Birds Optimization. Wang et al. [119] proposed Bumble Bees Mating Optimization as a method to solve VRP, with the inspiration coming from Honey Bees Mating Optimization, but the differences between these two schemes are quite a few.

Trying to make a simpler version of PSO, [120] came up with Weightless Swarm Algorithm. Pigeons have been an inspiration for [121] to develop a method for air robot path planning. The migration strategy of animals born the homonym algorithm [122]. Wang et al. [123] proposed Monarch Butterfly Optimization. Cheng et al. [124] proposed Cockroach Colony Optimization as a new version of Cockroach Swarm Optimization implementing robot navigation.

One of the most common authors of nature-inspired algorithms is S. Mirjalili, which has published two aforementioned Swarm Intelligent techniques; Grey Wolf Optimizer and Whale Optimization Algorithm. Algorithms proposed by him that belong in this subcategory are Moth Flame Optimization [125] and Dragonfly Algorithm [126], while also he was co-author of Salp Swarm Algorithm [127] and Grasshopper Optimization Algorithm [128]. The majority of these algorithms are applied on engineering design optimization problems in their initial work.

In 2017, [129] proposed Selfish Herd Optimizer, while in 2018 four more algorithms were presented; Biology Migration Algorithm [130], Lion Pride Optimization Algorithm [131], Meerkat-inspired Algorithm [132] and Wildebeests Herd Optimization [133]. The same year, another cockroach-inspired method was presented, named Artificial Social Cockroaches [134], mainly for image detection.

The most recent techniques of this subcategory are Bison algorithm [135], Seagull Optimization Algorithm [136], Vocalization of Humpback Whale Optimization [137] and the upcoming Collective Animal Behavior [138].

15.2.2   Algorithms Inspired by Organisms

While Swarm Intelligent schemes are inspired by collective behavior, there are some other algorithms which are based on strategies and habits observed in individuals. Either these individuals are animals, fishes and birds, or belong in flower species, grasses and bushes, all algorithmic inspirations of this category come from organisms.

In the subcategory of Fauna-based methods belong the techniques that are inspired by species of animals (megafauna), birds (avifauna), fishes (piscifauna), while also based on bacterias (infauna and epifauna), microorganisms and viruses (microfauna), making it the largest of the subcategories. Flora-based methods include all algorithms that are inspired by plant life forms, such as trees and fungus.

The third subcategory consists of schemes that model operations of organs, like heart, or disease treatments, such as chemotherapy.

15.2.2.1   Fauna-Based Methods

The first algorithm met in this subcategory is Marriage in Honey Bees [139], which was initially proposed to solve the satisfiability problem. The next one was proposed one year after, named Bacterial Chemotaxis [140], and implemented to solve an airfoil design problem. Fauna includes also viruses, so algorithms based on them are categorized here. These would be Viral Systems [141], Swine Influenza Models Based Optimization [142], Virulence Optimization Algorithm [143] and Virus Colony Search [144].

Although that its name refers to Swarm Intelligence (SI), Cat Swarm Optimization [145] is the best example of the mistaken comprehension of what SI is and the confusion with the population-based methods. The same year, Shuffle Frog-leaping Algorithm [146] and Honey-bees Mating Optimization [147] were proposed. [148] were inspired by animal searching behavior and group living theory and proposed Group Search Optimizer.

In 2007, the first of the two schemes that were based on monkeys was presented [149] to solve Potential Problem. The second one was proposed one year later [150]. One more algorithm inspired by bees, named Bee Collecting Pollen Algorithm [151], was presented in 2008 and implemented for TSP, as the majority of bee-inspired algorithms. One of the most known algorithm was presented that year, named Biogeography-based Optimization [152] modeling the migration and mutation that was observed in animals based on their geographical particularities.

Two algorithms inspired by insects were proposed in 2011; Fruit Fly Optimization [153] and Mosquitos Oviposition [154]. Bats have been an inspiration topic once again, where [155] proposed Bat Sonar Algorithm. Aihara et al. [156] proposed a mathematical model of the calling method used by tree frogs, naming the algorithm Japanese Tree Frogs. Another bird, named Hoopoe, inspired [157] to come up with Hoopoe Heuristic Optimization. The chronologically first algorithm that was inspired by lions was the homonym algorithm proposed by Rajakumar [158]. The same year, The Great Salmon Run [159] was proposed, too.

Cuttlefish inspired [160] to propose the homonym algorithm. The echolocation of dolphins inspired [161] to propose an algorithm to solve engineering problems. Egyptian Vulture Optimization Algorithm was proposed by Sur et al. [162] to solve Knapsack problem. On the other hand, [163] proposed Green Heron Optimization Algorithm as a routing problems tackling method. Keshtel, a species of dabbling duck, was the inspiration for the homonym algorithm [164]. Jumper Firefly Algorithm was presented by [165] to solve the assignment problem. Tilahun and Ong [166] modeled the different behavior between preys and predators in their algorithm. Bacteria became famous inspiration topic through the years, something that made [167] propose Magnetotactic Bacteria Optimization Algorithm. What is more, [168] came up with Mussels Wandering Optimization.

In 2014 one more bird-based method made its appearance, named Bird Mating Optimizer [169]. Also, Coral Reefs Optimization Algorithm [170] was presented that year to cope with routing problems. The animals' way to divide food resources over their population brought up Competition over Resources [171]. On the other hand, the symbiosis of different species in a specific area inspired [172] to come up with Symbiotic Organism Search.

Mirjalili [173] makes his appearance in this category too, where he has proposed Ant Lion Optimizer. Artificial Algae Algorithm [174] and Jaguar Algorithm [175] were presented that year.

In 2016 there were six new techniques; Camel Algorithm [176], Crow Search Algorithm [177], Red Deer Algorithm [178], Mosquito Flying Optimization ([179], Shark Smell Optimization [180] and Sperm Whale Algorithm [181].

Butterflies [182] and beetles [183] inspired new algorithms in 2017. Fish Electrolocation Optimization was proposed by Haldar and Chakraborty [184] to solve reliability problems. Species of whales kept inspiring methods, where in 2017 Killer Whale Algorithm was proposed by Biyanto et al. [185]. Laying Chicken Algorithm [186] was presented the same year. Połap and Woz´niak [187] proposed Polar Bear Optimization Algorithm to solve engineering problems. Samareh Moosavi and

Khatibi Bardsiri [188] came up with the Satin Bowerbird Optimizer, inspired by the structures which are built by male satin bowerbirds.

As it has been seen in Fig. 15.1, 2018 was the year with the most published algorithms that were based on organisms. Inspired by barnacles that are met in sea, [189] presented Barnacles Mating Optimizer, applying it on Assignment Problem. Another dolphin-inspired scheme, named Dolphin Pod Optimization [190], was presented. Other algorithms appeared that year were Cheetah Based Optimization Algorithm [191], Circular Structures of Puffer Fish [192], Mouth Brooding Fish algorithm [193] and Pity Beetle Algorithm [194].

The same year, Beetle Swarm Optimization Algorithm [195] and Eagle Piercing Optimizer [196] were proposed as engineering optimization techniques. Owl Search Algorithm [197] was proposed to solve Heat Flow Exchange and Pontogammarus Maeoticus Swarm Optimization [198] has been proposed to solve Solar PV Arrays. At last, Rhinoceros Search Algorithm [199] was presented as flow-shop scheduling optimization method.

Andean Condor Algorithm [200] was presented to solve Cell Formation problems. From the same family of birds, [201] inspired the Bald Eagle Search. Donkey and Smuggler Optimization Algorithm was proposed by Shamsaldin et al. [202] to solve TSP problems. Falcon Optimization Algorithm [203], Flying Squirrel Optimizer [204], Harris Hawks Optimization [205] and Sooty Tern Optimization Algorithm [206] were all applied in engineering problems when they were proposed. [207] proposed Artificial Feeding Birds, applying it in various problems; TSP, training of Neural Networks etc.

Recent methods are also Raccoon Optimization Algorithm [208], Sea Lion Optimization Algorithm [209] and Squirrel Search Algorithm [210].

15.2.2.2   Flora-Based Methods

In 2006, Mehrabian and Lucas [211] presented Invasive Weed Optimization, being the first flora-based method. The same year, [212] proposed Saplings Growing up Algorithm. In 2008, Plant Growth Optimization [213] was proposed to solve speed reducer problem and in 2009 Paddy Field Algorithm [214] made its appearance.

Artificial Plant Optimization Algorithm [215] was proposed to solve engineering problems and Plant Propagation Algorithm [216] to cope with Chlorobenzene Purification problem. Ecology-inspired Optimization Algorithm [217] made its appearance also in 2011. Inspired by a mold, named physarum, [218] proposed Physarum Optimization. [219] came up with Flower Pollination Algorithm also as an engineering optimization tool.

In 2013, Root Mass Optimization was presented by Qi et al. [220] and one year later [221] proposed Root Growth Algorithm. The same year, Forest Optimization Algorithm [222] was proposed as a clustering method.

Roots and trees inspired also Runner-root Algorithm [223], Seed-based Plant Propagation Algorithm [224], Tree Seed Algorithm [225], Natural Forest Regeneration [226] and Root Tree Optimization Algorithm [227]. In 2016, [228] solved scheduling problems proposing Waterweeds Algorithm.

Two more tree-based methods were proposed in 2018, named Tree Based Optimization [229] and Tree Growth Algorithm [230]. Applied in engineering problems, Mushroom Reproduction Optimization [231] was presented also in 2018. Another physarum-inspired scheme, named Physarum-energy Optimization Algorithm [232], was proposed as a TSP tackling method.

15.2.2.3 Other Paradigms

The smallest subcategory consists of schemes that are based mostly on body organs. The first one was Clonal Selection Algorithm [233], proposed in 2000. If someone would like to be more generic, both Genetic Algorithm [17, 234] and Differential Evolution [235] would fit in this subcategory. However, both techniques are considered to be Evolutionary Algorithms, which is the parent category of Nature Inspired schemes.

Although this may rise a lot of discussion, the reincarnation concept inspired at first [236] and then [237] to propose nature-inspired optimization methods. The first algorithm was named Reincarnation Algorithm and was applied on TSP problems, while the second one was given the name of Zombie Survival Optimization and was proposed for image tracking.

Stem Cells Optimization Algorithm [238] was the first algorithm that was related to body parts, especially the stem cells. However, the first algorithm inspired by the functionality of a body part was Heart algorithm [239] and was presented for cluster analysis in 2014. One year later, Invasive Tumor Growth Optimization [240] was presented and later on, Kidney-inspired Algorithm [241] was proposed and applied on TSP problems. A method similar to neural networks, named Neuronal Communication [242], made recently its appearance as an engineering optimization method.

Based on the phenomenon that the human body is trying to stabilize a condition where something is not functioning in the right way, [243] presented Allostatic Optimization. Such a condition would be a disease, which is the inspiration of Artificial Infectious Disease Optimization [244]. The most recent method is called Chemotherapy Science Algorithm [245] and was proposed to solve TSP problems.

15.3 Application Areas and Open Problems for NII

Many researchers are quite skeptical about the existence of an actual physical analogue that inspired some of the authors to propose all these algorithms. However, in this work the usage of such schemes is investigated based on the application areas, in which each algorithm is tested or implemented.

In all Tables consisting the number of published works on each application area, the first number denotes the total number of published works, while the number in the parenthesis denotes how many of these works have been published in Journals.

Algorithms without any application were excluded from the Tables below. This case applies to the most recently published algorithms. Surprisingly, this is the case for some older techniques, too. In total, 31.3% of Swarm Intelligent (SI) schemes and 42.6% of Organisms-based algorithms have no implementations in optimization problems. These percentages are 57.8 and 73.1% in Clustering/Classification problems, whilst in other problem areas are 51.8 and 69.4% respectively. Overall, 27.7% of SI and 37% of Organisms-based methods have no application. One out of three algorithms has not been applied in any problem, from the time that it was proposed.

However, the majority of these algorithms were proposed recently, as it can be seen in Fig. 15.2. The algorithms that exist for over a decade and were not applied in a problem are Saplings Growing up Algorithm (2006), Bee Collecting Pollen Algorithm, Jumping Frogs Optimization (2008), Bumblebees algorithm, Dolphin Partner Optimization, Locust Swarms (2009), Hierarchical Swarm Model, Reincarnation Algorithm (2010). It would be interesting to investigate the performance of these algorithms in some well-known optimization problems.

Except for the established schemes, which were mentioned above, algorithms with multiple citations and over 100 applications are presented in Table 15.1. These algorithms can be considered nearly established schemes, because of their wide usage in multiple problem areas. The number of implementations of these schemes is growing daily, so we excluded them from the Tables below. However, they are collected in a separate Table to divide them from the established well-known techniques that exist for approximately two decades.

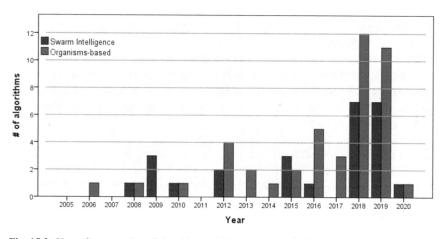

Fig. 15.2 Year of presentation of algorithms which have no applications

Table 15.1 Near-established algorithms based on their applications

	Citations	Applications	Category
Bat algorithm	≈2697	>150	Swarm Intelligence
Biogeography-based optimization	≈2280	>150	Organisms-based
Cuckoo search	≈3759	>150	Swarm Intelligence
Firefly Algorithm	≈1368	>100	Swarm Intelligence
Flower Pollination Algorithm	≈930	>100	Organisms-based
Grey Wolf Optimizer	≈2194	>100	Swarm Intelligence
Invasive Weed Optimization	≈959	>100	Organisms-based
Shuffled Frog Leaping Algorithm	≈768	>150	Organisms-based

Notable fact is that four of them are proposed by Professor Xin-She Yang, who has published a lot of work on the mathematical foundations of Nature Inspired Algorithms [246] and conducts also research on the applications of this kind of techniques [247–249].

15.3.1 Applications of Swarm Intelligent Methods

One detail, which can be extracted from Table 15.2, is that there are algorithms which seem to be very good optimization tools, such as Dragonfly Algorithm, Fish-swarm Algorithm, Moth-flame Optimization Algorithm, The Bees Algorithm and Whale Optimization Algorithm. These techniques are way more used in this area of problems than the other schemes.

Swarm Intelligent algorithms follow the trend of the nature-inspired family techniques, where optimization problems are the majority of implementations. The 41,7% of SI techniques are applied of clustering problems, where the majority of them has between 1-5 implementations.

What is more, the last rows of Table 15.2 provide a better interpretation of the application areas of Swarm Intelligence. No further results can be extracted from that Table. The strategy that is modeled in each algorithm (i.e. foraging, social behavior or other swarm behavior) seems not to be correlated with the type of problems that the algorithm can be applied to.

An interesting question should be if the specific characteristics met in the algorithms of a subcategory makes them better in a specific type of optimization problems. This information is given in Table 15.3. Apart from Fish-swarm Algorithm and Whale Optimization Algorithm, which are widely used in many problem areas, the rest of the techniques are not so often used.

Notable exceptions are Bee Colony Optimization (applied mostly on Vehicle Routing Problems), The Bees Algorithm which has many engineering applications,

Table 15.2 Areas of application of swarm intelligent algorithms

	Optimization	Classification/Clustering	Other
African Buffalo Optimization	10(8)	–	2(2)
African Wild Dog Algorithm	1(1)	–	–
Animal Migration Optimization	7(4)	3(3)	3(2)
Artificial Searching Swarm Algorithm	4(1)	–	–
Artificial Social Cockroaches	–	1(1)	–
Bacterial Swarming Algorithm	5(1)	–	1(1)
Bee Colony Optimization	16(15)	1(1)	4(2)
Bee Swarm Optimization	4(3)	–	2(1)
BeeHive	10(5)	–	–
Bird Swarm Algorithm	9(4)	3(3)	4(1)
Butterfly Optimization Algorithm	2(2)	1(1)	–
Bumble Bees Mating Optimization	2(2)	–	1(1)
Chicken Swarm Optimization	12(6)	5(1)	5(3)
Cockroach Swarm Optimization	3(3)	–	1(1)
Dolphin Swarm Algorithm	1(1)	–	1(1)
Dragonfly Algorithm	42(25)	4(3)	18(11)
Emperor Penguin Optimizer	–	1(1)	1(1)
Elephant Herding Optimization	15(8)	1(1)	2(1)
Elephant Search Algorithm	3(2)	2(1)	–
Fish-swarm Algorithm	86(50)	24(15)	38(22)
Glowworm Swarm Optimization	11(7)	–	10(8)
Hunting Search	13(7)	–	1(1)
Krill Herd	27(16)	8(5)	3(2)
Lion Optimization Algorithm	4(3)	2(2)	2(2)
Meerkat Clan Algorithm	1(1)	–	–
Migrating Birds Optimization	31(2)	3(1)	2(0)
Monarch Butterfly Optimization	5(3)	–	–
Moth-flame Optimization Algorithm	91(65)	10(6)	22(14)
OptBees	2(1)	3(1)	–
Penguins Search Optimization Algorithm	4(2)	2(2)	–
Pigeon-inspired Optimization	34(26)	8(8)	6(4)
Raven Roosting Optimization Algorithm	2(2)	–	–

(continued)

Table 15.2 (continued)

	Optimization	Classification/Clustering	Other
Roach Infestation Optimization	1(0)	–	–
Sailfish Optimizer	1(1)	–	–
Salp Swarm Algorithm	37(25)	5(5)	12(8)
See-See Partridge Chicks Optimization	–	2(2)	–
Seven-spot Ladybird Optimization	1(0)	–	–
Sheep Flocks Heredity Model Algorithm	8(7)	–	1(1)
Social Spider Optimization	32(21)	9(5)	10(8)
Spider Monkey Optimization Algorithm	28(21)	8(6)	2(1)
Spotted Hyena Optimizer	6(4)	–	1(0)
Swallow Swarm Optimization Algorithm	1(0)	–	–
Termite Colony Optimization	2(1)	–	1(0)
The Bees Algorithm	67(39)	9(3)	2(1)
The Bees Algorithm	51(27)	4(0)	16(5)
Weightless Swarm Algorithm	2(1)	–	–
Whale Optimization Algorithm	176(130)	26(22)	30(19)
Whale Swarm Algorithm	1(1)	–	–
Wolf Pack Algorithm	8(6)	–	6(5)
Wolf Search	2(2)	3(1)	2(1)
Foraging	317(217)	48(32)	66(34)
Social Behavior	229(144)	57(38)	78(53)
Other Swarm Behavior	287(194)	35(28)	78(49)
Total	833(555)	140(98)	222(136)

Chicken Swarm Optimization, Dragonfly Algorithm, Elephant Herding Optimization, Moth-flame Optimization, Salp Swarm Algorithm and Spider Monkey Optimization Algorithm, all with many energy applications, Krill Herd that has implemented in many engineering and energy problems and Pigeon-inspired Optimization (with many applications in unmanned vehicles, either on clustering problems or optimization).

The field with the least implementations is Finance. The plethora of applications is almost equally divided in Engineering and Energy problems.

Table 15.3 Number of Swarm Intelligence applications on optimization problems

	Engineering	Finance	Operational research	Energy problems	Other applications
African Buffalo Optimization	–	–	6(5)	3(2)	1(1)
African Wild Dog Algorithm	1(1)	–	–	–	–
Animal Migration Optimization	3(2)	–	2(1)	1(0)	1(1)
Artificial Searching Swarm Algorithm	3(1)	–	–	1(0)	–
Bacterial Swarming Algorithm	1(0)	–	–	4(1)	–
Bee Colony Optimization	3(3)	–	11(10)	–	2(2)
Bee Swarm Optimization	–	–	2(1)	2(2)	–
BeeHive	–	–	3(3)	1(0)	6(2)
Bird Swarm Algorithm	2(1)	–	1(1)	6(2)	–
Butterfly Optimization Algorithm	1(1)	–	1(1)	–	–
Bumble Bees Mating Optimization	–	–	2(2)	–	–
Chicken Swarm Optimization	–	–	2(1)	8(4)	2(1)
Cockroach Swarm Optimization	–	–	3(3)	–	–
Dolphin Swarm Algorithm	–	–	–	1(1)	–
Dragonfly Algorithm	5(5)	–	5(1)	28(16)	4(3)
Elephant Herding Optimization	1(1)	–	5(2)	8(4)	1(1)
Elephant Search Algorithm	–	–	2(2)	–	1(0)
Fish-swarm Algorithm	23(12)	1(1)	42(25)	21(8)	9(4)
Glowworm Swarm Optimization	–	–	5(2)	5(4)	1(1)

(continued)

Table 15.3 (continued)

	Engineering	Finance	Operational research	Energy problems	Other applications
Hunting Search	6(4)	–	3(2)	4(1)	–
Krill Herd	12(6)	1(1)	3(1)	10(7)	1(1)
Lion Optimization Algorithm	1(0)	–	2(2)	1(1)	–
Meerkat Clan Algorithm	–	–	1(1)	–	–
Migrating Birds Optimization	–	–	31(20)	–	–
Moth-flame Optimization Algorithm	15(11)	–	6(5)	69(48)	1(1)
Monarch Butterfly Optimization	1(0)	–	4(3)	–	–
OptBees	–	–	2(1)	–	–
Penguins Search Optimization Algorithm	–	–	3(1)	1(1)	–
Pigeon-inspired Optimization	6(4)	–	7(5)	6(5)	15(12)
Raven Roosting Optimization Algorithm	–	–	2(2)	–	–
Roach Infestation Optimization	1(0)	–	–	–	–
Sailfish Optimizer	1(1)	–	–	–	–
Salp Swarm Algorithm	6(3)	–	3(1)	24(18)	4(3)
Seven-spot Ladybird Optimization	–	–	–	1(0)	–
Sheep Flocks Heredity Model Algorithm	1(1)	–	7(6)	–	–
Social Spider Optimization	7(4)	–	8(6)	16(11)	1(0)
Spider Monkey Optimization Algorithm	4(3)	1(1)	4(4)	19(13)	–
Spotted Hyena Optimizer	4(3)	–	1(0)	1(1)	–

(continued)

Table 15.3 (continued)

	Engineering	Finance	Operational research	Energy problems	Other applications
Swallow Swarm Optimization Algorithm	–	–	1(0)	–	–
Termite Colony Optimization	–	–	1(0)	1(1)	–
The Bees Algorithm	32(17)	–	8(5)	5(1)	6(4)
Weightless Swarm Algorithm	1(1)	–	–	1(0)	–
Whale Optimization Algorithm	37(30)	–	35(26)	98(69)	6(5)
Whale Swarm Algorithm	1(1)	–	–	1(1)	–
Wolf Pack Algorithm	1(1)	–	6(4)	1(1)	–
Wolf Search	–	–	–	2(2)	–
Foraging	88(62)	–	78(58)	129(83)	22(14)
Social Behavior	50(27)	3(3)	85(53)	85(52)	16(9)
Other Swarm Behavior	44(29)	–	72(47)	144(96)	27(22)

15.3.2   Applications of Organisms-Inspired Algorithms

This category of algorithms includes four algorithms that are used in a big variety of problems; Ant Lion Optimizer, Fruit Fly Optimization and Symbiotic Organisms Search. While also, the aforementioned near-established schemes (Biogeography-based Optimization, Flower Pollination Algorithm, Invasive Weed Optimization and Shuffled Frog Leaping Algorithm) have numerous applications. What is more, quite a lot of implementations are observed in Cat Swarm Optimization, Crow Search Algorithm, Group Search Optimizer, Honey Bees Mating Optimization and Marriage in Honey Bees.

The percentage of Organisms-inspired algorithms that are applied on clustering or classification problems is 26,8%, which is smaller than the corresponding percentage of Swarm Intelligent schemes, as mentioned in (Table 15.4).

The big difference among the number of algorithms consisting each sub-category (i.e. fauna, flora and other paradigms), explains the divergence of total implementations in each problem field among these subcategories.

In Table 15.5 it is obvious that Organisms-based schemes are widely used in engineering and energy problems. The wide number of Fauna methods makes this

Table 15.4 Areas of application of organisms-inspired algorithms

	Optimization	Classification/Clustering	Other
Allostatic Optimization	–	–	1(1)
Ant Lion Optimizer	93(65)	5(4)	12(9)
Artificial Algae Algorithm	7(7)	–	1(1)
Artificial Coronary Circulation System	2(2)	–	–
Artificial Plant Optimization Algorithm	5(5)	–	–
Bacterial Chemotaxis	8(5)	–	6(3)
Barnacles Mating Optimizer	1(0)	–	–
Beetle Antennae Search Algorithm	5(4)	1(1)	1(1)
Bird Mating Optimizer	6(5)	6(6)	1(1)
Cat Swarm Optimization	38(21)	9(5)	8(8)
Cheetah Based Optimization Algorithm	1(1)	–	–
Chemotherapy Science Algorithm	1(1)	–	–
Competition over Resources	5(2)	–	–
Coral Reefs Optimization Algorithm	10(10)	1(0)	1(1)
Crow Search Algorithm	52(26)	7(6)	11(9)
Cuttlefish Algorithm	2(2)	2(2)	4(4)
Dolphin Echolocation	17(14)	–	–
Dolphin Pod Optimization	1(0)	–	–
Eco-inspired Evolutionary Algorithm	–	–	1(0)
Egyptian Vulture Optimization Algorithm	6(6)	1(1)	–
Fish Electrolocation Optimization	2(0)	–	–
Fish-school Search	7(3)	1(1)	7(2)
Forest Optimization Algorithm	5(2)	6(6)	2(2)
Fruit Fly Optimization	115(82)	13(7)	27(22)
Grasshopper Optimization Algorithm	10(9)	–	5(5)
Great Salmon Run	3(3)	1(1)	–
Green Herons Optimization Algorithm	1(0)	–	–

(continued)

Table 15.4 (continued)

	Optimization	Classification/Clustering	Other
Group Search Optimizer	72(62)	3(3)	5(4)
Harris Hawks Optimization	–	–	2(2)
Heart	2(2)	–	–
Honey-bees Mating Optimization	57(41)	6(5)	6(2)
Invasive Tumor Growth Optimization Algorithm	–	–	1(0)
Japanese Tree Frogs	1(0)	–	1(1)
Jumper Firefly Algorithm	2(1)	–	–
Keshtel Algorithm	1(1)	–	–
Kidney-inspired Algorithm	7(7)	1(1)	2(2)
Killer Whale Algorithm	2(1)	–	–
Lion Pride Optimizer	–	1(1)	1(0)
Marriage in Honey Bees	70(62)	3(3)	1(1)
Monkey Algorithm	18(15)	2(2)	1(1)
Monkey Search	3(2)	–	–
Mosquitoes Oviposition	2(0)	–	–
Mouth Brooding Fish Algorithm	1(1)	–	–
Mussels Wandering Optimization	5(4)	1(1)	–
Natural Forest Regeneration	1(1)	–	–
Paddy Field Algorithm	–	–	1(0)
Physarum Optimization	11(7)	1(1)	–
Plant Growth Optimization	20(14)	–	–
Plant Propagation Algorithm	3(3)	–	1(1)
Polar Bear Optimization Algorithm	2(2)	–	–
Prey-Predator Algorithm	4(4)	–	–
Root Tree Optimization Algorithm	8(2)	–	1(1)
Satin Bowerbird Optimizer	4(4)	–	–
Selfish Herd Optimizer	2(2)	–	–
Shark Smell Optimization	5(5)	–	1(1)
Sperm Whale Algorithm	1(1)	–	1(1)
Squirrel Search Algorithm	2(2)	–	–
Stem Cells Optimization Algorithm	1(0)	2(1)	–

(continued)

Table 15.4 (continued)

	Optimization	Classification/Clustering	Other
Swine Influenza Models Based Optimization	6(4)	–	1(0)
Symbiotic Organisms Search	112(85)	5(2)	12(10)
The Lion's Algorithm	3(3)	1(1)	3(3)
The Runner-Root Algorithm	2(2)	–	2(2)
Tree Growth Algorithm	2(2)	1(1)	–
Tree-seed algorithm	10(7)	3(2)	3(1)
Viral System	15(8)	–	–
Virus Colony Search	5(2)	–	–
Waterweeds Algorithm	1(0)	–	–
Fauna	1074(834)	75(58)	106(84)
Flora	298(253)	28(27)	14(10)
Other Paradigms	13(12)	3(2)	4(3)
Total	1385(1099)	106(87)	124(97)

subcategory the most used of Organisms-based subcategories. What is more, only Fauna-based methods have been used in finance optimization problems.

15.3.3 Comparison and Discussion

In Table 15.1, eight near established schemes were mentioned, namely Bat Algorithm, Biogeography-based optimization, Cuckoo Search, Firefly Algorithm, Flower Pollination Algorithm, Grey Wolf Optimizer, Invasive Weed Optimization and Shuffled Frog Leaping Algorithm. In this special group of techniques five more can be added, which were found to have more than 100 application works each. These algorithms are Ant Lion Optimizer, Moth-flame Optimization Algorithm and Whale Optimization Algorithm from Swarm Intelligent schemes, and Fruit Fly Optimization and Symbiotic Organisms Search from Organisms-based techniques.

Antagonists of Nature Inspired Intelligence usually argue for the performance of other Computational Intelligent methods, such as Neural Networks, in classification or clustering problems. As it can be seen in Tables 15.2 and 15.4, NIAs are more efficient in optimization problems. Only a few schemes have been proposed for other types of problems. Especially, 33,3% of all aforementioned algorithms have at least one implementation on clustering or classification problems and 38,5% are implemented for other problem areas.

What is more, the field of finance, where only a few algorithms have been used may provide challenging problems for these techniques. Portfolio optimization,

Table 15.5 Number of organisms-inspired algorithms' applications on optimization problems

	Engineering	Finance	Operational research	Energy problems	Other applications
Ant Lion Optimizer	19(13)	1(1)	21(15)	51(35)	1(1)
Artificial Algae Algorithm	1(1)	–	2(2)	4(4)	–
Artificial Coronary Circulation System	2(2)	–	–	–	–
Artificial Plant Optimization Algorithm	–	–	–	–	5(5)
Bacterial Chemotaxis	1(1)	–	–	6(3)	1(1)
Barnacles Mating Optimizer	–	–	–	1(0)	–
Beetle Antennae Search Algorithm	2(1)	–	–	–	3(3)
Bird Mating Optimizer	1(0)	–	2(2)	3(3)	–
Cat Swarm	2(1)	–	17(8)	15(11)	4(1)
Cheetah Based Optimization Algorithm	–	–	–	1(1)	–
Chemotherapy Science Algorithm	–	–	1(1)	–	–
Competition Over Resources	–	–	–	5(2)	–
Coral Reefs Optimization Algorithm	2(2)	–	2(2)	5(5)	1(1)
Crow Search Algorithm	7(3)	–	8(5)	35(17)	2(1)
Cuttlefish Algorithm	1(1)	–	–	–	1(1)
Dolphin Echolocation	15(13)	–	–	2(1)	–
Dolphin Pod Optimization	1(0)	–	–	–	–
Egyptian Vulture Optimization Algorithm	2(2)	–	3(3)	1(1)	–

(continued)

Table 15.5 (continued)

	Engineering	Finance	Operational research	Energy problems	Other applications
Fish Electrolocation Optimization	–	–	2(0)	–	–
Fish-school Search	2(1)	–	4(2)	1(0)	–
Forest Optimization Algorithm	–	–	4(2)	1(0)	–
Fruit Fly Optimization	37(30)	1(1)	44(28)	25(16)	8(7)
Grasshopper Optimization Algorithm	–	–	1(1)	8(7)	1(1)
Great Salmon Run	2(2)	–	–	1(1)	–
Green Herons Optimization Algorithm	–	–	1(0)	–	–
Group Search Optimizer	49(42)	–	2(2)	21(18)	–
Heart	–	–	2(2)	–	–
Honey-bees Mating Optimization	4(1)	–	16(12)	36(27)	1(1)
Japanese Tree Frogs	–	–	–	–	1(0)
Jumper Firefly Algorithm	–	–	1(0)	1(1)	–
Keshtel Algorithm	–	–	1(1)	–	–
Kidney-inspired Algorithm	1(1)	–	–	5(5)	1(1)
Killer Whale Algorithm	–	1(1)	–	1(0)	–
Marriage in Honey Bees	51(46)	–	5(4)	11(10)	3(2)
Monkey Algorithm	–	–	5(3)	3(2)	10(10)
Monkey Search	–	–	1(1)	2(1)	–
Mosquitoes Oviposition	–	–	–	2(0)	–
Mouth Brooding Fish Algorithm	–	1(1)	–	–	–

(continued)

Table 15.5 (continued)

	Engineering	Finance	Operational research	Energy problems	Other applications
Mussels Wandering Optimization	–	–	1(1)	1(1)	3(2)
Natural Forest Regeneration	1(1)	–	–	–	–
Physarum Optimization	–	–	5(4)	3(2)	3(1)
Plant Growth Optimization	–	–	1(1)	19(13)	–
Plant Propagation Algorithm	1(1)	–	1(1)	1(1)	–
Polar Bear Optimization Algorithm	–	–	–	2(2)	–
Prey-Predator Algorithm	1(1)	–	2(2)	1(1)	–
Queen-bee Evolution	8(7)	–	–	2(2)	–
Root Tree Optimization Algorithm	–	–	8(2)	–	–
Satin Bowerbird Optimizer	–	–	2(2)	2(2)	–
Selfish Herd Optimizer	–	–	–	1(1)	1(1)
Shark Smell Optimization	1(1)	–	1(1)	2(2)	1(1)
Sperm Whale Algorithm	–	–	–	1(1)	–
Stem Cells Optimization Algorithm	1(0)	–	–	–	–
Symbiotic Organisms Search	30(27)	1(1)	38(24)	43(33)	–
Swine Influenza Models Based Optimization	1(0)	–	1(1)	4(3)	–
The Lion's Algorithm	–	–	3(3)	–	–
The Runner-Root Algorithm	–	–	–	2(2)	–

(continued)

Table 15.5 (continued)

	Engineering	Finance	Operational research	Energy problems	Other applications
Tree Growth Algorithm	–	–	2(2)	–	–
Tree-seed Algorithm	4(3)	–	1(1)	3(2)	2(1)
Viral System	–	–	15(8)	–	–
Virus Colony Search	–	–	1(1)	4(1)	–
Waterweeds Algorithm	–	–	1(0)	–	–
Fauna	230(189)	5(5)	198(132)	293(206)	40(32)
Flora	6(5)	–	15(11)	37(26)	10(7)
Other Paradigms	4(3)	–	3(3)	5(5)	1(1)

index tracking, stock prediction and market forecasting are some of these problems. Although some of the aforementioned financial problems are included in the category entitled as "Other problems", the number of these works is equally low with the one mentioned in the "Finance" optimization category problems.

As it can be seen in Fig. 15.3, the majority of the methods mentioned in this work have less than 50 implementations. This figure refers to all kind of problem areas. This trend is observed too, when investigating apart the two nature-inspired categories that are described in this work. Based on these results, we can assume that 7 algorithms are already established, while 13 more tend to become established (exceed 100 implementations). This assumption may lead to the preliminary conclusion that approximately only 15% of the proposed schemes may provide a good method to

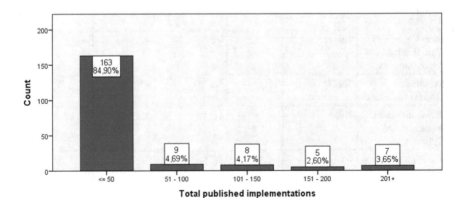

Fig. 15.3 Relation between algorithms and number of implementations

tackle problems. However, further research is needed. The relation between year and number of implementations can provide more accurate results.

15.3.4 Are All These Algorithms Actually Needed?

Looking more carefully in the existing schemes, someone can observe that there are eleven different algorithms inspired by bees, five based on cockroaches and dolphins and four lion algorithms. Also, three different techniques are based on bacteria, beetles, frogs, monkeys, penguins and wolves. And finally, there are a couple of techniques inspired by ants, bats and whales. The major question here is if the algorithms which share the same inspiration have common characteristics and the same optimization strategy.

Monkey Algorithm (MA) is based on a probabilistic climbing strategy of monkeys, while Monkey Search (MS) is a lot similar to that strategy; the monkeys climb on trees and randomly find new branches in their way up. Although these algorithms are inspired by the same strategy, MA is found to be way more useful in terms of applications with 18 published works on optimization problems, while MS has only 3. But the important detail here is that MA's majority of applications are on optimal sensor placement [250, 251]. What is more, MA has implementations on clustering problems [252, 253].

Neither Cockroach Swarm Optimization (CSO) nor Roach Infestation Algorithm seems to be useful optimization tools, even though both algorithms model the foraging strategy of cockroaches. Both techniques have only a few applications, where CSO is mostly applied in TSP and VRP problems. Artificial Social Cockroaches algorithm is applied only in information retrieval [134]. On the other hand, Cockroach Swarm Evolution and Cockroach Colony Optimization have no applications.

Bacteria Foraging Optimization (BFO) is considered an established method, thus it is not included in the Tables above. On the other hand, Bacterial Chemotaxis (BC) and Bacterial Swarming Algorithm are mostly applied on energy problems, but BC has also a few forecasting applications. However, both these schemes are far away from the number of published works that BFO has.

Among penguin-inspired algorithms, Penguins Search Optimization Algorithm is the most used. However, all three algorithms of this inspiration have very few implementations; 6 for Penguins Search Optimization Algorithm, 2 for Emperor Penguin Optimizer and none for Emperor Penguins Colony.

Another important thing is to point out the best of the same-inspiration techniques or at least come up with the application field that each algorithm is better than the others. In the group of bee-based algorithms, BeeHive algorithm was presented as a routing algorithm, which has been inspired by the communicative procedures of honey bees. This is a very good example of a method that shares inspiration with other nature-inspired algorithms, but solves a specific problem. The majority of its applications are in network problems [254, 255]. Such an example is also Bee Colony Optimization, which is mostly applied on TSP, assignment and scheduling problems.

The Bees Algorithm is one of the most-used bee-inspired techniques, counting 71 publications, where the 32 of them are on engineering optimization problems. Another such highly used algorithm is Marriage in Honey Bees with up to 70 implementations, where 51 of them are on engineering problems. Honey Bees Optimization is also good optimization method (counting 69 publications), used mostly on energy problems (36 of these papers). The well-known Artificial Bee Colony is considered established method and thus it cannot be compared with the other bee algorithms.

Contrary to the rest bee-inspired schemes, Bumblebees algorithm and Bee Pollen Collecting Algorithm have no application even though they were published in 2009 and 2008, respectively. Bee Swarm Optimization, OptBees and Bumblebees Mating Optimization are found only in a few papers.

Among the algorithms that are inspired by dolphins, Dolphin Echolocation (DE) has the most application by far. Dolphin Swarm Algorithm and Dolphin Pod Optimization cannot challenge the performance of DE, having one or two implementations each. Dolphin Partner Optimization and Dolphin Swarm Optimization Algorithm are not implemented in any problem area apart from their initial publication.

In a smaller scale, this is the case in beetle-inspired techniques. Pity Beetle Algorithm and Beetle Swarm Optimization Algorithm have no application, while Beetle Antennae Search Algorithm is found in 7 papers with 5 of them being on optimization problems.

Lion Optimization Algorithm and Lion's Algorithm have less than 5 applications each. Lion Pride Optimizer has one implementation in clustering and one on image segmentation [256], while Lion Pride Optimization has no applications being recently presented.

Although, they are based on the same strategy of frogs (but different species of them), Shuffled Frog Leaping Algorithm (SFLA), Japanese Tree Frogs (JTF) and Jumping Frogs Optimization (JFO) differ a lot in the number of applications. SFLA tends to be an established algorithm that solves mainly engineering problems, while JTF has one implementation on sensor networks [257] and one on graph coloring [258]. On the other hand, JFO has no applications although it has been proposed in 2008.

What is more, while Whale Optimization Algorithm is considered to become an established scheme counting 173 works where optimization problems are tackled, the rest of whale-inspired techniques do not follow this example. Whale Swarm Algorithm has only one implementation, Sperm Whale Algorithm and Killer Whale Algorithm have both two works, while Humpback Whale Optimization Algorithm is recently proposed and thus, has no applications.

A similar situation is observed between Grey Wolf Optimizer (GWO), Wolf Pack Algorithm (WPA) and Wolf Search (WS), where GWO is widely used in a lot of applications, whilst WPA and WS have been implemented only in a few. Furthermore, that is the case for Bat Algorithm (BA) and Bat Sonar Algorithm (BSA), where BA is established and BSA has no applications. The comparison between Ant Colony

Optimization and Ant Lion Optimization (ALO) wouldn't be fair. However, ALO can be found in up to 100 works, where many of them is tackling variations of the economic load dispatch problem, routing and scheduling problems, or engineering problems, usually related to design.

Although Spider Monkey Optimization has been proposed in 2014 (later on than the two other monkey-inspired schemes), it has been applied in numerous problems and has 38 published works. Monkey Algorithm is also popular with 21 implementations, but Monkey Search has only three implementations.

To conclude, all the above mentioned examples show that a new method should be proposed by authors, only if it solves a new problem or its strategy is proper to tackle a specific application.

15.4 Suggestions and Future Work

Each year, more and more algorithms that claim to be nature-inspired are presented. The majority of these algorithms, usually, does not solve a new problem or even worse does not provide a new strategy to cope with some problems. On the contrary, a lot of schemes are modifications of Particle Swarm Optimization and Ant Colony Optimization. The similarity of other methods with these two aforementioned schemes should get under research.

On undergoing work, authors are focusing on the conceptual blocks that form a useful Nature Inspired Algorithm for proposing a framework for developing such methods. However, the existence of all these schemes makes it impossible for someone to select the best algorithm for a specific kind of problem. Either, a metric should be presented or a web tool must be developed to report the best choices for each type of problem.

Something that could also provide useful information on the usage of Nature Inspired Algorithms may be to create separate survey reports in order to investigate which problems are solved more often in each problem field. Such a work could come up with useful findings on which optimization problems are being solved optimally and which can be new challenges for NIAs.

What is more, a wide percentage (near 30%) of the algorithms collected in this chapter has no application associated to them. Except for the recently published algorithms, the rest of the schemes should be tested on different problems to examine if they are useful for a specific problem or application area.

References

1. J. Jantzen, *Foundations of Fuzzy Control: A Practical Approach*, 2nd edn. (Wiley Publishing, 2013)
2. L.A. Zadeh, Fuzzy sets. Inf. Control **8**, 338–353 (1965)

3. Y. Kodratoff, R.S. Michalski (eds.), *Machine learning: an artificial intelligence approach*, vol. III (Morgan Kaufmann Publishers Inc., San Francisco, CA, USA, 1990)
4. R.S. Michalski, J.G. Carbonell, T.M. Mitchell (eds.), *Machine Learning: An Artificial Intelligence Approach* (Springer, Berlin Heidelberg, 1983)
5. S.R. Michalski, G.J. Carbonell, M.T. Mitchell (eds.), *Machine Learning an Artificial Intelligence Approach*, vol. II (Morgan Kaufmann Publishers Inc., San Francisco, CA, USA, 1986)
6. T.M. Mitchell, J.G. Carbonell, R.S. Michalski (eds.), *Machine Learning: A Guide to Current Research* (Springer, US, 1986)
7. F. Rosenblatt, Two theorems of statistical separability in the perceptron. Mechanisation of thought processes, in *Proceedings of a symposium held at the National Physical Laboratory*, ed. by DV Blake, Albert M. Uttley (Her Majesty's Stationery Office, London, 1959), pp. 419–456
8. P.J. Werbos, Beyond regression: new tools for prediction and analysis in the behavioral sciences. Ph.D Thesis, Harvard University (1974)
9. B. Widrow, M.E. Hoff, Adaptive switching circuits. Stanford University Ca Stanford Electronics Labs (1960)
10. L. Magdalena, Fuzzy Rule-Based Systems, in *Springer Handbook of Computational Intelligence*, ed. by J. Kacprzyk, W. Pedrycz (Springer, Berlin Heidelberg, 2015), pp. 203–218
11. H.R. Berenji, Fuzzy logic controllers, in *An Introduction to Fuzzy Logic Applications in Intelligent Systems*, ed. by R.R. Yager, L.A. Zadeh (Springer, US, Boston, MA, 1992), pp. 69–96
12. L.A. Zadeh, Knowledge representation in fuzzy logic, in *An Introduction to Fuzzy Logic Applications in Intelligent Systems*, ed. by R.R. Yager, L.A. Zadeh (Springer, US, Boston, MA, 1992), pp. 1–25
13. J.M. Keller, R. Krishnapuram, Fuzzy set methods in computer vision, in *An Introduction to Fuzzy Logic Applications in Intelligent Systems*, ed. by R.R. Yager, L.A. Zadeh (Springer, US, Boston, MA, 1992), pp. 121–145
14. S.K. Pal, Fuzziness, image information and scene analysis, in *An Introduction to Fuzzy Logic Applications in Intelligent Systems*, ed. by R.R. Yager, L.A. Zadeh (Springer, US, Boston, MA, 1992), pp. 147–183
15. V. Novák, Fuzzy sets in natural language processing, in *An Introduction to Fuzzy Logic Applications in Intelligent Systems*, ed. by R.R. Yager, L.A. Zadeh (Springer, US, Boston, MA, 1992), pp. 185–200
16. D. Michie, "Memo" functions and machine learning. Nature **218**, 19–22 (1968)
17. J.H. Holland, *Adaptation in natural and artificial systems: An introductory analysis with applications to biology, control, and artificial intelligence* (U Michigan Press, Oxford, England, 1975)
18. A. Tzanetos, G. Dounias, Nature inspired optimization algorithms related to physical phenomena and laws of science: a survey. Int. J. Artif. Intell. Tools **26**, 1750022 (2017). https://doi.org/10.1142/s0218213017500221
19. A. Chakraborty, A.K. Kar, Swarm Intelligence: a review of algorithms, in *Nature-Inspired Computing and Optimization: Theory and Applications*, ed. by S. Patnaik, X.-S. Yang, K. Nakamatsu (Springer International Publishing, Cham, 2017), pp. 475–494
20. R.S. Parpinelli, H.S. Lopes, New inspirations in swarm intelligence: a survey. Int. J. Bio-Inspired Comput. **3**, 1–16 (2011). https://doi.org/10.1504/ijbic.2011.0387
21. A. Slowik, H. Kwasnicka, Nature inspired methods and their industry applications—Swarm Intelligence Algorithms. IEEE Trans. Ind. Inf. **14**, 1004–1015 (2018). https://doi.org/10.1109/tii.2017.2786782
22. B Chawda, J Patel Natural Computing Algorithms–a survey. Int. J. Emerg. Technol. Adv. Eng. **6** (2016)
23. J. Del Ser, E. Osaba, D. Molina et al., Bio-inspired computation: where we stand and what's next. Swarm Evol. Comput. **48**, 220–250 (2019). https://doi.org/10.1016/j.swevo.2019.04.008

24. S. Roy, S. Biswas, S.S. Chaudhuri, Nature-inspired Swarm Intelligence and its applications. Int. J. Mod. Educ. Comput. Sci. (IJMECS) **6**, 55 (2014)
25. K.C.B. Steer, A. Wirth, S.K. Halgamuge, The rationale behind seeking inspiration from nature, in *Nature-Inspired Algorithms for Optimisation*, ed. by R. Chiong (Springer, Berlin Heidelberg, 2009), pp. 51–76
26. O.K. Erol, I. Eksin, A new optimization method: big bang–big crunch. Adv. Eng. Softw. **37**, 106–111 (2006)
27. A. Hatamlou, Black hole: a new heuristic optimization approach for data clustering. Inf. Sci. **222**, 175–184 (2013)
28. G. Beni, J. Wang, Swarm Intelligence in cellular robotic systems, in *Robots and Biological Systems: Towards a New Bionics?*, ed. by P. Dario, G. Sandini, P. Aebischer (Springer, Berlin, Heidelberg, 1993), pp. 703–712
29. JM Bishop, Stochastic searching networks, in *1989 First IEE International Conference on Artificial Neural Networks, (Conf. Publ. No. 313)* (IET, 1989), pp 329–331
30. V. Bhasin, P. Bedi, A. Singhal, Feature selection for steganalysis based on modified Stochastic Diffusion Search using Fisher score, in *2014 International Conference on Advances in Computing, Communications and Informatics (ICACCI)* (2014), pp. 2323–2330
31. J.M. Bishop, P. Torr, The stochastic search network, in *Neural Networks for Vision, Speech and Natural Language*, ed. by R. Linggard, D.J. Myers, C. Nightingale (Springer, Netherlands, Dordrecht, 1992), pp. 370–387
32. P.D. Beattie, J.M. Bishop, Self-localisation in the 'Senario' autonomous wheelchair. J. Intell. Rob. Syst. **22**, 255–267 (1998). https://doi.org/10.1023/a:1008033229660
33. M.M. al-Rifaie, A. Aber, Identifying metastasis in bone scans with Stochastic Diffusion Search, in *2012 International Symposium on Information Technologies in Medicine and Education* (2012), pp. 519–523
34. S. Hurley, R.M. Whitaker, An agent based approach to site selection for wireless networks, in *Proceedings of the 2002 ACM Symposium on Applied Computing* (ACM, New York, NY, USA, 2002), pp. 574–577
35. M. Dorigo, Optimization, learning and natural algorithms. Ph.D Thesis, Politecnico di Milano (1992)
36. M, Dorigo, T. Stützle, ACO algorithms for the traveling salesman problem, in *Evolutionary Algorithms in Engineering and Computer Science*, ed. by K. Miettinen (K. Miettinen, M. Makela, P. Neittaanmaki, J. Periaux, eds.) (Wiley, 1999), pp. 163–183
37. C. Fountas, A. Vlachos, Ant Colonies Optimization (ACO) for the solution of the Vehicle Routing Problem (VRP). J. Inf. Optim. Sci. **26**, 135–142 (2005). https://doi.org/10.1080/025 22667.2005.10699639
38. S. Fidanova, M. Durchova, Ant Algorithm for Grid Scheduling Problem, in *Large-Scale Scientific Computing*, ed. by I. Lirkov, S. Margenov, J. Waśniewski (Springer, Berlin Heidelberg, 2006), pp. 405–412
39. W.J. Gutjahr, M.S. Rauner, An ACO algorithm for a dynamic regional nurse-scheduling problem in Austria. Comput. Oper. Res. **34**, 642–666 (2007). https://doi.org/10.1016/j.cor. 2005.03.018
40. W. Wen, C. Wang, D. Wu, Y. Xie (2015) An ACO-based scheduling strategy on load balancing in cloud computing environment, in *2015 Ninth International Conference on Frontier of Computer Science and Technology*. pp. 364–369
41. T. Stützle, M. Dorigo, ACO algorithms for the quadratic assignment problem. New Ideas in Optimization (1999)
42. L. Lessing, I. Dumitrescu, T. Stützle, A Comparison Between ACO algorithms for the set covering problem, in *Ant Colony Optimization and Swarm Intelligence*, ed. by M. Dorigo, M. Birattari, C. Blum, et al. (Springer, Berlin Heidelberg, 2004), pp. 1–12
43. M.A.P. Garcia, O. Montiel, O. Castillo et al., Path planning for autonomous mobile robot navigation with Ant Colony Optimization and fuzzy cost function evaluation. Appl. Soft Comput. **9**, 1102–1110 (2009). https://doi.org/10.1016/j.asoc.2009.02.014

44. D. Martens, M. De Backer, R. Haesen et al., Classification With Ant Colony Optimization. IEEE Trans. Evol. Comput. **11**, 651–665 (2007). https://doi.org/10.1109/tevc.2006.890229
45. A. Shmygelska, H.H. Hoos, An ant colony optimisation algorithm for the 2D and 3D hydrophobic polar protein folding problem. BMC Bioinform. **6**, 30 (2005). https://doi.org/10.1186/1471-2105-6-30
46. R. Eberhart, J. Kennedy, A new optimizer using particle swarm theory, in *MHS'95, Proceedings of the Sixth International Symposium on Micro Machine and Human Science.* (IEEE, 1995), pp. 39–43
47. H. Wu, J. Geng, R. Jin et al., An improved comprehensive learning Particle Swarm Optimization and Its application to the semiautomatic design of antennas. IEEE Trans. Antennas Propag. **57**, 3018–3028 (2009). https://doi.org/10.1109/tap.2009.2028608
48. R.C. Eberhart, X. Hu, Human tremor analysis using Particle Swarm Optimization, in *Proceedings of the 1999 Congress on Evolutionary Computation-CEC99 (Cat. No. 99TH8406)*, vol. 3 (1999), pp. 1927–1930
49. M. Shen, Z. Zhan, W. Chen et al., Bi-velocity discrete Particle Swarm Optimization and its application to multicast routing problem in communication networks. IEEE Trans. Ind. Electron. **61**, 7141–7151 (2014). https://doi.org/10.1109/tie.2014.2314075
50. J. Nenortaite, R. Simutis, Adapting Particle Swarm Optimization to stock markets, in *5th International Conference on Intelligent Systems Design and Applications (ISDA'05)* (2005), pp. 520–525
51. A.A.A. Esmin, G. Lambert-Torres, Loss power minimization using Particle Swarm Optimization, in *The 2006 IEEE International Joint Conference on Neural Network Proceedings* (2006) pp. 1988–1992
52. Y. Zhang, D. Gong, J. Zhang, Robot path planning in uncertain environment using multiobjective particle swarm optimization. Neurocomputing **103**, 172–185 (2013). https://doi.org/10.1016/j.neucom.2012.09.019
53. C.-J. Liao, Chao-Tang Tseng, P. Luarn, A discrete version of Particle Swarm Optimization for flowshop scheduling problems. Comput. Oper. Res. **34**, 3099–3111 (2007). https://doi.org/10.1016/j.cor.2005.11.017
54. B. Xue, M. Zhang, W.N. Browne, Particle Swarm Optimization for feature selection in classification: a multi-objective approach. IEEE Trans. Cybern. **43**, 1656–1671 (2013). https://doi.org/10.1109/tsmcb.2012.2227469
55. Y. Wang, J. Lv, L. Zhu, Y. Ma, Crystal structure prediction via Particle-Swarm Optimization. Phys. Rev. B **82**, 094116 (2010). https://doi.org/10.1103/physrevb.82.094116
56. I.-H. Kuo, S.-J. Horng, T.-W. Kao et al., An improved method for forecasting enrollments based on fuzzy time series and Particle Swarm Optimization. Expert Syst. Appl. **36**, 6108–6117 (2009). https://doi.org/10.1016/j.eswa.2008.07.043
57. I. Fister Jr., X-S. Yang, I. Fister et al., A brief review of nature-inspired algorithms for optimization (2013). arXiv:13074186
58. K.M. Passino, Biomimicry of bacterial foraging for distributed optimization and control. IEEE Control Syst. Mag. **22**, 52–67 (2002). https://doi.org/10.1109/mcs.2002.1004010
59. H.F. Wedde, M. Farooq, Y. Zhang, BeeHive: an efficient fault-tolerant routing algorithm inspired by Honey Bee behavior, in *Ant Colony Optimization and Swarm Intelligence*, ed. by M. Dorigo, M. Birattari, C. Blum, et al. (Springer, Berlin Heidelberg, 2004), pp. 83–94
60. D. Teodorovic, M. Dell'Orco, Bee colony optimization–a cooperative learning approach to complex transportation problems. Adv. OR AI Methods Trans. **51**, 60 (2005)
61. D.T. Pham, A. Ghanbarzadeh, E. Koç et al., The Bees Algorithm—a novel tool for complex optimisation problems, in *Intelligent Production Machines and Systems*, ed. by D.T. Pham, E.E. Eldukhri, A.J. Soroka (Elsevier Science Ltd, Oxford, 2006), pp. 454–459
62. D. Karaboga, B. Basturk, A powerful and efficient algorithm for numerical function optimization: artificial bee colony (ABC) algorithm. J. Glob. Optim. **39**, 459–471 (2007). https://doi.org/10.1007/s10898-007-9149-x
63. X.-S. Yang, A new metaheuristic bat-inspired algorithm, in *Nature Inspired Cooperative Strategies for Optimization (NICSO 2010)*, ed. by J.R. González, D.A. Pelta, C. Cruz, et al. (Springer, Berlin Heidelberg, 2010), pp. 65–74

64. R. Akbari, A. Mohammadi, K. Ziarati, A novel bee swarm optimization algorithm for numerical function optimization. Commun. Nonlinear Sci. Numer. Simul. **15**, 3142–3155 (2010). https://doi.org/10.1016/j.cnsns.2009.11.003
65. R. Oftadeh, M.J. Mahjoob, M. Shariatpanahi, A novel meta-heuristic optimization algorithm inspired by group hunting of animals: hunting search. Comput. Math Appl. **60**, 2087–2098 (2010). https://doi.org/10.1016/j.camwa.2010.07.049
66. C. Zhaohui, T. Haiyan, Cockroach swarm optimization for vehicle routing problems. Energy Proc. **13**, 30–35 (2011). https://doi.org/10.1016/j.egypro.2011.11.007
67. M.T.M.H. Shirzadi, M.H. Bagheri, A novel meta-heuristic algorithm for numerical function optimization: Blind, Naked Mole-Rats (BNMR) algorithm. SRE **7**, 3566–3583 (2012). https://doi.org/10.5897/sre12.514
68. F. Ahmadi, H. Salehi, K. Karimi, Eurygaster algorithm: a new approach to optimization. Int. J. Comput. Appl. **57**, 9–13 (2012)
69. R.D. Maia, L.N. de Castro, W.M. Caminhas, Bee colonies as model for multimodal continuous optimization: The OptBees algorithm, in *2012 IEEE Congress on Evolutionary Computation* (2012), pp. 1–8
70. R,. Tang, S. Fong, X. Yang, S. Deb, Wolf search algorithm with ephemeral memory, in *Seventh International Conference on Digital Information Management (ICDIM 2012)* (2012), pp. 165–172
71. C. Subramanian, A. Sekar, K. Subramanian, A new engineering optimization method: African wild dog algorithm. Int. J. Soft Comput. **8**, 163–170 (2013). https://doi.org/10.3923/ijscomp.2013.163.170
72. Y. Gheraibia, A. Moussaoui, Penguins Search Optimization Algorithm (PeSOA), in *Recent Trends in Applied Artificial Intelligence*, ed. by M. Ali, T. Bosse, K.V. Hindriks, et al. (Springer, Berlin Heidelberg, 2013), pp. 222–231
73. P. Wang, Z. Zhu, S. Huang, Seven-Spot Ladybird Optimization: a novel and efficient meta-heuristic algorithm for numerical optimization. Sci. World J. (2013). https://doi.org/10.1155/2013/378515
74. S. Mirjalili, S.M. Mirjalili, A. Lewis, Grey Wolf Optimizer. Adv. Eng. Softw. **69**, 46–61 (2014). https://doi.org/10.1016/j.advengsoft.2013.12.007
75. X. Meng, Y. Liu, X. Gao, H. Zhang, A new bio-inspired algorithm: Chicken Swarm Optimization, in *International Conference in Swarm Intelligence* (Springer, 2014) pp. 86–94
76. S.-J. Wu, C.-T. Wu, A bio-inspired optimization for inferring interactive networks: Cockroach swarm evolution. Expert Syst. Appl. **42**, 3253–3267 (2015). https://doi.org/10.1016/j.eswa.2014.11.039
77. S.-J. Wu, C.-T. Wu, Computational optimization for S-type biological systems: Cockroach Genetic Algorithm. Math. Biosci. **245**, 299–313 (2013). https://doi.org/10.1016/j.mbs.2013.07.019
78. S. Arora, S. Singh, Butterfly algorithm with Lèvy Flights for global optimization, in *2015 International Conference on Signal Processing, Computing and Control (ISPCC)* (2015), pp. 220–224
79. S. Arora, S. Singh, Butterfly optimization algorithm: a novel approach for global optimization. Soft. Comput. **23**, 715–734 (2019). https://doi.org/10.1007/s00500-018-3102-4
80. W. Yong, W. Tao, Z. Cheng-Zhi, H. Hua-Juan, A new stochastic optimization approach—Dolphin Swarm Optimization Algorithm. Int. J. Comput. Intell. Appl. **15**, 1650011 (2016). https://doi.org/10.1142/s1469026816500115
81. A. Brabazon, W. Cui, M. O'Neill, The raven roosting optimisation algorithm. Soft. Comput. **20**, 525–545 (2016). https://doi.org/10.1007/s00500-014-1520-5
82. S. Mirjalili, A. Lewis, The whale optimization algorithm. Adv. Eng. Softw. **95**, 51–67 (2016)
83. X.-B. Meng, X.Z. Gao, L. Lu et al., A new bio-inspired optimisation algorithm: Bird Swarm Algorithm. J. Exp. Theor. Artif. Intell. **28**, 673–687 (2016). https://doi.org/10.1080/0952813x.2015.1042530
84. G. Dhiman, V. Kumar, Spotted hyena optimizer: a novel bio-inspired based metaheuristic technique for engineering applications. Adv. Eng. Softw. **114**, 48–70 (2017). https://doi.org/10.1016/j.advengsoft.2017.05.014

85. B. Zeng, L. Gao, X. Li, Whale Swarm Algorithm for function optimization, in *Intelligent Computing Theories and Application* ed. by D.-S. Huang, V. Bevilacqua, P. Premaratne, P. Gupta (Springer International Publishing, 2017), pp. 624–639
86. D. Zaldívar, B. Morales, A. Rodríguez et al., A novel bio-inspired optimization model based on Yellow Saddle Goatfish behavior. Biosystems **174**, 1–21 (2018). https://doi.org/10.1016/j.biosystems.2018.09.007
87. ATS Al-Obaidi, HS Abdullah, O. Ahmed Zied, Meerkat Clan Algorithm: a New Swarm Intelligence Algorithm. Indones. J. Electric. Eng. Comput. Sci. **10**, 354–360 (2018). https://doi.org/10.11591/ijeecs.v10.i1
88. S. Shadravan, H.R. Naji, V.K. Bardsiri, The Sailfish Optimizer: a novel nature-inspired meta-heuristic algorithm for solving constrained engineering optimization problems. Eng. Appl. Artif. Intell. **80**, 20–34 (2019). https://doi.org/10.1016/j.engappai.2019.01.001
89. X.L. Li, Z.J. Shao, J.X. Qian, An optimizing method based on autonomous animates: Fish-swarm Algorithm. Syst. Eng. Theory Pract. **22**, 32–38 (2002). https://doi.org/10.12011/1000-6788(2002)11-32
90. K.N. Krishnanand, D. Ghose, Glowworm swarm based optimization algorithm for multimodal functions with collective robotics applications. Multiagent Grid Syst. **2**, 209–222 (2006)
91. T.C. Havens, C.J. Spain, N.G. Salmon, J.M. Keller, Roach Infestation Optimization, in *2008 IEEE Swarm Intelligence Symposium* (2008), pp. 1–7
92. F. Comellas, J. Martinez-Navarro, Bumblebees: a multiagent combinatorial optimization algorithm inspired by social insect behaviour, in *Proceedings of the First ACM/SIGEVO Summit on Genetic and Evolutionary Computation* (ACM, 2009), pp. 811–814
93. X. Yang, S. Deb, Cuckoo Search via Lévy flights, in *2009 World Congress on Nature & Biologically Inspired Computing (NaBIC)* (2009), pp. 210–214
94. X.-S. Yang, Firefly Algorithms for multimodal optimization, in *Stochastic Algorithms: Foundations and Applications*, ed. by O. Watanabe, T. Zeugmann (Springer, Berlin Heidelberg, 2009), pp. 169–178
95. Y. Shiqin, J. Jianjun, Y. Guangxing, A Dolphin Partner Optimization, in *2009 WRI Global Congress on Intelligent Systems* (IEEE, 2009), pp. 124–128
96. S Chen, Locust Swarms-a new multi-optima search technique, in *2009 IEEE Congress on Evolutionary Computation* (2009), pp. 1745–1752
97. R. Hedayatzadeh, F.A. Salmassi, M. Keshtgari et al., Termite colony optimization: a novel approach for optimizing continuous problems, in *2010 18th Iranian Conference on Electrical Engineering* (2010), pp. 553–558
98. A.H. Gandomi, A.H. Alavi, Krill herd: a new bio-inspired optimization algorithm. Commun. Nonlinear Sci. Numer. Simul. **17**, 4831–4845 (2012)
99. E. Cuevas, M. Cienfuegos, D. Zaldívar, M. Pérez-Cisneros, A swarm optimization algorithm inspired in the behavior of the social-spider. Expert Syst. Appl. **40**, 6374–6384 (2013). https://doi.org/10.1016/j.eswa.2013.05.041
100. M. Neshat, G. Sepidnam, M. Sargolzaei, Swallow swarm optimization algorithm: a new method to optimization. Neural Comput. Appl. **23**, 429–454 (2013). https://doi.org/10.1007/s00521-012-0939-9
101. J.C. Bansal, H. Sharma, S.S. Jadon, M. Clerc, Spider Monkey Optimization Algorithm for numerical optimization. Memet. Comput. **6**, 31–47 (2014). https://doi.org/10.1007/s12293-013-0128-0
102. J.B. Odili, M.N.M. Kahar, African buffalo optimization (ABO): a new meta-heuristic algorithm. J. Adv. Appl. Sci. **3**, 101–106 (2015)
103. S. Deb, S. Fong, Z. Tian, Elephant Search Algorithm for optimization problems, in *2015 Tenth International Conference on Digital Information Management (ICDIM)* (2015), pp. 249–255
104. G-G. Wang, S. Deb, L.D.S. Coelho, Elephant herding optimization, in *2015 3rd International Symposium on Computational and Business Intelligence (ISCBI)* (IEEE, 2015) pp. 1–5
105. R. Omidvar, H. Parvin, F. Rad, SSPCO optimization algorithm (See-See Partridge Chicks Optimization), in *2015 Fourteenth Mexican International Conference on Artificial Intelligence (MICAI)* (2015), pp. 101–106

106. G.-G. Wang, S. Deb, L.D.S. Coelho, Earthworm optimization algorithm: a bio-inspired meta-heuristic algorithm for global optimization problems. Int. J. Bio-Inspired Comput. **7**, 1–23 (2015)
107. M. Yazdani, F. Jolai, Lion optimization algorithm (LOA): a nature-inspired metaheuristic algorithm. J. Comput. Des. Eng. **3**, 24–36 (2016)
108. T. Wu, M. Yao, J. Yang, Dolphin Swarm Algorithm. Front. Inf. Technol. Electron. Eng. **17**, 717–729 (2016). https://doi.org/10.1631/fitee.1500287
109. J. Pierezan, L. Dos Santos Coelho, Coyote optimization algorithm: a new metaheuristic for global optimization problems, in *2018 IEEE Congress on Evolutionary Computation (CEC)* (2018), pp 1–8
110. G. Dhiman, V. Kumar, Emperor penguin optimizer: a bio-inspired algorithm for engineering problems. Knowl. Based Syst. **159**, 20–50 (2018). https://doi.org/10.1016/j.knosys.2018.06.001
111. S. Harifi, M. Khalilian, J. Mohammadzadeh, S. Ebrahimnejad, Emperor Penguins Colony: a new metaheuristic algorithm for optimization. Evol. Intel. **12**, 211–226 (2019). https://doi.org/10.1007/s12065-019-00212-x
112. K. Hyunchul, A. Byungchul, A new evolutionary algorithm based on sheep flocks heredity model, in *2001 IEEE Pacific Rim Conference on Communications, Computers and Signal Processing*, vol. 2 (IEEE Cat. No.01CH37233) (2001), pp. 514–517
113. W.J. Tang, Q.H. Wu, J.R. Saunders, A bacterial swarming algorithm for global optimization, in *2007 IEEE Congress on Evolutionary Computation* (2007), pp. 1207–1212
114. F.J.M. Garcia, J.A.M. Pérez, Jumping frogs optimization: a new swarm method for discrete optimization. Documentos de Trabajo del DEIOC 3 (2008)
115. T Chen, A simulative bionic intelligent optimization algorithm: artificial searching Swarm Algorithm and its performance analysis, in *2009 International Joint Conference on Computational Sciences and Optimization* (2009), pp 864–866
116. C.J.A.B. Filho, F.B. de Lima Neto, A.J.C.C. Lins et al., Fish School search, in *Nature-Inspired Algorithms for Optimisation*, ed. by R. Chiong (Springer, Berlin Heidelberg, 2009), pp. 261–277
117. H. Chen, Y. Zhu, K. Hu, X. He, Hierarchical Swarm Model: a new approach to optimization. Discret. Dyn. Nat. Soc. (2010). https://doi.org/10.1155/2010/379649
118. E. Duman, M. Uysal, A.F. Alkaya, Migrating Birds Optimization: a new meta-heuristic approach and its application to the quadratic assignment problem, in *Applications of Evolutionary Computation*, ed. by C. Di Chio, S. Cagnoni, C. Cotta, et al. (Springer, Berlin Heidelberg, 2011), pp. 254–263
119. Y. Marinakis, M. Marinaki, Bumble bees mating optimization algorithm for the vehicle routing problem, in *Handbook of Swarm Intelligence: Concepts, Principles and Applications*, ed. by B.K. Panigrahi, Y. Shi, M.-H. Lim (Springer, Berlin Heidelberg, 2011), pp. 347–369
120. T.O. Ting, K.L. Man, S.-U. Guan et al., Weightless Swarm Algorithm (WSA) for dynamic optimization problems, in *Network and Parallel Computing*, ed. by J.J. Park, A. Zomaya, S.-S. Yeo, S. Sahni (Springer, Berlin Heidelberg, 2012), pp. 508–515
121. P. Qiao, H. Duan, Pigeon-inspired optimization: a new swarm intelligence optimizer for air robot path planning. Int. J. Intell. Comput. Cyber **7**, 24–37 (2014). https://doi.org/10.1108/ijicc-02-2014-0005
122. X. Li, J. Zhang, M. Yin, Animal migration optimization: an optimization algorithm inspired by animal migration behavior. Neural Comput. Appl. **24**, 1867–1877 (2014)
123. G.-G. Wang, S. Deb, Z. Cui, Monarch butterfly optimization. Neural Comput. Appl. (2015). https://doi.org/10.1007/s00521-015-1923-y
124. L. Cheng, L. Han, X. Zeng et al., Adaptive Cockroach Colony Optimization for rod-like robot navigation. J. Bionic Eng. **12**, 324–337 (2015). https://doi.org/10.1016/s1672-6529(14)60125-6
125. S. Mirjalili, Moth-flame optimization algorithm: a novel nature-inspired heuristic paradigm. Knowl. Based Syst. **89**, 228–249 (2015)

126. S. Mirjalili, Dragonfly algorithm: a new meta-heuristic optimization technique for solving single-objective, discrete, and multi-objective problems. Neural Comput. Appl. **27**, 1053–1073 (2016). https://doi.org/10.1007/s00521-015-1920-1

127. S. Mirjalili, A.H. Gandomi, S.Z. Mirjalili et al., Salp Swarm Algorithm: a bio-inspired optimizer for engineering design problems. Adv. Eng. Softw. **114**, 163–191 (2017). https://doi.org/10.1016/j.advengsoft.2017.07.002

128. S.Z. Mirjalili, S. Mirjalili, S. Saremi et al., Grasshopper optimization algorithm for multi-objective optimization problems. Appl. Intell. **48**, 805–820 (2018). https://doi.org/10.1007/s10489-017-1019-8

129. F. Fausto, E. Cuevas, A. Valdivia, A. González, A global optimization algorithm inspired in the behavior of selfish herds. Biosystems **160**, 39–55 (2017). https://doi.org/10.1016/j.biosystems.2017.07.010

130. Q. Zhang, R. Wang, J. Yang et al., Biology migration algorithm: a new nature-inspired heuristic methodology for global optimization. Soft. Comput. (2018). https://doi.org/10.1007/s00500-018-3381-9

131. A. Kaveh, S. Mahjoubi, Lion pride optimization algorithm: a meta-heuristic method for global optimization problems. Scientia Iranica **25**, 3113–3132 (2018). https://doi.org/10.24200/sci.2018.20833

132. C.E. Klein, L. dos Santos Coelho, Meerkats-inspired algorithm for global optimization problems, in *26th European Symposium on Artificial Neural Networks, ESANN 2018* (Bruges, Belgium, 25–27 Apr 2018)

133. M.M. Motevali, A.M. Shanghooshabad, R.Z. Aram, H. Keshavarz, WHO: a new evolutionary algorithm bio-inspired by Wildebeests with a case study on bank customer segmentation. Int. J. Pattern Recogn. Artif. Intell. **33**, 1959017 (2018). https://doi.org/10.1142/s0218001419590171

134. HA Bouarara, RM Hamou, A Abdelmalek, Enhanced Artificial Social Cockroaches (EASC) for modern information retrieval, in *Information Retrieval and Management: Concepts, Methodologies. Tools, and Application* (2018), pp. 928–960. https://doi.org/10.4018/978-1-5225-5191-1.ch040

135. A. Kazikova, M. Pluhacek, R. Senkerik, A. Viktorin, Proposal of a new swarm optimization method inspired in bison behavior, in *Recent Advances in Soft Computing*, ed. by R Matoušek (Springer International Publishing, 2019), pp. 146–156

136. G. Dhiman, V. Kumar, Seagull optimization algorithm: theory and its applications for large-scale industrial engineering problems. Knowl. Based Syst. **165**, 169–196 (2019). https://doi.org/10.1016/j.knosys.2018.11.024

137. R. Masadeh, A. Sharieh, B. Mahafzah, Humpback whale optimization algorithm based on vocal behavior for task scheduling in cloud computing. Int. J. Adv. Sci. Technol. **13**, 121–140 (2019)

138. E. Cuevas, F. Fausto, A. González, A Swarm Algorithm inspired by the collective animal behavior, in *New Advancements in Swarm Algorithms: Operators and Applications*, ed. by E. Cuevas, F. Fausto, A. González (Springer International Publishing, Cham, 2020), pp. 161–188

139. H.A. Abbass, MBO: marriage in honey bees optimization-a Haplometrosis polygynous swarming approach, in *Proceedings of the 2001 Congress on Evolutionary Computation*, vol. 1 (IEEE Cat. No.01TH8546, 2001), pp. 207–214

140. S.D. Muller, J. Marchetto, S. Airaghi, P. Kournoutsakos, Optimization based on bacterial chemotaxis. IEEE Trans. Evol. Comput. **6**, 16–29 (2002). https://doi.org/10.1109/4235.985689

141. P. Cortés, J.M. García, J. Muñuzuri, L. Onieva, Viral systems: a new bio-inspired optimisation approach. Comput. Oper. Res. **35**, 2840–2860 (2008). https://doi.org/10.1016/j.cor.2006.12.018

142. S.S. Pattnaik, K.M. Bakwad, B.S. Sohi et al., Swine Influenza Models Based Optimization (SIMBO). Appl. Soft Comput. **13**, 628–653 (2013). https://doi.org/10.1016/j.asoc.2012.07.010

143. M. Jaderyan, H. Khotanlou, Virulence Optimization Algorithm. Appl. Soft Comput. **43**, 596–618 (2016). https://doi.org/10.1016/j.asoc.2016.02.038
144. M.D. Li, H. Zhao, X.W. Weng, T. Han, A novel nature-inspired algorithm for optimization: Virus colony search. Adv. Eng. Softw. **92**, 65–88 (2016). https://doi.org/10.1016/j.advengsoft.2015.11.004
145. S.-C. Chu, P. Tsai, J.-S. Pan, Cat Swarm Optimization, in *PRICAI 2006: Trends in Artificial Intelligence*, ed. by Q. Yang, G. Webb (Springer, Berlin Heidelberg, 2006), pp. 854–858
146. M. Eusuff, K. Lansey, F. Pasha, Shuffled frog-leaping algorithm: a memetic meta-heuristic for discrete optimization. Eng. Optim. **38**, 129–154 (2006). https://doi.org/10.1080/030521 50500384759
147. O.B. Haddad, A. Afshar, M.A. Mariño, Honey-Bees Mating Optimization (HBMO) Algorithm: a new heuristic approach for water resources optimization. Water Resour. Manag. **20**, 661–680 (2006). https://doi.org/10.1007/s11269-005-9001-3
148. S. He, Q. H. Wu, J.R. Saunders, A novel Group Search Optimizer inspired by animal behavioural ecology, in *2006 IEEE International Conference on Evolutionary Computation* (2006), pp. 1272–1278
149. A. Mucherino, O. Seref, Monkey search: a novel metaheuristic search for global optimization. AIP Conf. Proc. **953**, 162–173 (2007). https://doi.org/10.1063/1.2817338
150. R. Zhao, W. Tang, Monkey algorithm for global numerical optimization. J. Uncertain Syst. **2**, 165–176 (2008)
151. X. Lu, Y. Zhou, A novel global convergence algorithm: Bee Collecting Pollen Algorithm, in *Advanced Intelligent Computing Theories and Applications. With Aspects of Artificial Intelligence*, ed. by D.-S. Huang, D.C. Wunsch, D.S. Levine, K.-H. Jo (Springer, Berlin Heidelberg, 2008), pp. 518–525
152. D. Simon, Biogeography-based optimization. IEEE Trans. Evol. Comput. **12**, 702–713 (2008)
153. W.T. Pan, A new evolutionary computation approach: fruit fly optimization algorithm, in *Proceedings of the Conference on Digital Technology and Innovation Management* (2011)
154. A.A. Minhas F ul, M. Arif, MOX: A novel global optimization algorithm inspired from Oviposition site selection and egg hatching inhibition in mosquitoes. App. Soft Comput. **11**, 4614–4625 (2011). https://doi.org/10.1016/j.asoc.2011.07.020
155. M.A. Tawfeeq, Intelligent Algorithm for Optimum Solutions Based on the Principles of Bat Sonar (2012). arXiv:12110730
156. I. Aihara, H. Kitahata, K. Yoshikawa, K. Aihara, Mathematical modeling of frogs' calling behavior and its possible application to artificial life and robotics. Artif. Life Robot. **12**, 29–32 (2008). https://doi.org/10.1007/s10015-007-0436-x
157. M. El-Dosuky, A. El-Bassiouny, T. Hamza, M. Rashad, New Hoopoe Heuristic Optimization (2012). CoRR arXiv:abs/1211.6410
158. B.R. Rajakumar, The Lion's Algorithm: a new nature-inspired search algorithm. Proc. Technol. **6**, 126–135 (2012). https://doi.org/10.1016/j.protcy.2012.10.016
159. A. Mozaffari, A. Fathi, S. Behzadipour, The great salmon run: a novel bio-inspired algorithm for artificial system design and optimisation. IJBIC **4**, 286 (2012). https://doi.org/10.1504/ijbic.2012.049889
160. A.S. Eesa, A.M.A. Brifcani, Z. Orman, Cuttlefish Algorithm–a novel bio-inspired optimization algorithm. Int. J. Sci. Eng. Res. **4**, 1978–1986 (2013)
161. A. Kaveh, N. Farhoudi, A new optimization method: dolphin echolocation. Adv. Eng. Softw. **59**, 53–70 (2013). https://doi.org/10.1016/j.advengsoft.2013.03.004
162. C. Sur, S. Sharma, A. Shukla, Egyptian vulture optimization algorithm–a new nature inspired meta-heuristics for knapsack problem, in *The 9th International Conference on Computing and Information Technology (IC2IT2013)* (Springer, 2013), pp. 227–237
163. C. Sur, A. Shukla, New bio-inspired meta-heuristics-Green Herons Optimization Algorithm-for optimization of travelling salesman problem and road network, in Swarm, Evolutionary, and Memetic Computing ed. by B.K. Panigrahi, P.N. Suganthan, S. Das, S.S. Dash (Springer International Publishing, 2013), pp. 168–179

164. M. Hajiaghaei-Keshteli, M. Aminnayeri, Keshtel Algorithm (KA): a new optimization algorithm inspired by Keshtels' feeding, in *Proceeding in IEEE Conference on Industrial Engineering and Management Systems* (IEEE, Rabat, Morocco, 2013), pp. 2249–2253
165. M. Bidar, H. Rashidy Kanan, Jumper firefly algorithm, in *ICCKE 2013* (2013), pp. 267–271
166. S.L. Tilahun, H.C. Ong, Prey-Predator Algorithm: a new metaheuristic algorithm for optimization problems. Int. J. Info. Tech. Dec. Mak. **14**, 1331–1352 (2013). https://doi.org/10.1142/s021962201450031x
167. H. Mo, L. Xu, Magnetotactic bacteria optimization algorithm for multimodal optimization, in *2013 IEEE Symposium on Swarm Intelligence (SIS)* (2013), pp. 240–247
168. J. An, Q. Kang, L. Wang, Q. Wu, Mussels Wandering Optimization: an ecologically inspired algorithm for global optimization. Cogn. Comput. **5**, 188–199 (2013). https://doi.org/10.1007/s12559-012-9189-5
169. A. Askarzadeh, Bird mating optimizer: an optimization algorithm inspired by bird mating strategies. Commun. Nonlinear Sci. Numer. Simul. **19**, 1213–1228 (2014)
170. S. Salcedo-Sanz, J. Del Ser, I. Landa-Torres et al., The Coral Reefs Optimization Algorithm: a novel metaheuristic for efficiently solving optimization problems. Sci. World J. (2014). https://doi.org/10.1155/2014/739768
171. S. Mohseni, R. Gholami, N. Zarei, A.R. Zadeh, Competition over Resources: A new optimization algorithm based on animals behavioral ecology, in *2014 International Conference on Intelligent Networking and Collaborative Systems* (2014), pp. 311–315
172. M.-Y. Cheng, D. Prayogo, Symbiotic organisms search: A new metaheuristic optimization algorithm. Comput. Struct. **139**, 98–112 (2014). https://doi.org/10.1016/j.compstruc.2014.03.007
173. S. Mirjalili, The Ant Lion Optimizer. Adv. Eng. Softw. **83**, 80–98 (2015). https://doi.org/10.1016/j.advengsoft.2015.01.010
174. S.A. Uymaz, G. Tezel, E. Yel, Artificial algae algorithm (AAA) for nonlinear global optimization. Appl. Soft Comput. **31**, 153–171 (2015). https://doi.org/10.1016/j.asoc.2015.03.003
175. C. Chen, Y. Tsai, I. Liu, et al., A novel metaheuristic: Jaguar Algorithm with learning behavior, in *2015 IEEE International Conference on Systems, Man, and Cybernetics* (2015), pp. 1595–1600
176. M.K. Ibrahim, R.S. Ali, Novel optimization algorithm inspired by camel traveling behavior. Iraqi J. Electric. Electron. Eng. **12**, 167–177 (2016)
177. A. Askarzadeh, A novel metaheuristic method for solving constrained engineering optimization problems: Crow search algorithm. Comput. Struct. **169**, 1–12 (2016). https://doi.org/10.1016/j.compstruc.2016.03.001
178. A.F. Fard, M. Hajiaghaei-Keshteli, Red Deer Algorithm (RDA); a new optimization algorithm inspired by Red Deers' mating (2016), pp. 33–34
179. M. Alauddin, Mosquito flying optimization (MFO), in *2016 International Conference on Electrical, Electronics, and Optimization Techniques (ICEEOT)* (2016), pp. 79–84
180. O. Abedinia, N. Amjady, A. Ghasemi, A new metaheuristic algorithm based on shark smell optimization. Complexity **21**, 97–116 (2016). https://doi.org/10.1002/cplx.21634
181. A. Ebrahimi, E. Khamehchi, Sperm whale algorithm: an effective metaheuristic algorithm for production optimization problems. J. Nat. Gas Sci. Eng. **29**, 211–222 (2016). https://doi.org/10.1016/j.jngse.2016.01.001
182. X. Qi, Y. Zhu, H. Zhang, A new meta-heuristic butterfly-inspired algorithm. J. Comput. Sci. **23**, 226–239 (2017). https://doi.org/10.1016/j.jocs.2017.06.003
183. X. Jiang, S. Li, BAS: Beetle Antennae Search Algorithm for optimization problems (2017). CoRR arXiv:abs/1710.10724
184. V. Haldar, N. Chakraborty, A novel evolutionary technique based on electrolocation principle of elephant nose fish and shark: fish electrolocation optimization. Soft. Comput. **21**, 3827–3848 (2017). https://doi.org/10.1007/s00500-016-2033-1
185. T.R. Biyanto, Irawan S. Matradji et al., Killer Whale Algorithm: an algorithm inspired by the life of Killer Whale. Procedia Comput. Sci. **124**, 151–157 (2017). https://doi.org/10.1016/j.procs.2017.12.141

186. E. Hosseini, Laying chicken algorithm: a new meta-heuristic approach to solve continuous programming problems. J App. Comput. Math. **6**, 10–4172 (2017). https://doi.org/10.4172/2168-9679.1000344
187. D. Połap, M. Woz´niak, Polar Bear Optimization Algorithm: meta-heuristic with fast population movement and dynamic birth and death mechanism. Symmetry **9** (2017). https://doi.org/10.3390/sym9100203
188. S.H. Samareh Moosavi, V. Khatibi Bardsiri, Satin bowerbird optimizer: A new optimization algorithm to optimize ANFIS for software development effort estimation. Eng. Appl. Artif. Intell. **60**, 1–15 (2017). https://doi.org/10.1016/j.engappai.2017.01.006
189. M.H. Sulaiman, Z. Mustaffa, M.M. Saari et al., Barnacles Mating Optimizer: an evolutionary algorithm for solving optimization, in *2018 IEEE International Conference on Automatic Control and Intelligent Systems (I2CACIS)* (IEEE, 2018), pp. 99–104
190. A. Serani, M. Diez, Dolphin Pod Optimization, in *Machine Learning, Optimization, and Big Data* ed. by G. Nicosia, P. Pardalos, G. Giuffrida, R. Umeton (Springer International Publishing, 2018), pp. 50–62
191. C.E. Klein, V.C. Mariani, L.D.S. Coelho, Cheetah Based Optimization Algorithm: a novel swarm intelligence paradigm, in *26th European Symposium on Artificial Neural Networks, ESANN 2018, UCL Upcoming Conferences for Computer Science & Electronics* (Bruges, Belgium, 25–27 Apr 2018), pp. 685–690
192. M.C. Catalbas, A. Gulten, Circular structures of puffer fish: a new metaheuristic optimization algorithm, in *2018 Third International Conference on Electrical and Biomedical Engineering, Clean Energy and Green Computing (EBECEGC)*. (IEEE, 2018), pp. 1–5
193. E. Jahani, M. Chizari, Tackling global optimization problems with a novel algorithm–Mouth Brooding Fish algorithm. Appl. Soft Comput. **62**, 987–1002 (2018). https://doi.org/10.1016/j.asoc.2017.09.035
194. N.A. Kallioras, N.D. Lagaros, D.N. Avtzis, Pity Beetle Algorithm–a new metaheuristic inspired by the behavior of bark beetles. Adv. Eng. Softw. **121**, 147–166 (2018). https://doi.org/10.1016/j.advengsoft.2018.04.007
195. T. Wang, L. Yang, Q. Liu, Beetle Swarm Optimization Algorithm: theory and application (2018). arXiv:180800206
196. A.T. Khan, S. Li, P.S. Stanimirovic, Y. Zhang, Model-free optimization using eagle perching optimizer (2018). CoRR arXiv:abs/1807.02754
197. M. Jain, S. Maurya, A. Rani, V. Singh, Owl search algorithm: A novel nature-inspired heuristic paradigm for global optimization. J. Intell. Fuzzy Syst. **34**, 1573–1582 (2018). https://doi.org/10.3233/jifs-169452
198. B. Ghojogh, S. Sharifian, Pontogammarus Maeoticus Swarm Optimization: a metaheuristic optimization algorithm (2018). CoRR arXiv:abs/1807.01844
199. S. Deb, Z. Tian, S. Fong et al., Solving permutation flow-shop scheduling problem by rhinoceros search algorithm. Soft. Comput. **22**, 6025–6034 (2018). https://doi.org/10.1007/s00500-018-3075-3
200. B. Almonacid, R. Soto, Andean Condor Algorithm for cell formation problems. Nat. Comput. **18**, 351–381 (2019). https://doi.org/10.1007/s11047-018-9675-0
201. H.A. Alsattar, A.A. Zaidan, B.B. Zaidan, Novel meta-heuristic bald eagle search optimisation algorithm. Artif. Intell. Rev. (2019). https://doi.org/10.1007/s10462-019-09732-5
202. A.S. Shamsaldin, T.A. Rashid, R.A. Al-Rashid Agha et al., Donkey and Smuggler Optimization Algorithm: a collaborative working approach to path finding. J. Comput. Design Eng. (2019). https://doi.org/10.1016/j.jcde.2019.04.004
203. E.H. de Vasconcelos Segundo, V.C. Mariani, L. dos Santos Coelho, Design of heat exchangers using falcon optimization algorithm. Appl. Thermal Eng. (2019). https://doi.org/10.1016/j.applthermaleng.2019.04.038
204. G, Azizyan, F. Miarnaeimi, M. Rashki, N. Shabakhty, Flying Squirrel Optimizer (FSO): a novel SI-based optimization algorithm for engineering problems. Iran. J. Optim. (2019)
205. A.A. Heidari, S. Mirjalili, H. Faris et al., Harris hawks optimization: algorithm and applications. Future Gener. Comput. Syst. **97**, 849–872 (2019). https://doi.org/10.1016/j.future.2019.02.028

206. G. Dhiman, A. Kaur, STOA: A bio-inspired based optimization algorithm for industrial engineering problems. Eng. Appl. Artif. Intell. **82**, 148–174 (2019). https://doi.org/10.1016/j.eng appai.2019.03.021
207. J.-B. Lamy, Artificial Feeding Birds (AFB): a new metaheuristic inspired by the behavior of pigeons, in *Advances in Nature-Inspired Computing and Applications*, ed. by S.K. Shandilya, S. Shandilya, A.K. Nagar (Springer International Publishing, Cham, 2019), pp. 43–60
208. S. Zangbari Koohi, N.A.W. Abdul Hamid, M. Othman, G. Ibragimov, Raccoon Optimization Algorithm. IEEE Access **7**, 5383–5399 (2019). https://doi.org/10.1109/access.2018.2882568
209. R. Masadeh, Sea Lion Optimization Algorithm. Int. J. Adv. Comput. Sci. Appl. **10**, 388–395 (2019)
210. M. Jain, V. Singh, A. Rani, A novel nature-inspired algorithm for optimization: Squirrel search algorithm. Swarm Evol. Comput. **44**, 148–175 (2019). https://doi.org/10.1016/j.swevo.2018. 02.013
211. A.R. Mehrabian, C. Lucas, A novel numerical optimization algorithm inspired from weed colonization. Ecol. Inf. **1**, 355–366 (2006). https://doi.org/10.1016/j.ecoinf.2006.07.003
212. A. Karci, B. Alatas, Thinking capability of Saplings Growing up Algorithm, in *Intelligent Data Engineering and Automated Learning–IDEAL 2006*, ed. by E. Corchado, H. Yin, V. Botti, C. Fyfe (Springer, Berlin Heidelberg, 2006), pp. 386–393
213. W. Cai, W. Yang, X. Chen, A global optimization algorithm based on plant growth theory: Plant Growth Optimization, in *2008 International Conference on Intelligent Computation Technology and Automation (ICICTA)* (2008), pp. 1194–1199
214. U. Premaratne, J. Samarabandu, T. Sidhu, A new biologically inspired optimization algorithm, in *2009 International Conference on Industrial and Information Systems (ICIIS)* (IEEE, 2009), pp. 279–284
215. Z. Zhao, Z. Cui, J. Zeng, X. Yue, Artificial plant optimization algorithm for constrained optimization problems, in *2011 Second International Conference on Innovations in Bio-inspired Computing and Applications* (2011), pp. 120–123
216. A. Salhi, E.S. Fraga, Nature-inspired optimisation approaches and the new Plant Propagation Algorithm (Yogyakarta, Indonesia, 2011), pp. K2?1–K2?8
217. R.S. Parpinelli, H.S. Lopes, An eco-inspired evolutionary algorithm applied to numerical optimization, in *2011 Third World Congress on Nature and Biologically Inspired Computing*, pp. 466–471 (2011)
218. Y. Song, L. Liu, H. Ma, A.V. Vasilakos, Physarum Optimization: a new heuristic algorithm to minimal exposure problem, in *Proceedings of the 18th Annual International Conference on Mobile Computing and Networking* (ACM, New York, NY, USA, 2012), pp. 419–422
219. X.-S. Yang, Flower Pollination Algorithm for global optimization, in *Unconventional Computation and Natural Computation*, ed. by J. Durand-Lose, N. Jonoska (Springer, Berlin Heidelberg, 2012), pp. 240–249
220. X. Qi, Y. Zhu, H. Chen et al., An idea based on plant root growth for numerical optimization, in *Intelligent Computing Theories and Technology*, ed. by D.-S. Huang, K.-H. Jo, Y.-Q. Zhou, K. Han (Springer, Berlin Heidelberg, 2013), pp. 571–578
221. H. Zhang, Y. Zhu, H. Chen, Root growth model: a novel approach to numerical function optimization and simulation of plant root system. Soft. Comput. **18**, 521–537 (2014). https:// doi.org/10.1007/s00500-013-1073-z
222. M. Ghaemi, M.-R. Feizi-Derakhshi, Forest Optimization Algorithm. Expert Syst. Appl. **41**, 6676–6687 (2014). https://doi.org/10.1016/j.eswa.2014.05.009
223. F. Merrikh-Bayat, The runner-root algorithm: a metaheuristic for solving unimodal and multimodal optimization problems inspired by runners and roots of plants in nature. Appl. Soft Comput. **33**, 292–303 (2015). https://doi.org/10.1016/j.asoc.2015.04.048
224. M. Sulaiman, A. Salhi, A seed-based Plant Propagation Algorithm: the feeding station model. Sci. World J. (2015). https://doi.org/10.1155/2015/904364
225. M.S. Kiran, TSA: Tree-seed algorithm for continuous optimization. Expert Syst. Appl. **42**, 6686–6698 (2015). https://doi.org/10.1016/j.eswa.2015.04.055

226. H. Moez, A. Kaveh, N. Taghizadieh, Natural forest regeneration algorithm: a new meta-heuristic. Iran. J. Sci. Technol. Trans. Civil Eng. **40**, 311–326 (2016). https://doi.org/10.1007/s40996-016-0042-z

227. Y. Labbi, D.B. Attous, H.A. Gabbar et al., A new rooted tree optimization algorithm for economic dispatch with valve-point effect. Int. J. Electr. Power Energy Syst. **79**, 298–311 (2016). https://doi.org/10.1016/j.ijepes.2016.01.028

228. L. Cheng, Q. Zhang, F. Tao et al., A novel search algorithm based on waterweeds reproduction principle for job shop scheduling problem. Int. J. Adv. Manuf. Technol. **84**, 405–424 (2016). https://doi.org/10.1007/s00170-015-8023-0

229. B. Ghojogh, S. Sharifian, H. Mohammadzade, Tree-based optimization: a meta-algorithm for metaheuristic optimization (2018). CoRR arXiv:abs/1809.09284

230. A. Cheraghalipour, M. Hajiaghaei-Keshteli, M.M. Paydar, Tree Growth Algorithm (TGA): A novel approach for solving optimization problems. Eng. Appl. Artif. Intell. **72**, 393–414 (2018). https://doi.org/10.1016/j.engappai.2018.04.021

231. M. Bidar, H.R. Kanan, M. Mouhoub, S. Sadaoui, Mushroom Reproduction Optimization (MRO): a novel nature-inspired evolutionary algorithm, in *2018 IEEE Congress on Evolutionary Computation (CEC)* (2018), pp. 1–10

232. X. Feng, Y. Liu, H. Yu, F. Luo, Physarum-energy optimization algorithm. Soft. Comput. **23**, 871–888 (2019). https://doi.org/10.1007/s00500-017-2796-z

233. L.N. de Castro, F.J. von Zuben, The clonal selection algorithm with engineering applications, in *GECCO'00, Workshop on Artificial Immune Systems and their Applications* (2000), pp. 36–37

234. J.H. Holland, Genetic algorithms. Sci. Am. **267**, 66–73 (1992)

235. R. Storn, K. Price, Differential evolution–a simple and efficient heuristic for global optimization over continuous spaces. J. Global Optim. **11**, 341–359 (1997)

236. A. Sharma, A new optimizing algorithm using reincarnation concept, in *2010 11th International Symposium on Computational Intelligence and Informatics (CINTI)* (2010), pp. 281–288

237. H.T. Nguyen, B. Bhanu, Zombie Survival Optimization: a Swarm Intelligence Algorithm inspired by zombie foraging, in *Proceedings of the 21st International Conference on Pattern Recognition (ICPR2012)* (IEEE, 2012), pp. 987–990

238. M. Taherdangkoo, M. Yazdi, M.H. Bagheri, Stem Cells Optimization Algorithm, in *Bio-Inspired Computing and Applications*, ed. by D.-S. Huang, Y. Gan, P. Premaratne, K. Han (Springer, Berlin Heidelberg, 2012), pp. 394–403

239. A. Hatamlou, Heart: a novel optimization algorithm for cluster analysis. Prog. Artif. Intell. **2**, 167–173 (2014). https://doi.org/10.1007/s13748-014-0046-5

240. D. Tang, S. Dong, Y. Jiang et al., ITGO: Invasive Tumor Growth Optimization Algorithm. Appl. Soft Comput. **36**, 670–698 (2015). https://doi.org/10.1016/j.asoc.2015.07.045

241. N.S. Jaddi, J. Alvankarian, S. Abdullah, Kidney-inspired algorithm for optimization problems. Commun. Nonlinear Sci. Numer. Simul. **42**, 358–369 (2017). https://doi.org/10.1016/j.cnsns.2016.06.006

242. S. Asil Gharebaghi, M. Ardalan Asl, New meta-heuristic optimization algorithm using neuronal communication. Int. J. Optim. Civil Eng. **7**, 413–431 (2017)

243. V. Osuna-Enciso, E. Cuevas, D. Oliva et al., A bio-inspired evolutionary algorithm: allostatic optimisation. IJBIC **8**, 154–169 (2016)

244. G. Huang, Artificial infectious disease optimization: a SEIQR epidemic dynamic model-based function optimization algorithm. Swarm Evol. Comput. **27**, 31–67 (2016). https://doi.org/10.1016/j.swevo.2015.09.007

245. M.H. Salmani, K. Eshghi, A metaheuristic algorithm based on chemotherapy science: CSA. J. Opti. (2017). https://doi.org/10.1155/2017/3082024

246. X.-S. Yang, X.-S. He, Mathematical analysis of algorithms: part I, in *Mathematical Foundations of Nature-Inspired Algorithms*, ed. by X.-S. Yang, X.-S. He (Springer International Publishing, Cham, 2019), pp. 59–73

247. X.-S. Yang, Nature-inspired mateheuristic algorithms: success and new challenges. J. Comput. Eng. Inf. Technol. **01** (2012). https://doi.org/10.4172/2324-9307.1000e101
248. X.-S. Yang, Metaheuristic optimization: nature-inspired algorithms and applications, in *Artificial Intelligence, Evolutionary Computing and Metaheuristics* (Springer, 2013), pp. 405–420
249. X.-S. Yang, X.-S. He, Applications of nature-inspired algorithms, in *Mathematical Foundations of Nature-Inspired Algorithms*, ed. by X.-S. Yang, X.-S. He (Springer International Publishing, Cham, 2019), pp. 87–97
250. T.-H. Yi, H.-N. Li, M. Gu, X.-D. Zhang, Sensor placement optimization in structural health monitoring using niching monkey algorithm. Int. J. Str. Stab. Dyn. **14**, 1440012 (2014). https://doi.org/10.1142/s0219455414400124
251. T.-H. Yi, H.-N. Li, G. Song, X.-D. Zhang, Optimal sensor placement for health monitoring of high-rise structure using adaptive monkey algorithm. Struct. Control Health Monit. **22**, 667–681 (2015). https://doi.org/10.1002/stc.1708
252. I, Hodashinsky, S. Samsonov, Design of fuzzy rule based classifier using the monkey algorithm. Bus. Inform. 61–67 (2017). https://doi.org/10.17323/1998-0663.2017.1.61.67
253. J. Zhang, Y. Zhang, J. Sun, Intrusion detection technology based on Monkey Algorithm–《Computer Engineering》2011年14期. Comput. Eng. **37**, 131–133 (2011)
254. K. Kiran, P.D. Shenoy, K.R. Venugopal, L.M. Patnaik, Fault tolerant BeeHive routing in mobile ad-hoc multi-radio network, in *2014 IEEE Region 10 Symposium* (2014), pp. 116–120
255. X. Wang, Q. Chen, R. Zou, M. Huang, An ABC supported QoS multicast routing scheme based on BeeHive algorithm, in *Proceedings of the 5th International ICST Conference on Heterogeneous Networking for Quality, Reliability, Security and Robustness. ICST (Institute for Computer Sciences, Social-Informatics and Telecommunications Engineering)* (ICST, Brussels, Belgium, 2008), pp. 23:1–23:7
256. W. Li, M. Jiang, Fuzzy-based lion pride optimization for grayscale image segmentation, in *2018 IEEE International Conference of Safety Produce Informatization (IICSPI)* (2018), pp. 600–604
257. H. Hernández, C. Blum, Implementing a model of Japanese Tree Frogs' calling behavior in sensor networks: a study of possible improvements, in *Proceedings of the 13th Annual Conference Companion on Genetic and Evolutionary Computation.* (ACM, New York, NY, USA, 2011), pp. 615–622
258. H. Hernández, C. Blum, Distributed graph coloring: an approach based on the calling behavior of Japanese tree frogs. Swarm Intell. **6**, 117–150 (2012). https://doi.org/10.1007/s11721-012-0067-2

Chapter 16
Detecting Magnetic Field Levels Emitted by Tablet Computers via Clustering Algorithms

Alessia Amelio and Ivo Rumenov Draganov

Abstract In this chapter, we study the problem of detecting ranges of extremely low frequency magnetic field emitted by the tablet computers. Accordingly, the measurement of the magnetic field from a set of tablets is carried out. Furthermore, the measurement results are analyzed and clustered according to seven well-known algorithms: (i) K-Means, (ii) K-Medians, (iii) Self-Organizing Map, (iv) Fuzzy C-Means, (v) Expectation Maximization with Gaussian Mixture Models, (vi) Agglomerative hierarchical clustering, and (vii) DBSCAN, to obtain different magnetic field ranges from each algorithm. The obtained ranges are compared and evaluated according to the reference level proposed by the Tjänstemännens Centralorganisation (TCO) standard in order to find dangerous areas on the tablet surface, which are established during the typical work with the tablet computers. The analysis shows differences and similarities between the ranges found by the seven algorithms, and suggests the most suitable algorithm to use in the case of tablet measurement. Then, the magnetic field ranges are used to compute the factor of exposure which determines the real level of tablets' users exposure to the extremely low frequency magnetic field. The study proves that dangerous areas on the tablet surface correspond to specific inner components, and gives suggestions for a safe use of the tablet.

Keywords Pattern recognition · Clustering · Magnetic field · Tablet computer · TCO standard

A. Amelio
DIMES University of Calabria, Via Pietro Bucci 44, 87036 Rende, CS, Italy
e-mail: aamelio@dimes.unical.it

I. R. Draganov (✉)
Department of Radio Communications and Video Technologies,
Technical University of Sofia, 1756 Sofia, Bulgaria
e-mail: idraganov@tu-sofia.bg

379

16.1 Introduction

The measurement of the magnetic field represents a process where the collection of data is immanent. The variety of data depends on the positions where and from the time when the measurement is performed. Hence, all data incorporate various elements and characteristics which should be adequately processed and further analyzed. Due to data variability, the creation of data groups with similar characteristics can make the data analysis easier. Hence, clustering can be one of the most suitable data mining choices for a solid and meaningful data analysis.

The humans can be exposed to a variety of magnetic field constantly during their lives. The magnetic field can be divided according to its characteristics, which means by its source as well as its frequency. If we take into consideration the magnetic field sources, then only natural and artificial magnetic field exist. A natural magnetic field is created by the Earth. It is characterized by a constant amplitude and frequency around 8 Hz. On the contrary, the artificial magnetic field is man-made and it is a consequence of the technological development. It also spreads over the Extremely Low Frequency (ELF), i.e. between 30 and 300 Hz. Furthermore, it includes spikes and peaks especially around 50 or 60 Hz.

A tablet represents a mobile computer including a touchscreen from the top, a battery from the bottom, and a motherboard with the circuitry inside. Furthermore, it is equipped with a microphone, loudspeaker, cameras, and a variety of sensors. It has a screen of at least 7" wide. If it is not supplied with the keyboard, then it is typically called booklet. Otherwise, it is called convertible tablet. Due to the different electronic circuitry included, it radiates a frequency dependent magnetic field.

Many standards have been proposed in order to safely use the electronic devices. The TCO standard prescribes the measurement geometry and test procedure. It proposes a safety limit ≤ 200 nT measured at 0.30 m in front of and around the tablet or laptop computer in the ELF frequency range, i.e. in the extended ELF range between 5 Hz and 2 kHz [28]. In our approach, we extend the TCO standard by proposing a new measurement geometry, which is more accustomed to the way of working with the tablets. Furthermore, many researchers in their studies pointed out the dangers of the emitted magnetic field to the human health [1, 18]. Accordingly, they have proposed different safety limits of the magnetic field, such as 1 μT [20], 0.4 μT [1], 0.2 μT [18]. In this paper, we use the reference limit proposed by the TCO standard, which is 0.2 μT, above which there exists a danger to the human health.

Many studies have investigated the destructive consequences of the human exposure (short or long term) to the ELF magnetic field. It is of high importance, because the users' skin, blood, lymph, and bones are under constant exposure to the ELF magnetic field. Hence, some concern about an adequate use of the tablet is mandatory [32]. Specifically, the risk of exposure of the laptop users to the ELF magnetic field has been investigated [2]. It states that the changes in engineering practice to reduce the ELF exposure from equipment or devices should be more explored due to the hazardous effects to the human health of a long exposure to the ELF magnetic field. Furthermore, some research has been also conducted on the influence of the magnetic field exposure to the biological systems, such as birds, which confirms general changes in their behavior, reproductive success, growth and development, physiology and endocrinology, and oxidative stress [15]. At the end, a wide exploration about the impact of the ELF magnetic field exposure has been considered in [4]. It shows many opposed reports about the effects on the DNA (27 research works reporting an effect, while 14 research works reporting no significant effect). Furthermore, leukemia and tumors have been under special consideration. Some analysis has shown that the exposure to an ELF magnetic field, which is higher than the level of 0.4 μT can be considered as extremely dangerous, because it doubles the risk of childhood leukemia [10]. Also, the exposure to such a level of ELF magnetic field is very risky for pregnant women and the fetus [2]. Hence, at the present time the most common practice is to avoid an excessive exposure over a long period of time (usually called Prudent Avoidance) [29]. It is worth noting that recent opinions by experts have documented some deficiencies in the current exposure standards [4]. Hence, some new standardization concerning the ELF magnetic field is precautional.

In this chapter, we explore the elements of the ELF magnetic field initiated by the tablet devices. Although some researchers have examined the dangerous effects of the laptop on the human health [2, 6–9, 22], to the very best of our knowledge, nobody has explored the problem of the tablet computers. It represents a very important topic, due to the tablet widespreading and tendency to be in close contact with the body of the users, as well as its high portability. For this reason, researching in this direction is of great importance for human health preservation.

The focus of this chapter is on detecting ELF magnetic field ranges emitted by the tablet computers, corresponding to dangerousness classes. To pursue this goal, we employ seven well-known clustering algorithms: (i) K-Means, (ii) K-Medians, (iii) Self-Organizing Map, (iv) Fuzzy C-Means, (v) Expectation Maximization with Gaussian Mixture Models, (vi) Agglomerative hierarchical clustering, and (vii) DBSCAN, determining clusters of measurement positions of the tablet, corresponding to ELF magnetic field ranges. Obtained results by the seven algorithms are analyzed and compared, according to the safety limit of the TCO standard, in order to understand what is the most suitable algorithm for this task. In this context, a clustering analysis is more useful than a "flat" analysis of measured data for different reasons. The first one is that clustering allows to automatically detect magnetic field ranges associated with dangerousness classes, which is not possible to find by a simple data observation. The second one is that dangerousness classes envelope different measurements associated with the tablet computers' positions. Hence, they are important to make a general evaluation of high magnetic field emission points of the tablet computer. A similar evaluation can be easily performed on a few tablet computers by observing the measured data at each position. However, an accurate evaluation would require a given number of tablet computer's positions and observation of their measurements. Accordingly, clustering is suitable for the automatization and generalization of such a process. Finally, the application of clustering for detecting magnetic field ranges emitted by a tablet computer is an alternative way to perform the measurement process. Basically, it reduces the errors affecting the measurement of single values, because the measurement is now conceived as measurement intervals. The TCO standard is our choice for the measurement and comparison of the ELF magnetic field ranges because of its strict rules connected with the proposed geometry and reference limits. Its widespread acceptance from the manufacturers' side plays an important role, too. Specifically, we take into account 9 different tablets in order to measure their ELF magnetic field. Hence, we cluster the ELF magnetic field values associated with the measurement positions of the tablets by using K-Means, K-Medians, Self-Organizing Map, Fuzzy C-Means, Expectation Maximization with Gaussian Mixture Models, Agglomerative hierarchical clustering, and DBSCAN algorithms. Then, we compare the obtained results to designate the best magnetic field ranges associated with the dangerous areas of the tablet. The best magnetic field ranges are used to compute a new factor of exposure. It is specifically linked with the real ELF emission of the tablet to which the tablets' users are exposed during different modes of working with the tablet. At the end, its level is compared to the safety reference limit determined by the TCO standard in order to differentiate the safe and dangerous mode of tablet use. Hence, the clustering and factor of exposure along with the finer mesh of the measurement geometry determine an advanced and finer evaluation level of the ELF magnetic field to which the tablet users are exposed.

All aforementioned gives a special attention to the following elements: (i) application of different clustering techniques in order to detect high-risk areas of ELF magnetic field that the tablets emit, (ii) proposition of a new user-centric measurement geometry of the tablets, (iii) measurement of the ELF magnetic field that the tablets emit in their neighborhood, (iv) obtained results that are valuable to a safe use of the tablets, (v) obtained results that are immanent to the tablet designers in order to develop new circuits emitting a lower magnetic field as well as to produce new materials with a smaller magnetic field permeability, and (vi) introduction of a new measure called factor of exposure which evaluates the real exposure of the tablet users to the ELF magnetic field during the use of the tablet in different modes.

The chapter is organized as follows. Section 16.2 explains the measurement procedure and the typical way of working with the tablet. Section 16.3 describes the different methods which are used for data clustering. Section 16.4 evaluates the tablet user's exposure to the ELF magnetic field. Section 16.5 presents the results of the measurement process, clustering methods and factor of exposure, and discusses them. Finally, Sect. 16.6 draws the conclusions and outlines future work directions.

16.2   Measurement of the Tablet Magnetic Field

16.2.1   Magnetic Field

Each electronic circuit supplied by a current I induces a magnetic field. Hence, any electronic circuit inside a tablet radiates a magnetic field. If the current I is steady and constant in the uniform wire, then the induced magnetic field can be represented by the Biot-Savart law [16]:

$$\mathbf{B}(\mathbf{r}) = \frac{\mu_0 I}{4\pi} \int_{wire} \frac{d\mathbf{l} \cdot \mathbf{r}}{r^2}, \tag{16.1}$$

where the integral sums over the wire length, the vector $d\mathbf{l}$ is the vector line element with direction as the current I, μ_0 is the magnetic permeability constant, r is the distance between the location of $d\mathbf{l}$ and the location where the magnetic field is calculated, and $\mathbf{r}$ is a unit vector in the direction of r.

Furthermore, the magnetic field $\mathbf{B}$ is characterized by its magnitude and direction. It can be calculated as [2]:

$$\mathbf{B}(\mathbf{r}) = B_x \cdot \mathbf{x} + B_y \cdot \mathbf{y} + B_z \cdot \mathbf{z}, \tag{16.2}$$

where $\mathbf{x}$, $\mathbf{y}$ and $\mathbf{z}$ are positional vectors orthogonal to each other, and B_x, B_y and B_z are the scalar intensity values of the magnetic induction in the direction of these vectors, respectively. The measuring devices measure the aforementioned scalar components of the magnetic induction $\mathbf{B}$. Using these values, the Root Mean Square (RMS) of the magnetic induction $\mathbf{B}$ can be calculated as:

$$B = \sqrt{B_x{}^2 + B_y{}^2 + B_z{}^2}. \tag{16.3}$$

16.2.2 Measuring Devices

The ELF magnetic field is measured by the devices Lutron 3D EMF-828 and Aaronia Spectran NF-5030.

Lutron 3D EMF-828 tester with external probe is designed to determine the magnitude of the magnetic field generated by power lines, computers, gadgets, electrical appliances and any other similar device that emits a magnetic field in the neighbourhood [24]. It measures the level of the magnetic field at the direction of all three axes (X, Y, Z directions), i.e. B_x, B_y and B_z. The measurement is performed at the frequency bandwidth between 30 and 300 Hz, which is commonly called ELF interval. The magnetic field can be measured in three measurement intervals: 20 μT, 200 μT and 2000 μT with a resolution of 0.01 μT, 0.1 μT and 1 μT, respectively.

The other measurement device is SPECTRAN NF Spectrum Analyzer NF-5030 [27]. It measures the values of minimum, maximum, average and RMS of the magnetic field B, given by its directional versions B_x, B_y and B_z, or the total one B. The measurement extends from 1 pT to 2 mT in the ELF interval between 1 Hz and 1 MHz. It has 6 pre-programmed frequency intervals for the measurement. Still, the user can define its own frequency interval, where the measurement will take place. Figure 16.1 illustrates the measuring devices.

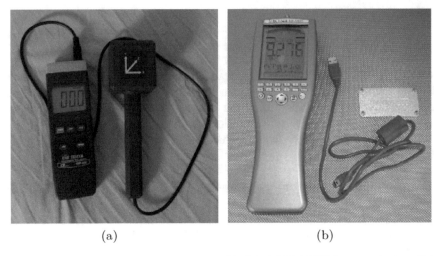

(a)   (b)

Fig. 16.1 Measuring devices: **a** Lutron 3D EMF-828, **b** SPECTRAN NF Spectrum Analyzer NF-5030

16.2.3   TCO Standard

TCO'05 proposed a measurement methodology of the magnetic field RMS value in two frequency ranges: (i) band 1 with a frequency range between 50 Hz and 2 kHz, and (ii) band II with a frequency range between 2 and 400 kHz [28].

The TCO measuring geometry includes measurement positions, which are 30 cm away from the emitter of the low frequency magnetic field. Typically, the emitters represent portable computers like a laptop or tablet. Furthermore, the proposed measurement geometry has a universal approach. Hence, it can be used also for magnetic field emitters such as AC adapters.

Unfortunately, the body position of the portable computer user may be closer than the proposed measurement positions. Hence, an investigation of the ELF magnetic field value at a smaller distance or in its centre is mandatory from the users' point of view. The measuring positions proposed by the TCO standard for the objects emitting the magnetic field are given in Fig. 16.2.

16.2.4   The Realized Experiment

The experiment includes the measurement of the magnetic field level at designated measurement positions or around the tablets. It is conducted by measuring the magnetic field on each tablet at its top and bottom in 9 different positions (a total of 18 measurement positions). Figure 16.3a, b illustrates the tablet's top and bottom measurement positions. In contrast to the TCO measurement geometry, we propose a

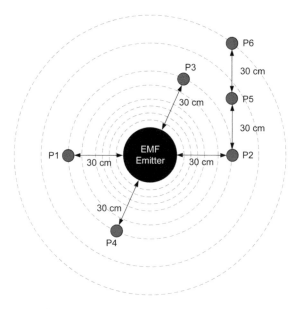

Fig. 16.2 Measuring positions proposed by the TCO standard for the objects that emit a magnetic field (P1, ..., P6)

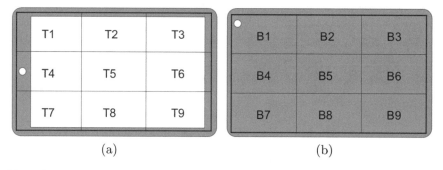

Fig. 16.3 Tablet measurement positions: **b** at the top part (T1,...,T9), **c** at the bottom part (B1,...,B9)

new measurement geometry, which is more accustomed to the way of using portable computers [9] or tablets, including 18 measurement positions instead of only 6 proposed by the TCO standard. Furthermore, this new proposed positions are mutually distant around 5-6 cm instead of 30 cm as in the TCO standard, which is 5–6 times finer than the proposed TCO. It is worth noting that the users' fingers and hands are all the time in close contact with the top and bottom area of the tablet. Hence, the parts of both hands are exposed to a direct ELF magnetic field at the full extent.

The measurement is performed on 9 tablets of different manufacturer, named as Tablet 1,...,9. The first one, Tablet 1, has the motherboard inside the body of the tablet and it is equipped with an external keyboard (convertible tablet). The other ones, Tablet 2,...,9, have also the motherboard inside the body of the tablet, but they don't have an external keyboard. In our case, the keyboard of the Tablet 1 is not considered, because it emits a negligible magnetic field.

Figure 16.4 shows the measurement of the tablet's ELF magnetic field emission at each measurement position (measurement geometry) and some high-risk areas on the tablet surface corresponding to specific inner components. Furthermore, the emission of the ELF magnetic field at the top and bottom of the tablet in the different

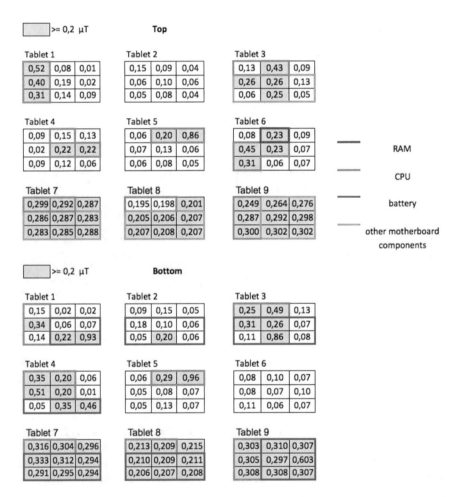

Fig. 16.4 Measurement results of the tablet's ELF magnetic field emission (in μT) at the top and bottom parts. Values above the safety reference limit of 0.2 μT are marked in grey. The regions corresponding to specific inner components of the tablet are differently coloured

modes when the screen is OFF and ON reduces approximately of 5%. The reason for such a small reduction of ELF magnetic field emission is because the screen has a low power consumption and it has an integrated protected gorilla glass at its top, which has lower magnetic permeability.

It is worth noting that all tablets are commonly used in typical working conditions. It means that they are used for Internet browsing or Skype calling, picture viewing and reading documents and typing. Also, it can be noticed that the tablets are typically used as a battery powered device, only.

16.2.4.1   Pin-Point Results of the ELF Magnetic Field Measurement

In this experiment, we introduce a pin-point measurement of a higher number of points where the ELF magnetic field is emitted. Figure 16.5 illustrates the level of the ELF magnetic field at the top of each tablet (Tablet 1,...,9).

Similarly, Fig. 16.6 illustrates the level of the ELF magnetic field at the bottom of each tablet (Tablet 1,...,9).

From all above results, it is clear that 18 measurement positions are enough to qualify and quantify the real exposure of the tablet users to the ELF magnetic field emitted by the tablet itself.

16.2.5   A Typical Way of Working with the Tablet

To measure the real ELF magnetic field exposure of the tablet users, it is needed to explore the typical way of working with the tablet as well as the humans' body positions in these circumstances. The tablet can be used in the so-called portrait and landscape orientations. In the portrait orientation, the tablet is commonly used for reading documents, for browsing the Internet, for watching pictures, etc. Furthermore, the tablet in the landscape orientation is also used for reading documents, for typing documents, for browsing the Internet, for watching pictures, for taking pictures, etc. The tablet orientation the user will exploit mainly depends on the users' preferences. Hence, depending on the typical way of working with the tablet, a different level of tablet user's exposure to an ELF magnetic field will be measured.

Firstly, the typical tablet landscape orientation use is presented. Figure 16.7 illustrates typical positions of the users' hands which are in close contact with the tablet, i.e. at its top and bottom parts. Secondly, the typical tablet portrait orientation use is presented. Figure 16.8 illustrates typical positions of the users' hands which are in close contact with the tablet, i.e. at its top and bottom parts. From Figs. 16.7 and 16.8, it is worth noting that the users' hands and fingers are not in close contact with some areas of the tablet, while there exist specific parts which the users are in close contact with. Obviously, in some parts of the tablet which the users' hands and

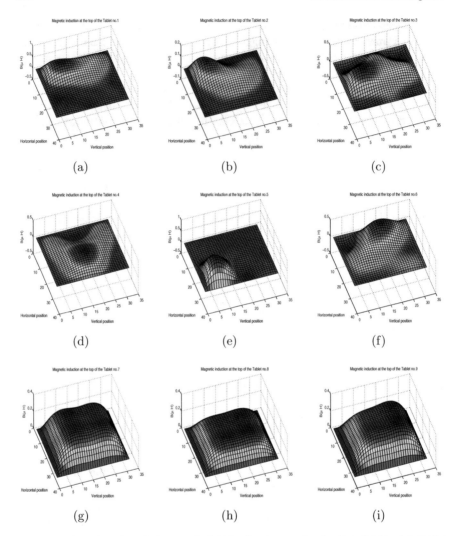

Fig. 16.5 Illustration of the ELF magnetic field level at the top of each tablet: **a** Tablet 1, **b** Tablet 2, **c** Tablet 3, **d** Tablet 4, **e** Tablet 5, **f** Tablet 6, **g** Tablet 7, **h** Tablet 8, **i** Tablet 9

fingers are in close contact with, the exposure to the ELF magnetic field emission is stronger and more dangerous. On the contrary, the other parts of the tablet which the users' hands and fingers are not in close contact with, represent a lower risk to the high exposure of the ELF magnetic field emission, because the ELF magnetic field rapidly reduces its amplitude with the distance.

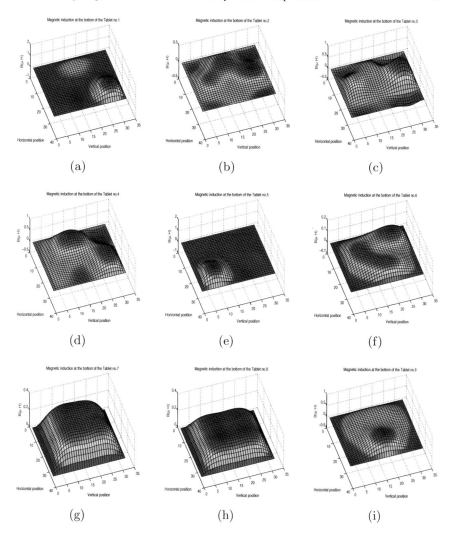

Fig. 16.6 Illustration of the ELF magnetic field level at the bottom of each tablet: **a** Tablet 1, **b** Tablet 2, **c** Tablet 3, **d** Tablet 4, **e** Tablet 5, **f** Tablet 6, **g** Tablet 7, **h** Tablet 8, **i** Tablet 9

16.3 Magnetic Field Clustering

For each of the 9 tablets, each measurement position at the top part, Ti i = 1...9, and at the bottom part, Bi i = 1...9, is associated with a single feature. It represents the RMS of the measured magnetic field induction at that position, that we called B. Consequently, each tablet determines 9 features at the top part and 9 features at the bottom part. Hence, two separate datasets are created from the measurement positions of the tablets. The first one collects 81 features from the top positions and

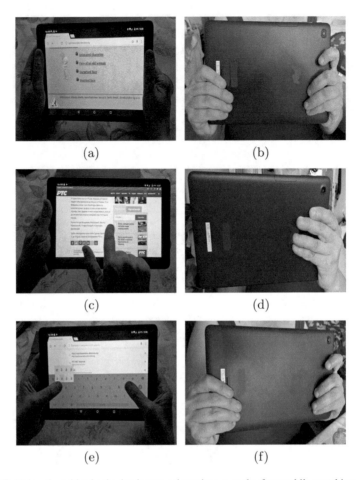

Fig. 16.7 Using the tablet in the landscape orientation: **a** at the front while watching, **b** at the bottom while watching, **c** at the front while browsing, **d** at the back while browsing, **e** at the front while typing, and **f** at the back while typing

it is named as *top*. The second one contains 81 features from the bottom positions and it is named as *bottom*.

The measurement positions are grouped into clusters for detecting the emission ranges of ELF magnetic field on the tablet top and bottom surface. It is achieved by considering the magnetic field feature at each position. Then, clustering is performed on the positions according to their feature representation. It determines the ELF magnetic field ranges which are emitted in different high-risk areas of the tablet surface.

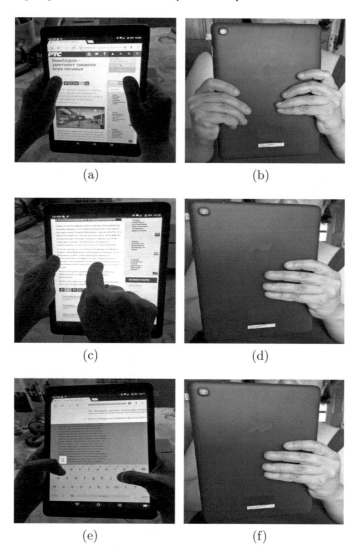

(a)   (b)

(c)   (d)

(e)   (f)

Fig. 16.8 Using a tablet in portrait orientation: **a** at the front while watching, **b** at the bottom while watching, **c** at the front while browsing, **d** at the back while browsing, **e** at the front while typing, and **f** at the back while typing

To pursue this goal, each dataset is partitioned by using seven well-known clustering algorithms of different typology: (i) K-Means [17, 25, 30], (ii) K-Medians [5, 8, 9, 23, 31], (iii) Self-Organizing Map [21, 30, 34], (iv) Expectation Maximization with Gaussian Mixture Models [12, 19], (v) Agglomerative hierarchical clustering [26], (vi) Fuzzy-C-Means [3, 13], and (vii) DBSCAN [14].

16.3.1 K-Means Clustering

K-Means algorithm is a partitional center-based clustering method [25], employed in the literature for discretizing numerical values in ranges, based on the natural characteristics of the data [17, 30]. The number K of the clusters (ranges) is prior set as an input parameter. The algorithm is based on the fact that each cluster is represented by a *centroid*, which is the mean of the data points in that cluster.

The procedure is started by selecting K random initial centroids. Since the location of the initial centroids influences the final clustering solution, the centroids should be carefully positioned, i.e. as much as possible far to each other. The first step of the algorithm consists of assigning each data point to its nearest centroid, in terms of the Euclidean distance. In the second step, the new K centroids are re-computed from the previous assignments. These two steps are iterated, while the centroids progressively change their location, until no more changes are required.

The aim of the algorithm is to minimize the squared error function, which is the following objective function:

$$J = \sum_{m=1}^{K} \sum_{i=1}^{z_m} ||x_i - c_m||^2 \qquad (16.4)$$

where $||x_i - c_m||^2$ is the square Euclidean distance between the data point x_i in cluster m and the centroid c_m in cluster m. z_m denotes the number of data points in cluster m.

16.3.2 K-Medians Clustering

K-Medians algorithm [5] is a well-known representative of the center-based clustering methods, adopted in some contexts for clustering one-dimensional data, such as the measurements [8, 9, 23, 31]. The number of clusters K is prior fixed and it is an input parameter of the method. Similarly to K-Means, its center-based characteristic derives from the concept of *centroid*, which is the representative of the cluster in the algorithm. For each cluster, its centroid is computed as the median value of the data points belonging to that cluster. Its advantage relies on the robustness to outliers and ability to generate more compact clusters than K-Means [25]. The algorithm is composed of three main phases:

1. Initial centroid selection,
2. Data points assignment, and
3. Centroid re-computation.

In the first phase, the algorithm selects K initial centroids. It is a critical aspect, because their choice can influence the final clustering result. In the second phase, each data point is assigned to the centroid which is the nearest in terms of L_1 distance between the corresponding features. In the last phase, the K centroids are re-computed from the new obtained clusters. Phases 2 and 3 are iterated multiple times, until the centroids do not modify anymore their position.

K-Medians aims to minimize the following function:

$$J = \sum_{m=1}^{K} \sum_{i=1}^{z_m} ||x_i - c_m|| \tag{16.5}$$

where $||x_i - c_m||$ is the L_1 distance between the data point x_i in cluster m and the centroid c_m in cluster m. z_m denotes the number of data points in cluster m.

16.3.3 Self-Organizing Map Clustering

Kohonen's Self-Organizing Map (SOM) [21, 34] is a classification method successfully adopted in some contexts for discretizing numerical values [30]. It is based on a neural network, which is learned following the "shape" of the training data for determining the output. It is characterized by a total of N neurons located on a regular one or two-dimensional grid. The number n of the input neurons is equal to the dimensions of the data point. Output neurons are related to the clusters in output, hence their number K may be equal to the number of clusters. In the network, each output neuron is connected with all input neurons. Accordingly, an output neuron i is associated with a n-dimensional vector, $w_i = [w_{i1}, ..., w_{in}]$, called *prototype vector*, which is the weight of the i-th neuron.

In order to train the network, an algorithm composed of multiple steps is iteratively run, starting from the first epoch. Initially, all the weights are initialized to small random values. In the first step, the algorithm randomly selects from the training set a data point $x = \{x^1, ..., x^n\}$. Then, the output neuron i^* having the minimum distance from x, called Best-Matching Unit (BMU), is selected as:

$$i^* = \arg\min_i \ d^2(x, w_i) \tag{16.6}$$

where $d^2(x, w_i)$ is the L_2 (Euclidean) distance computed between x and the prototype vector w_i of the output neuron i:

Layer

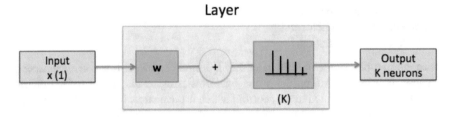

Fig. 16.9 Adopted one-dimensional SOM

$$d^2(x, w_i) = \sum_{k=1}^{n} (x^k - w_{ik})^2 \tag{16.7}$$

In the second step, all the prototype vectors within a neighbourhood distance from BMU are updated by employing the following weight update rule:

$$w_i(t+1) = w_i(t) + \eta(t)(x - w_i(t)) \tag{16.8}$$

where η is the learning rate, decreasing monotonically over time. After that, the epoch is incremented and the procedure is repeated until computational bounds do not exceed. At the end of this procedure, the SOM network is ready for clustering. Hence, given the input data point x, its corresponding BMU defines the cluster which the point belongs to.

In our case, the SOM network is one-dimensional, because we manage ELF magnetic field values. Hence, we have that $n = 1$, because the input data point is one-dimensional. Accordingly, a prototype vector w_i reduces to a single weight. Furthermore, the number K of output neurons corresponds to the number of ELF magnetic field ranges. The advantage of the SOM network is the dimensionality reduction of high-dimensional input data into a one or two dimensional grid. It makes easier to understand and interpret the data. In this case, the input data are one-dimensional. Also, the output neurons are located on a $1 \times K$ one-dimensional grid. Hence, the conversion of dimensions is implicitly realized. Figure 16.9 shows the employed SOM model.

16.3.4 DBSCAN Clustering

DBSCAN [14] comes from Density-Based Spatial Clustering of Applications with Noise. It is an algorithm that locates high density areas with regions of low density in-between. The density is defined as the number of points covered by a given radius (*Eps*). Three types of points are being introduced for the purposes of analysis: (i) core points, such that their number (*MinPts*) is over a certain number within *Eps* so they form the cluster itself, (ii) border points, they are fewer than *MinPts* encapsulated by *Eps* and still are in the neighbourhood of the core points, and (iii) noise points, all the other points from the dataset. Core points separated by a distance smaller than *Eps* belong to the same cluster. Border points close enough to the core points are also part of the same cluster while the noise points are not. A set of values need to be defined by the user. These are the physical distance *Eps* and the desired minimum cluster size *MinPts*. The first one can be selected by using a k-distance graph but too small values lead to the absence of clustering and too large ones to the formation of only one cluster. The *MinPts* parameter is derived from the number of dimensions D following the condition $MinPts \geq D + 1$. Larger values are preferable especially with the growth of the dataset. Thus, the data points within a radius of *Eps* from a reference data point form the so called *Eps*-Neighbourhood and the *Eps*-Neighbourhood of a data point containing no less than *MinPts* data points form the so called core points. A data point q is considered as directly density-reachable from a data point p if q is within the *Eps*-Neighbourhood of p and p is a core point. Also, a data point p is density-connected to a data point q with respect to *Eps* and *MinPts* if there is a data point o such that both p and q are density-reachable from o with respect to *Eps* and *MinPts*. As an input N data points are passed with the globally defined *Eps* and *MinPts*. The output consists of clusters of data points. The following steps are performed for that purpose:

1. Random selection of a point P,
2. All points around P that are density-reachable with respect to *Eps* and *MinPts* are found,
3. If P is a core point, the formation of a cluster is implemented,
4. If P is a border point, continue to the next data point from the dataset,
5. Continue processing until the end of the dataset.

The algorithm is resistant to noise and copes with clusters varying in shape and size. The advantages of DBSCAN include that: (i) it is not necessary to have prior knowledge of the number of clusters in the data, and various types of clusters in shape could be found, (ii) robustness to outliers, (iii) insensitiveness to the data arrangement in the dataset, (iii) acceleration of region queries, (iv) only two input parameters are needed. The disadvantages are that it is not entirely deterministic and the results depend on a distance measure when querying the regions while a good knowledge of the scale of the input data is required.

16.3.5 Expectation-Maximization with Gaussian Mixture Models

The Expectation-Maximization (EM) algorithm [12, 19] is applied over statistical models for estimation of maximum likelihood parameters without the possibility of finding a direct solution. Partial data observations are the input, typically accompanied by latent variables and unknown parameters. Assuming the existence of unregistered samples, the statistical model can be simplified which in the case of a mixture suggests the presence of a corresponding unobserved data point to each observed one defining the mixture component. Having the likelihood function, a maximum could be found as a solution from the derivatives in relation to all unknown values, latent variables, and the parameters. The presence of latent variables makes it impossible, in most cases, to have simultaneous solutions to all the equations that emerge from the so stated problem. The solution may be found numerically. Arbitrary values can be selected for one of the sets of unknowns in order to obtain the ones for the other set and then get better approximation for the first set. The switching of calculation between the two sets of values continues until they reach a stationary condition. The derivative of the likelihood at this stage is zero corresponding to a maximum or a saddle point [33]. The marginal likelihood of the set of input samples X from which the maximum likelihood estimate of the unknown parameters is derived has the form:

$$L(\theta, X) = p(X|\theta) = \int p(X, Z|\theta)dZ, \qquad (16.9)$$

where Z is the set of unobserved latent data or missing values, θ vector of unknown parameters, and the very likelihood function is $L(\theta; X, Z) = p(X, Z|\theta)$. The practical application of this form is difficult since Z as a sequence of events leads to exponential growth of processed values with the increase of the length of the sequence. Thus, two steps are iteratively performed as a more substantial approach:

1. Expectation step, calculation of the expected value of the log likelihood function where the conditional distribution of Z with known X during the current approximation of $\theta(t)$:

$$Q(\theta|\theta^{(t)}) = E_{Z|X,\theta^{(t)}}[logL(\theta; X, Z)]. \qquad (16.10)$$

2. Maximization step, estimating the parameter maximizing the following value:

$$\theta^{(t+1)} = argmax_\theta Q(\theta|\theta^{(t)}). \qquad (16.11)$$

Often, in practice, the set of input samples is formed by independent observations containing mixture of two or more multivariate normal distributions of a given dimension in which case the algorithm is defined as EM with Gaussian Mixture Models. Despite the increase of the observed data likelihood function it is not guaranteed that a convergence will occur leading to a maximum likelihood which means reaching a local maximum. Additional techniques, mainly based on heuristics, are applied in order to escape it.

16.3.6 Hierarchical Clustering

Hierarchical clustering [26] is a method of cluster analysis aiming the hierarchical representation of clusters. There are two types of hierarchical clustering:

1. Agglomerative, a bottom-up approach where pairs of clusters are being merged starting from all the clusters at the bottom level and moving up the hierarchy.
2. Divisive, a top-down approach at which all the data points initially fall into one cluster and by splitting applied recursively, a movement is achieved down the hierarchy.

The resulting hierarchy is typically represented as a dendrogram. The decision of merging two clusters is made based on a measure of dissimilarity between the input data points. Pr−+oper metrics are introduced for measuring the distance between data points pairs along with a linkage criterion defining the dissimilarity of groups as pairwise distances of the data points in them. Frequently used metrics are the Euclidean distance, Manhattan distance, Maximum distance, Mahalanobis distance and others. Some of the popular linkage criteria include complete-linkage clustering, single-linkage clustering, average linkage clustering, centroid linkage clustering, minimum energy clustering, etc. A particular advantage of this type of clustering is that any valid distance measure can be applied with it. Then, a matrix of distances between the input data points initiates the clustering process instead of the data points themselves.

An example of agglomerative clustering is given in Fig. 16.10. Let the Euclidean distance be the employed metric. Partitioning clustering could be achieved after any height of the resulting tree, e.g. after the second level the accumulated clusters are $\{x_0\}, \{x_{12}\}, \{x_{34}\}, \{x_5\}, \{x_6\}$. With the increase of the tree level the clustering becomes grosser, e.g. after level 3 the data points fall within the clusters $\{x_0\}, \{x_{12}\}, \{x_{3456}\}$.

The process relies on progressively merging clusters. It starts from uniting the two closest elements laying in a distance less than a chosen limit from one another. A more straightforward approach is to build a distance matrix in which the value in the i-th row and j-th column is the distance between the i-th and j-th data points. Getting higher into the tree level leads to merging of rows and columns with updated distances. The distance between two clusters Y and Z may be selected from one of the following:

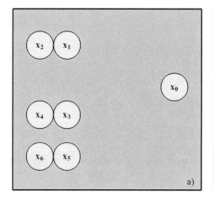 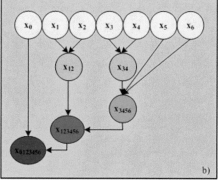

Fig. 16.10 Agglomerative clustering: **a** input data points, **b** resulting dendrogram

1. Complete-linkage clustering, the maximum distance among data points from analyzed clusters: $max\{d(y, z) : y \in Y, z \in Z\}$.
2. Single-linkage clustering, the minimum distance among data points from analyzed clusters: $min\{d(y, z) : y \in Y, z \in Z\}$.
3. The mean distance among elements of each cluster (average linking clustering): $(1/|Y||Z|) \times \sum_{y \in Y} \sum_{z \in Z} d(y, z)$.
4. The accumulated value of the intra-cluster variance.
5. The decrease of variance of the merged cluster.
6. The probability of the new formed clusters derived from the same distribution function.

The number criterion for clustering termination supposes the reach of a predefined relatively small number of clusters. Another possibility for stopping the clustering is the distance criterion. There, the agglomeration occurs at a greater distance relative to a previous iteration.

16.3.7 Fuzzy-C-Means Clustering

Fuzzy C-Means clustering was proposed by Dunn [13] and then further developed by Bezdek [33]. The input data points taking part into the clustering process could belong to more than one cluster. In order a convergence to occur, the minimization has to be provided for the following objective function:

$$J_m = \sum_{i=1}^{D} \sum_{j=1}^{K} \lambda_{ij}^{m} ||x_i - c_j||^2, \tag{16.12}$$

where D is the number of data points, K the number of clusters, m a factor influencing the fuzzy overlap represented by a real number greater than 1, x_i the i-th data point, c_j the center of the j-th cluster, λ_{ij} the level of belonging of x_i to the j-th cluster. It is true that:

$$\sum_{j=1}^{K} \lambda_{ij} = 1. \tag{16.13}$$

The following steps are performed during the clustering:

1. Random initialization of all λ_{ij},
2. Find the cluster centers according to the equation:

$$c_j = \frac{\sum_{i=1}^{D} \lambda_{ij}^{m} x_i}{\sum_{i=1}^{D} \lambda_{ij}^{m}}. \tag{16.14}$$

3. λ_{ij} are updated by the application of the expression:

$$\lambda_{ij} = \frac{1}{\sum_{k=1}^{K} \frac{||x_i - c_j||}{||x_i - c_k||}^{\frac{2}{(m-1)}}}. \tag{16.15}$$

4. The objective function J_m is found using (16.12).
5. Repeat steps 2–4 until J_m undergoes a change less than a preliminary defined threshold or a given number of iterations is reached.

The main difference with the K-Means algorithm is the introduction of the degree of membership for each data point to different clusters and the factor determining the fuzzy overlap. The intra-cluster variance is minimized as a result. Nevertheless, the achieved minimum of the objective function is local and the partitioning depends on the initial choice of weights which is a disadvantage also typical for the K-Means.

16.4   Evaluation of the Tablet User Exposure to ELF Magnetic Field

Taking into account all given facts, we introduce an additional measure for evaluation of the real exposure to the ELF magnetic field emission. It consists of the differences between the danger and safe level obtained by the clustering tools. In this way, we obtain a measure which is more accustomed to the real circumstances of the ELF magnetic field exposure. This is achieved by introducing a corrective factor cf_1 and cf_2 set to 1 or 2/3, i.e. 0.66 (which is a value that is empirically obtained, due to the distance which reduces the level of the ELF magnetic field exposure), respectively. In the areas where the users' hands and fingers are in close contact with the tablet areas, the correction factor cf is set to 1, while in other areas it is set to 0.66.

Also, we introduce a different factor of exposure (*foe*) for the calculation of the users' exposure to the ELF magnetic field emission, when they are working in different modes. Hence, the following *foe* are introduced: (i) foe_l for the tablet in landscape mode, and (ii) foe_p for the tablet in portrait mode. Due to a different exposure of the tablet users in each tablet orientation, we introduce a sub-differentiation as working in the watching, browsing and typing mode, which is given as (i) landscape use: (a) $foe_{l,w}$ for watching mode, (b) $foe_{l,b}$ for browsing mode, and (c) $foe_{l,t}$ for typing mode, respectively, as well as (ii) portrait use: (a) $foe_{p,w}$ for watching mode, (b) $foe_{p,b}$ for browsing mode, and (c) $foe_{p,t}$ for typing mode, respectively.

To elaborate these *foe* factors, a further exploration of danger and safe regions according to the Figs. 16.7 and 16.8 will be given. After that, the determination and calculation of different *foe* factors will be understandable and clear. Firstly, we shall explore the dangerous (given the corrective factor 1) and safe (given the corrective factor 0.66) parts of the tablet, when using in the portrait mode. If we have a total of 18 measurement positions which include 9 measurement positions at the top and 9

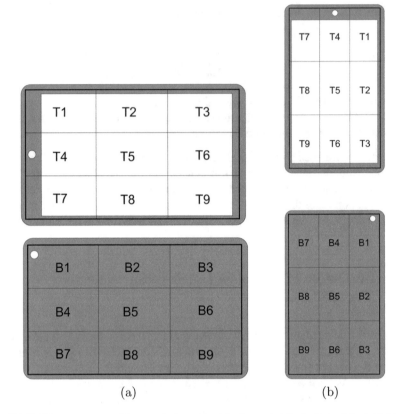

Fig. 16.11 Measurement positions equivalence: **a** at the top and at the bottom of the tablet in the landscape orientation, **b** equivalence made at the top and at the bottom of the tablet in the portrait orientation

measurement positions at the bottom of the tablet, then we have the circumstances for landscape and portrait orientation given as in Fig. 16.11.

Furthermore, all measurement positions in different working conditions with the tablet are given below. The positions in which the users are in close contact with the tablet are labeled in red. Taking into account the red labeled positions which are linked with the correction factor 1 (in which the users' hands and fingers are in close contact with the body of the tablet), the grey labeled positions are linked with the correction factor 0.66 (in which the users' hands and fingers are not always in close contact with the body of the tablet, but are in the near neighbourhood).

Figure 16.12 shows the use of the tablet in the landscape orientation in the following modes: (a) watching mode, (b) browsing mode, and (c) typing mode (see Fig. 16.7 for reference).

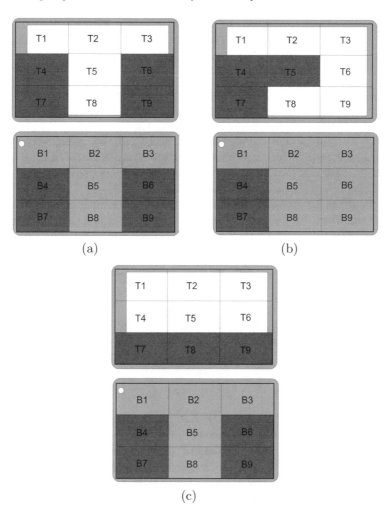

Fig. 16.12 The use of the tablet in the landscape orientation: **a** watching mode, **b** browsing mode, and **c** typing mode (the red labeled measurement areas represent the positions in which the users' hands and fingers are in close contact with the body of the tablet)

According to the Fig. 16.12, firstly we define a separate factor of exposure *foe* for top and bottom part in different ways of working with the tablets in landscape mode.

In this way, the following *foe* factors for the landscape operation mode at the top positions of the tablet: (a) $foe_{top,lw}$ for watching mode, (b) $foe_{top,lb}$ for browsing mode, and (c) $foe_{top,lt}$ for typing mode, are calculated as:

$$foe_{top,lw} = \frac{\sum_{t=4,6,7,9} c_t \cdot cf_1 + \sum_{t=1-3,5,8} c_t \cdot cf_2}{n_{top}}, \qquad (16.16)$$

$$foe_{top,lb} = \frac{\sum_{t=4,5,7} c_t \cdot cf_1 + \sum_{t=1-3,6,8,9} c_t \cdot cf_2}{n_{top}}, \quad (16.17)$$

$$foe_{top,lt} = \frac{\sum_{t=7-9} c_t \cdot cf_1 + \sum_{t=1-6} c_t \cdot cf_2}{n_{top}}, \quad (16.18)$$

where cf_1 and cf_2 represent a corrective factor set to 1 and 0.66, respectively. Furthermore, n_{top} is the number of measurement points at the top of the tablet equal to 9. At the end, c_t represents the ELF magnetic field range number obtained by clustering at the top positions (landscape orientation mode).

Furthermore, the *foe* factors for the landscape operation mode at the bottom positions of the tablet: (a) $foe_{bottom,lw}$ for watching mode, (b) $foe_{bottom,lb}$ for browsing mode, and (c) $foe_{bottom,lt}$ for typing mode, are calculated as:

$$foe_{bottom,lw} = \frac{\sum_{b=4,6,7,9} c_b \cdot cf_1 + \sum_{b=1-3,5,8} c_b \cdot cf_2}{n_{bottom}}, \quad (16.19)$$

$$foe_{bottom,lb} = \frac{\sum_{b=4,7} c_b \cdot cf_1 + \sum_{b=1-3,5,6,8,9} c_b \cdot cf_2}{n_{bottom}}, \quad (16.20)$$

$$foe_{bottom,lt} = \frac{\sum_{b=4,6,7,9} c_b \cdot cf_1 + \sum_{b=1-3,5,8} c_b \cdot cf_2}{n_{bottom}}, \quad (16.21)$$

where cf_1 and cf_2 represent a corrective factor set to 1 and 0.66, respectively. Furthermore, n_{bottom} is the number of measurement points at the bottom of the tablet equal to 9. At the end, c_b represents the ELF magnetic field range number obtained by clustering at the bottom positions (landscape orientation mode).

Accordingly, Fig. 16.13 shows the use of the tablet in the portrait orientation in the following modes: (a) watching mode, (b) browsing mode, and (c) typing mode (see Fig. 16.8 for reference).

According to the Fig. 16.13, firstly we define a separate factor of exposure *foe* for top and bottom part in different ways of working with the tablets in portrait mode.

Hence, the following *foe* factors for the portrait operation mode at the top positions of the tablet: (a) $foe_{top,pw}$ for watching mode, (b) $foe_{top,pb}$ for browsing mode, and (c) $foe_{top,pt}$ for typing mode, are calculated as:

$$foe_{top,pw} = \frac{\sum_{t=2,8} c_t \cdot cf_1 + \sum_{t=1,3-7,9} c_t \cdot cf_2}{n_{top}}, \quad (16.22)$$

$$foe_{top,pb} = \frac{\sum_{t=5,8} c_t \cdot cf_1 + \sum_{t=1-4,6,7,9} c_t \cdot cf_2}{n_{top}}, \quad (16.23)$$

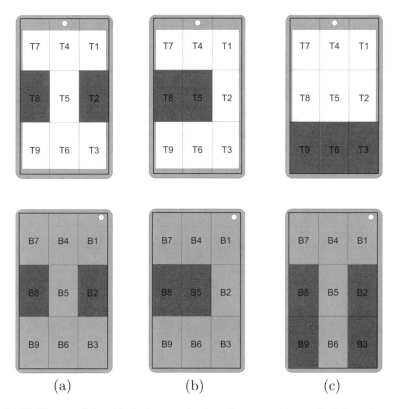

Fig. 16.13 The use of the tablet in the portrait orientation: **a** watching mode, **b** browsing mode, and **c** typing mode (the red labeled measurement areas represent the positions in which the users' hands and fingers are in close contact with the body of the tablet)

$$foe_{top,pt} = \frac{\sum_{t=3,6,9} c_t \cdot cf_1 + \sum_{t=1,2,4,5,7,8} c_t \cdot cf_2}{n_{top}}, \tag{16.24}$$

where cf_1 and cf_2 represent a corrective factor set to 1 and 0.66, respectively. Furthermore, n_{top} is the number of measurement points at the top of the tablet equal to 9. At the end, c_t represents the ELF magnetic field range number obtained by clustering at the top positions (portrait orientation mode).

Then, the *foe* factors for the portrait operation mode at the bottom positions of the tablet: (a) $foe_{bottom,pw}$ for watching mode, (b) $foe_{bottom,pb}$ for browsing mode, and (c) $foe_{bottom,pt}$ for typing mode, are calculated as:

$$foe_{bottom,pw} = \frac{\sum_{b=2,8} c_b \cdot cf_1 + \sum_{b=1,3-7,9} c_b \cdot cf_2}{n_{bottom}}, \tag{16.25}$$

$$foe_{bottom,pb} = \frac{\sum_{b=5,8} c_b \cdot cf_1 + \sum_{b=1-4,6,7,9} c_b \cdot cf_2}{n_{bottom}}, \tag{16.26}$$

$$foe_{bottom,pt} = \frac{\sum_{b=2,3,8,9} c_b \cdot cf_1 + \sum_{b=1,4-7} c_b \cdot cf_2}{n_{bottom}}, \quad (16.27)$$

where cf_1 and cf_2 represent a corrective factor set to 1 and 0.66, respectively. Furthermore, n_{bottom} is the number of measurement points at the bottom of the tablet equal to 9. At the end, c_b represents the ELF magnetic field range number obtained by clustering at the bottom positions (portrait orientation mode).

At the end, we consider both the top and bottom positions into integrated or so-called total *foe* factors. These total factors best reflect the users' exposure to the emitted ELF magnetic field during different ways of working with the tablets. The total factors of exposure *foe* in the landscape tablet orientation are: (i) for watching mode $foe_{total,lw}$, (ii) for browsing mode $foe_{total,lb}$, and (iii) for typing mode $foe_{total,lt}$. They are calculated as follows:

$$foe_{total,lw} = \frac{\sum_{t=4,6,7,9} c_t \cdot cf_1 + \sum_{t=1-3,5,8} c_t \cdot cf_2 + \sum_{b=4,6,7,9} c_b \cdot cf_1 + \sum_{b=1-3,5,8} c_b \cdot cf_2}{n_{total}}, \quad (16.28)$$

$$foe_{total,lb} = \frac{\sum_{t=4,5,7} c_t \cdot cf_1 + \sum_{t=1-3,6,8,9} c_t \cdot cf_2 + \sum_{b=4,7} c_b \cdot cf_1 + \sum_{b=1-3,5,6,8,9} c_b \cdot cf_2}{n_{total}}, \quad (16.29)$$

$$foe_{total,lt} = \frac{\sum_{t=7-9} c_t \cdot cf_1 + \sum_{t=1-6} c_t \cdot cf_2 + \sum_{b=4,6,7,9} c_b \cdot cf_1 + \sum_{b=1-3,5,8} c_b \cdot cf_2}{n_{total}}, \quad (16.30)$$

where cf_1 and cf_2 represent a corrective factor set to 1 and 0.66, respectively. Furthermore, n_{total} is the number of measurement points equal to 18 (9 at the top plus 9 at the bottom). At the end, c_t and c_b represent the ELF magnetic field range number obtained by clustering at the top and bottom positions (landscape orientation mode), respectively.

In a similar manner, the factors of exposure *foe* in the portrait tablet orientation are: (i) for watching mode $foe_{total,pw}$, (ii) for browsing mode $foe_{total,pb}$, and (iii) for typing mode $foe_{total,pt}$. Furthermore, they are calculated as follows:

$$foe_{total,pw} = \frac{\sum_{t=2,8} c_t \cdot cf_1 + \sum_{t=1,3-7,9} c_t \cdot cf_2 + \sum_{b=2,8} c_b \cdot cf_1 + \sum_{b=1,3-7,9} c_b \cdot cf_2}{n_{total}},$$
$$(16.31)$$

$$foe_{total,pb} = \frac{\sum_{t=5,8} c_t cf_1 + \sum_{t=1-4,6,7,9} c_t cf_2 + \sum_{b=5,8} c_b cf_1 + \sum_{b=1-4,6,7,9} c_b cf_2}{n_{total}}, \quad (16.32)$$

$$foe_{total,pt} = \frac{\sum_{t=3,6,9} c_t cf_1 + \sum_{t=1,2,4,5,7,8} c_t cf_2 + \sum_{b=2,3,8,9} c_b cf_1 + \sum_{b=1,4-7} c_b cf_2}{n_{total}},$$
$$(16.33)$$

where cf_1 and cf_2 represent a corrective factor set to 1 and 0.66, respectively. Furthermore, n_{total} is the number of measurement points equal to 18 (9 at the top plus 9 at the bottom). At the end, c_t and c_b represent the ELF magnetic field range number obtained by clustering at the top and bottom positions (portrait orientation mode), respectively.

At the end, the factor of exposure *foe* is compared with the reference safety level foe_{safe} determined by the classification level and taking into account the reference level emission given by the TCO standard, i.e. 0.2 μT. As a consequence, the tablets can be classified in two groups: (i) safe tablet for use, and (ii) not fully safe tablets for use. The division is established according to the level of the factor of exposure *foe*. If the *foe* value is below or nearby the *foe_safe*, then the tablet can be treated as a safe one for use. In the contrast, if the *foe* value is higher than the *foe_safe*, then the tablet needs additional measures to be safely used.

16.5 Results and Discussion

Next, we present the results derived from the measurement of the ELF magnetic field at the typical measurement positions of the top and bottom parts of the tablets. Then, we apply the clustering algorithms on the obtained results for detecting the ELF magnetic field ranges on the top and bottom tablet surface. Hence, we discuss and compare the ranges obtained by the seven clustering algorithms for finding the best results. At the end, we use the best ELF magnetic field ranges for computing the factor of exposure and evaluate the obtained results.

16.5.1 Measurement Results

The statistical processing of the measured values has been performed in MS Excel version 14.0.0. Tables 16.1 and 16.2 report the statistical measures of minimum, maximum, average and median of the ELF magnetic field measured respectively at the top (T1...T9) and bottom positions (B1...B9) of the 9 tablets.

A first important observation is that the tablet emits peaks of ELF magnetic field at the different top and bottom positions which are much higher than the safety reference limit of 0.2 μT. It is visible for the top part of the tablet at positions T1-T9, and for the bottom part of the tablet at positions B1-B9. It can be also noticed by observing the average values, which are above 0.2 μT at positions T2–T5 for the top part of the tablet, and at positions B1–B4 and B8–B9 for the bottom part of the tablet. Again, it is visible from the median values, which are above 0.2 μT at position T4, T5 and T7 for the top part of the tablet and at positions B1, B2, B4, B8 and B9 for the bottom part of the tablet.

Specifically, we can observe that the absolute maximum for the top part of the tablet is obtained at position T3, with a value of 0.8629 μT. On the contrary, the absolute maximum is reached for the bottom part of the tablet at position B3, with a value of 0.9571 μT. Also, the absolute minimum for the top part of the tablet has a value of 0.0100 μT at position T3, while for the bottom part of the tablet it has a value of 0.0141 μT at position B6. Furthermore, if we observe the average values for the top and bottom parts of the tablet, we can notice that the bottom part emits some higher peaks which are not obtained by the top part. They are at positions B2–B4, B8, and B9, with values respectively of 0.2310 μT, 0.2341 μT, 0.2575 μT, 0.2926

Table 16.1 Statistical measures of the emitted ELF magnetic field at the top part of the tablets

Measurement pos.	Minimum	Maximum	Average	Median
T1	0.0574	0.5167	0.1959	0.1513
T2	0.0812	0.4329	0.2150	0.1980
T3	0.0100	0.8629	0.2217	0.1300
T4	0.0224	0.4456	0.2254	0.2557
T5	0.0990	0.2920	0.2121	0.2202
T6	0.0224	0.2980	0.1499	0.1304
T7	0.0510	0.3145	0.1864	0.2070
T8	0.0592	0.3020	0.1698	0.1364
T9	0.0412	0.3020	0.1285	0.0700

Table 16.2 Statistical measures of the emitted ELF magnetic field at the bottom part of the tablets

Measurement pos.	Minimum	Maximum	Average	Median
B1	0.0616	0.3501	0.2017	0.2130
B2	0.0245	0.4949	0.2310	0.2090
B3	0.0173	0.9571	0.2341	0.1288
B4	0.0490	0.5101	0.2575	0.3050
B5	0.0574	0.3120	0.1751	0.2000
B6	0.0141	0.6030	0.1654	0.0735
B7	0.0510	0.3080	0.1471	0.1145
B8	0.0640	0.8592	0.2926	0.2229
B9	0.0640	0.9327	0.2774	0.2080

μT and 0.2774 μT. It is also confirmed by the median values for the bottom part of the tablet, with a higher peak of 0.3050 μT at position B4. From this analysis, we can conclude that the tablet emits a higher ELF magnetic field at its bottom part than at its top part.

This trend is more clearly visible in Fig. 16.14, illustrating the same computed statistical measures at the top positions of the tablet, T1...T9, (see Fig. 16.14a) and at the bottom positions of the tablet, B1...B9 (see Fig. 16.14b). It is worth noting that the highest peaks of maximum ELF magnetic field are reached at the bottom positions B3, B8 and B9, outperforming 0.80 μT. On the contrary, at the top part of the tablet, the measured ELF magnetic field only outperforms 0.80 μT at position T3. Furthermore, at the bottom part of the tablet the average value overcomes 0.2 μT at positions B2–B4, and B8–B9. On the contrary, at the top part of the tablet the average value overcomes 0.2 μT at positions T2–T4. Again, the median value is above 0.2 μT at positions T4 and T5 at the top part of the tablet, and at positions B1–B2, B4

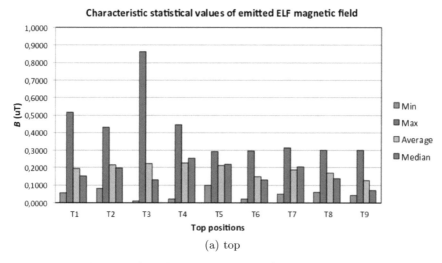

(a) top

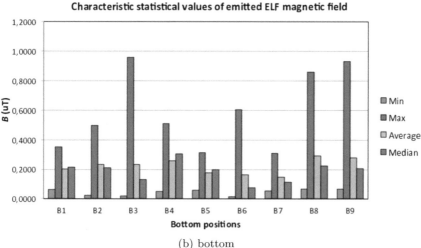

(b) bottom

Fig. 16.14 Statistical measures of the measured emitted ELF magnetic field: **a** at the top measurement positions of the tablet (T1...T9), **b** at the bottom measurement positions of the tablet (B1...B9)

and B8–B9 at the bottom part of the tablet. It confirms that the bottom part of the tablet emits a stronger ELF magnetic field in terms of the statistical measures.

The result of the statistical analysis derives from the punctual values of the measured ELF magnetic field at the different tablet positions (top and bottom), for each analyzed tablet. For this reason, in order to deepen the analysis, we show in Fig. 16.15 the distribution of the punctual measured values for each tablet at the top positions, T1...T9, (see Fig. 16.15a) and at the bottom positions, B1...B9, (see Fig. 16.15b). We

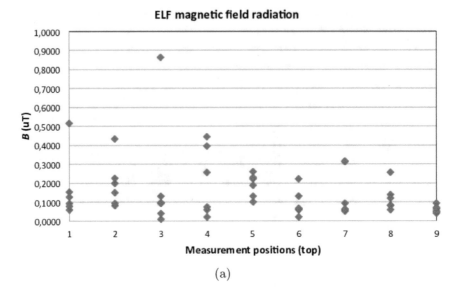

(a)

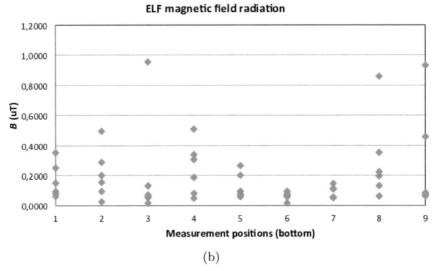

(b)

Fig. 16.15 Distribution of the measured ELF magnetic field: **a** at the top measurement positions of the tablet (1...9 represents T1...T9), **b** at the bottom measurement positions of the tablet (1...9 represents B1...B9)

can observe that more than 1/3 of the measurement positions at the top part of the tablet emit high values above the safety reference limit of 0.2 μT (see Fig. 16.15a). In particular, the distribution of the measured ELF magnetic field values is mostly concentrated up to 0.30 μT at positions T2 and T5, up to 0.2 μT at positions T1 and

T8, up to 0.10 µT at positions T3, T4, T6, T7 and T9. The highest peaks of ELF magnetic field are in correspondence of positions T1, T2, T3 and T4, determining the highest maximum values in Table 16.1. Also, these high peaks increase the average values, especially at positions T2, T3 and T4. In correspondence of positions T4 and T5, around a half of the values overcome 0.2 µT. For this reason, the positions T4 and T5 exhibit the highest median values of 0.2557 and 0.2202 µT. In Fig. 16.15b (bottom positions) we can notice that more data are distributed along higher ELF magnetic field values w.r.t. Figure 16.15a (top positions). It is especially visible at positions B2, B4, B8 and B9. Furthermore, we can observe that most of the measured values are concentrated up to 0.40 µT at positions B1, B4 and B8, up to 0.30 µT at positions B2 and B5, up to 0.2 µT at positions B3 and B7, and up to 0.10 µT at positions B6 and B9. The highest peaks of the measured ELF magnetic field are obtained in correspondence of the positions B3, B8 and B9, influencing the average values to be above 0.2 µT (see Table 16.2). Again, at position B4, the measured values are mostly distributed between 0.10 and 0.40 µT, with a high peak just above 0.50 µT. Because it is the most balanced distribution around 0.25 µT, it determines the highest median value of 0.3050 µT (see Table 16.2). Again, it confirms that the bottom positions have a stronger emission than the top positions of the tablet.

16.5.2 Clustering Results

Clustering analysis has been performed in Matlab R2017a. Each clustering algorithm has been separately applied on the *top* and *bottom* datasets, for obtaining 5 ranges of top and bottom measurement positions, which was found as a suitable number for laptop analysis [9]. Each range is represented in terms of minimum and maximum ELF magnetic field values of the measurement positions included in that range. Clustering prototypes are not informative in this context. Also, top and bottom ranges are sorted in increasing order (from the lowest to the highest) based on their minimum and maximum values, and differently coloured and numbered, from 1 to 5. Then, dangerousness maps are built on the 9 top and bottom tablet representations involved in the clustering. They show the measurement positions at the top and bottom parts of the tablets emitting ELF magnetic field values in a given top or bottom range. According to the minimum and maximum ELF magnetic field values in each range, the following ranges are obtained:

1. Very high,
2. High,
3. Middle,
4. Low, and
5. Very low.

For each clustering algorithm, the obtained ranges, which are the clusters of the *top* and *bottom* datasets, are considered as dangerousness classes for the top and bottom parts of the tablet.

A trial and error procedure has been employed for determining the parameters' value of each clustering algorithm. Then, the parameters combination determining the best solution has been selected for each algorithm.

Consequently, K-Means and K-Medians algorithms have been run 10 times on each dataset (*top* and *bottom*). Then, the clustering result determining the minimum value of the optimization function has been selected as the final clustering solution. For each run, the centroids have been initialized by using the most efficient approach which randomly selects K data points from the dataset as initial centroids. Also, the K input parameter of K-Means and K-Medians is set to 5, in order to obtain cluster classes of positions associated with the 5 ELF magnetic field ranges.

SOM builds a neural network with $K = 5$ output neurons, for determining 5 ELF magnetic field ranges. The epoch number is equal to 200 and the dimension of a neuron layer is 1×5. The number of training steps for initial covering of the input space is 100 and the initial size of the neighbourhood is 3. The distance between two neurons is calculated as the number of steps separating one from the other. A preliminary test has been performed for varying the number of the output neurons from 2 to 10. From the obtained results, we observed that a number between 2 and 4 neurons is too low, resulting in underfitting of the training data, with the consequence that the model classifies high values as middle range. On the other hand, a number of neurons above 6 resulted in overfitting of the training data, with the consequence that the model is not able to well-generalize when it processes new samples. At the end, we realized that a number of neurons between 5 and 6 was suitable for this task. Consequently, we selected 5 output neurons in order to compare the classification results with those of the other algorithms.

In the DBSCAN, the *Eps* distance value is set to 0.02 and the minimum number of points (*MinPts*) for the *Eps*-neighbourhood is set to 1.5.

In EM, the number of the clusters is set to 5, too.

Hierarchical clustering uses a bottom-up agglomerative method based on average linkage with Euclidean distance. The obtained dendrogram is horizontally cut in order to have 5 clusters.

Finally, in the Fuzzy-C-Means algorithm, the number of the clusters is set to 5 and each value is assigned to the cluster with the highest membership probability, in order to obtain a hard partitioning of the values.

16.5.2.1 K-Means Results

Fig. 16.16 shows the ELF magnetic field ranges detected by K-Means algorithm (see Fig. 16.16a) and the corresponding top and bottom dangerousness maps for the 9 tablets (see Fig. 16.16b).

We can observe that the top ranges exhibit a minimum difference with their corresponding bottom ranges. It is especially visible for very low, middle and high ranges. Hence, the obtained ranges do not capture the real condition of the ELF magnetic field values, which are visibly higher at the bottom part than at the top part of the tablet. Furthermore, we can notice that the very high top range is composed of a

	Top ranges	min.	max.	Bottom ranges	min.	max.
1	very low	0,0100	0,0949	very low	0,0141	0,0964
2	low	0,0990	0,1513	low	0,1068	0,1530
3	middle	0,1879	0,2640	middle	0,1841	0,2627
4	high	0,2760	0,5167	high	0,2867	0,6030
5	very high	0,8629	0,8629	very high	0,8592	0,9571

(a)

Top

Tablet 1

4	1	1
4	3	1
4	2	1

Tablet 2

2	1	1
1	2	1
1	1	1

Tablet 3

2	4	1
3	3	2
1	3	1

Tablet 4

1	2	2
1	3	3
1	2	1

Tablet 5

1	3	5
1	2	1
1	1	1

Tablet 6

1	3	1
4	3	1
4	1	1

Tablet 7

4	4	4
4	4	4
4	4	4

Tablet 8

3	3	3
3	3	3
3	3	3

Tablet 9

3	3	4
4	4	4
4	4	4

Bottom

Tablet 1

2	1	1
4	1	1
2	3	5

Tablet 2

1	2	1
3	1	1
1	3	1

Tablet 3

3	4	2
4	3	1
2	5	1

Tablet 4

4	3	1
4	3	1
1	4	4

Tablet 5

1	4	5
1	1	1
1	2	1

Tablet 6

1	1	1
1	1	1
2	1	1

Tablet 7

4	4	4
4	4	4
4	4	4

Tablet 8

3	3	3
3	3	3
3	3	3

Tablet 9

4	4	4
4	4	4
4	4	4

(b)

Fig. 16.16 ELF magnetic field ranges (cluster classes) obtained by K-Means algorithm (**a**), and corresponding top and bottom dangerousness maps (**b**)

single value which is $0.8629\ \mu T$, while the other high values are included in the high range. In any case, for the top and bottom parts, we can notice that middle, high and very high ranges are associated with values above the reference limit of $0.2\ \mu T$. The middle range is borderline, because it also includes values which are just below the reference limit. Finally, the very low and low ranges are safe.

Comparing Fig. 16.16b with Fig. 16.4, it is worth noting that middle, high and very high ranges are mostly emitted at the positions where the CPU is located for the top part, and the battery is located for the bottom part. However, some positions emitting a middle range correspond to other inner components integrated on the motherboard (see Tablet 3 top) or to the RAM (see Tablet 6 top). Also, we can observe that the very high range is only emitted in correspondence of the highest peak of $0.8629\ \mu T$ at the top part (see Tablet 5) where the CPU is located and in correspondence of the highest peaks at the bottom part (see Tablets 1, 3, and 5) where the battery and the CPU are located. All the other positions of the CPU and battery areas emit high and middle ranges, although some of them are particularly dangerous (high emission values).

16.5.2.2 K-Medians Results

Figure 16.17 illustrates the ELF magnetic field ranges detected by K-Medians algorithm (see Fig. 16.17a) and the corresponding top and bottom dangerousness maps for the 9 tablets (see Fig. 16.17b).

We can notice that any range is composed of a single value. Also, the bottom ranges are much higher than their corresponding top ranges. Hence, the difference between the ELF magnetic field values measured at top and bottom parts is visible at range level, too. In fact, K-Medians distributes the ELF magnetic field values in two top ranges below 0.2 μT, which are very low and low, in two top ranges above 0.2 μT, which are high and very high, and in one borderline top range, which is middle, including values below and above the reference limit. For the bottom part, the algorithm finds one safe range, which is very low, three dangerous ranges, which are middle, high and very high, and one borderline range, which is low. It contributes to capture the difference between the top and bottom parts' emission.

Observing Figs. 16.4 and 16.17b, we can notice that K-Medians algorithm is accurate in identifying the positions above the reference limit of 0.2 μT, which are classified as middle, high and very high according to their value. In particular, we can observe that a very high range is always emitted at the CPU area (see Tablets 1, 3, 5 and 6 top and Tablets 3–5 bottom) and at the battery area (see Tablets 1, 3, 4,

	Top ranges	min.	max.	Bottom ranges	min.	max.
1	very low	0,0100	0,0735	very low	0,0141	0,1304
2	low	0,0768	0,1364	low	0,1432	0,2229
3	middle	0,1503	0,2291	middle	0,2506	0,2970
4	high	0,2490	0,3145	high	0,3030	0,3501
5	very high	0,3957	0,8629	very high	0,4601	0,9571

(a)

Top

Tablet 1			Tablet 2			Tablet 3		
5	2	1	3	2	1	2	5	2
5	3	1	1	2	1	4	4	2
4	2	2	1	2	1	1	4	1

Tablet 4			Tablet 5			Tablet 6		
2	3	2	1	3	5	2	3	2
1	3	3	1	2	1	5	3	1
2	2	1	1	2	1	4	1	1

Tablet 7			Tablet 8			Tablet 9		
4	4	4	3	3	3	4	4	4
4	4	4	3	3	3	4	4	4
4	4	4	3	3	3	4	4	4

Bottom

Tablet 1			Tablet 2			Tablet 3		
2	1	1	1	2	1	3	5	1
4	1	1	2	1	1	4	3	1
2	2	5	1	2	1	1	5	1

Tablet 4			Tablet 5			Tablet 6		
4	2	1	1	3	5	1	1	1
5	2	1	1	1	1	1	1	1
1	4	5	1	1	1	1	1	1

Tablet 7			Tablet 8			Tablet 9		
4	4	3	2	2	2	4	4	4
4	4	3	2	2	2	4	3	5
3	3	3	2	2	2	4	4	4

(b)

Fig. 16.17 ELF magnetic field ranges (cluster classes) obtained by K-Medians algorithm (**a**), and corresponding top and bottom dangerousness maps (**b**)

and 9 bottom). On the contrary, the RAM emits a middle range, which is correctly associated with positions whose value is just above 0.2 μT (see Tablet 6 top).

16.5.2.3 SOM Results

The clustering results obtained by the SOM algorithm are reported in Fig. 16.18 depicting the ELF magnetic field ranges (see Fig. 16.18a) and the corresponding top and bottom dangerousness maps for the 9 tablets (see Fig. 16.18b).

We can observe that SOM algorithm determines well-spaced top ranges from 1 to 4 (from very low to high, with a min-max difference between 0.06 and 0.24, see Fig. 16.18a). However, it has the problem to reduce the very high range to a single value which is 0.8629 μT. This is not the case for the bottom ranges. Also, it is visible that the difference between top and bottom emission of the tablet is not captured by the ranges. In fact, the top ranges are quite similar in minimum and maximum values to their corresponding bottom ranges. In any case, the algorithm finds two top ranges below the reference limit of 0.2 μT, which are very low and low, two top ranges above 0.2 μT, which are high and very high, and a middle top range which is slightly borderline. The same is for the bottom ranges.

	Top ranges	min.	max.	Bottom ranges	min.	max.
1	very low	0,0100	0,0735	very low	0,0141	0,1304
2	low	0,0768	0,1364	low	0,1432	0,2229
3	middle	0,1503	0,2291	middle	0,2506	0,2970
4	high	0,2490	0,3145	high	0,3030	0,3501
5	very high	0,3957	0,8629	very high	0,4601	0,9571

(a)

Top

Tablet 1
5	2	1
5	3	1
4	2	2

Tablet 2
3	2	1
1	2	1
1	2	1

Tablet 3
2	5	2
4	4	2
1	4	1

Tablet 4
2	3	2
1	3	3
2	2	1

Tablet 5
1	3	5
1	2	1
1	2	1

Tablet 6
2	3	2
5	3	1
4	1	1

Tablet 7
4	4	4
4	4	4
4	4	4

Tablet 8
3	3	3
3	3	3
3	3	3

Tablet 9
4	4	4
4	4	4
4	4	4

Bottom

Tablet 1
2	1	1
4	1	1
2	2	5

Tablet 2
1	2	1
2	1	1
1	2	1

Tablet 3
3	5	1
4	3	1
1	5	1

Tablet 4
4	2	1
5	2	1
1	4	5

Tablet 5
1	3	5
1	1	1
1	1	1

Tablet 6
1	1	1
1	1	1
1	1	1

Tablet 7
4	4	3
4	4	3
3	3	3

Tablet 8
2	2	2
2	2	2
2	2	2

Tablet 9
4	4	4
4	3	5
4	4	4

(b)

Fig. 16.18 ELF magnetic field ranges (cluster classes) obtained by SOM algorithm (**a**), and corresponding top and bottom dangerousness maps (**b**)

The dangerousness maps in Fig. 16.18b and the tablet emission in Fig. 16.4 show that a single top position corresponding to the CPU area emits a very high range (see Tablet 5 top). On the contrary, the other positions where the CPU is located are classified as middle and high. Also, the RAM area emits a middle range (see Tablet 6 top). At the bottom part, the highest range is emitted by the CPU and the battery areas (see Tablets 1, 3 and 5 bottom). However, most of the positions where the CPU and the battery are located are classified as middle and high.

16.5.2.4   DBSCAN Results

The clustering results of the DBSCAN algorithm are reported in Fig. 16.19 which shows the ELF magnetic field ranges obtained by the algorithm (see Fig. 16.19a) and the corresponding top and bottom dangerousness maps for the 9 tablets (see Fig. 16.19b).

We can observe that the top ranges are lower than their corresponding bottom ranges. Hence, the DBSCAN preserves the characteristic that the emission at the top part is lower than the emission at the bottom part that we observed in the statistical analysis. However, at the top part most of the dangerous values above the reference limit of 0.2 μT are included in the middle and very high ranges, while the high range

	Top ranges	min.	max.	Bottom ranges	min.	max.
1	very low	0,0100	0,0990	very low	0,0141	0,2506
2	low	0,1204	0,1513	low	0,2627	0,3501
3	middle	0,1879	0,3145	middle	0,4601	0,5101
4	high	0,4329	0,4456	high	0,6030	0,8592
5	very high	0,3957	0,8629	very high	0,9327	0,9571

(a)

(b)

Fig. 16.19 ELF magnetic field ranges (cluster classes) obtained by DBSCAN algorithm **a**, and corresponding top and bottom dangerousness maps **b**

is very strict (with a min-max difference of 0.01, see Fig. 16.19a). Another problem is that the high and very high ranges are overlapped, in particular the high range is a sub-range of the very high range. At the bottom part, we can observe that the very high range is very strict (with a min-max difference of 0.02, see Fig. 16.19a). Also, there is only one safe range, the very low range, which is borderline since it includes values below and above 0.2 μT.

This condition is visible in the dangerousness maps of the top part in Fig. 16.19b. We can observe that most of the positions with a high emission value are classified as middle, while a few of them are classified as high and very high. In particular, the highest peaks corresponding to the CPU area emit high and very high ranges (see Tablets 1, 3, 5, and 6 and Fig. 16.4). However, some other high peaks at the CPU area are classified as middle (see Tablet 1 and Fig. 16.4). All the other inner components, including the RAM, are classified as middle (see Tablet 6). At the bottom part, we can observe that the only positions with the highest peaks corresponding to the CPU and battery areas emit high and very high ranges (see Tablets 1, 3, 5, and 9 and Fig. 16.4). All the other dangerous positions above 0.2 μT are classified as low and middle (see Tablets 3, 4, and 7 and Fig. 16.4).

16.5.2.5 EM Results

Figure 16.20 shows the clustering results obtained by EM algorithm in terms of the ELF magnetic field ranges (see Fig. 16.20a) and the corresponding top and bottom dangerousness maps (see Fig. 16.20b) for the 9 tablets.

At the top part, it is visible that the algorithm detects only one safe range (very low), three dangerous ranges above the safety reference limit of 0.2 μT (middle, high and very high), and one borderline range (low). A similar situation is at the bottom part, where the middle and high ranges are overlapped, in particular the middle range is a sub-range of the high range. Also, we can observe that the minimum and maximum values of the bottom ranges are quite similar to the values of the corresponding top ranges. Hence, the algorithm does not capture the difference between the top and bottom emission.

From the top dangerousness maps and Fig. 16.4, we can notice that most of the CPU positions emit high and very high ranges (see Tablets 1, 3, 5, and 6). On the contrary, the RAM emits a low range (see Tablet 6). About the other inner components, they do not have a clear classification, because they are sometimes classified as high (see Tablets 7 and 9), sometimes as middle (see Tablet 3), and sometimes as low (see Tablet 8). At the bottom part, high and very high ranges are associated with the CPU and the battery (see Tablets 1, 3, 4, and 5). However, because the middle and high ranges are overlapped, some positions which are classified as middle could be classified as high (e.g. in Tablet 7).

	Top ranges	min.	max.	Bottom ranges	min.	max.
1	very low	0,0100	0,1513	very low	0,0141	0,1530
2	low	0,1879	0,2291	low	0,1841	0,2229
3	middle	0,2490	0,2640	middle	0,2910	0,3160
4	high	0,2760	0,3145	high	0,2506	0,3501
5	very high	0,3957	0,8629	very high	0,4601	0,9571

(a)

Top   **Bottom**

Tablet 1

5	1	1
5	2	1
4	1	1

Tablet 2

1	1	1
1	1	1
1	1	1

Tablet 3

1	5	1
3	3	1
1	3	1

Tablet 1

1	1	1
4	1	1
1	2	5

Tablet 2

1	1	1
2	1	1
1	2	1

Tablet 3

4	5	1
3	4	1
1	5	1

Tablet 4

1	1	1
1	2	2
1	1	1

Tablet 5

1	2	5
1	1	1
1	1	1

Tablet 6

1	2	1
5	2	1
4	1	1

Tablet 4

4	2	1
5	2	1
1	4	5

Tablet 5

1	4	5
1	1	1
1	1	1

Tablet 6

1	1	1
1	1	1
1	1	1

Tablet 7

4	4	4
4	4	4
4	4	4

Tablet 8

2	2	2
2	2	2
2	2	2

Tablet 9

3	3	4
4	4	4
4	4	4

Tablet 7

3	3	3
4	3	3
3	3	3

Tablet 8

2	2	2
2	2	2
2	2	2

Tablet 9

3	3	3
3	3	3
3	3	3

(b)

Fig. 16.20 ELF magnetic field ranges (cluster classes) obtained by EM algorithm **a**, and corresponding top and bottom dangerousness maps **b**

16.5.2.6   Hierarchical Clustering Results

Figure 16.21a reports the ELF magnetic field ranges found by the agglomerative hierarchical clustering, and Fig. 16.21b shows the corresponding top and bottom dangerousness maps for the 9 tablets.

It is worth noting that the top ranges exhibit a minimum and maximum value which is similar to the values of the corresponding bottom ranges. The only two exceptions are middle and high ranges, which are higher at the bottom part. Hence, the difference between the top and bottom emission is partially visible at range level. At the top part, the algorithm detects one safe range below 0.2 μT (very low), three ranges above 0.2 μT (middle, high and very high), and a range which is borderline (low), because it also includes some values below 0.2 μT. However, the high and very high ranges are composed of a single value, which is respectively 0.5167 μT and 0.8629 μT. At the bottom part, we find a similar condition, with the difference that only the high range is characterized by a single value, which is 0.6030 μT.

Observing the dangerousness maps and Fig. 16.4, we can notice that at the top part only two high peaks corresponding to a part of the CPU area are classified as high ad very high (see Tablets 1 and 5). On the other hand, the middle range is emitted by a part of the CPU area (see Tablets 3 and 6). All the other tablet positions which are associated with values above 0.2 μT are classified as low and very low, including the

		Top ranges	min.	max.	Bottom ranges	min.	max.
	1	very low	0,0100	0,1513	very low	0,0141	0,1145
	2	low	0,1879	0,3145	low	0,1288	0,3501
	3	middle	0,3957	0,4456	middle	0,4601	0,5101
	4	high	0,5167	0,5167	high	0,6030	0,6030
	5	very high	0,8629	0,8629	very high	0,8592	0,9571

(a)

Top   **Bottom**

Tablet 1   Tablet 2   Tablet 3   Tablet 1   Tablet 2   Tablet 3

4	1	1		1	1	1		1	3	1		2	1	1		1	2	1		2	3	2
3	2	1		1	1	1		2	2	1		2	1	1		2	1	1		2	2	1
2	1	1		1	1	1		1	2	1		2	2	5		1	2	1		1	5	1

Tablet 4   Tablet 5   Tablet 6   Tablet 4   Tablet 5   Tablet 6

1	1	1		1	2	5		1	2	1		2	2	1		1	2	5		1	1	1
1	2	2		1	1	1		3	2	1		3	2	1		1	1	1		1	1	1
1	1	1		1	1	1		2	1	1		1	2	3		1	2	1		1	1	1

Tablet 7   Tablet 8   Tablet 9   Tablet 7   Tablet 8   Tablet 9

2	2	2		2	2	2		2	2	2		2	2	2		2	2	2		2	2	2
2	2	2		2	2	2		2	2	2		2	2	2		2	2	2		2	2	4
2	2	2		2	2	2		2	2	2		2	2	2		2	2	2		2	2	2

(b)

Fig. 16.21 ELF magnetic field ranges (cluster classes) obtained by agglomerative hierarchical clustering algorithm (**a**), and corresponding top and bottom dangerousness maps (**b**)

RAM position (see Tablet 6). At the bottom part, only a few positions corresponding to a part of the CPU and battery area are classified as very high and high (see Tablets 1, 3, 5, and 9). Other few positions of CPU and battery with high values are classified as middle. Finally, all the other positions with values above 0.2 μT are classified as very low and low.

16.5.2.7   Fuzzy-C-Means Results

The clustering results detected by Fuzzy-C-Means together with the corresponding top and bottom dangerousness maps for the 9 tablets are depicted in Fig. 16.22.

We can observe that the top and bottom ranges do not have specific problems (see Fig. 16.22a). At the top part, two ranges include the values below the reference limit of 0.2 μT (very low and low), two ranges include the values above 0.2 μT (high and very high), and the middle range is borderline, including a few values slightly below 0.2 μT. At the bottom part, the very low range is the only below 0.2 μT. On the contrary, middle, high and very high ranges include values above 0.2 μT. Finally, the low range is borderline. With the only exception of the very low range, the other ranges show a difference between the top and bottom part. In particular, the middle, high and very high top ranges have minimum and maximum values which are lower

	Top ranges	min.	max.	Bottom ranges	min.	max.
1	very low	0,0100	0,0911	very low	0,0141	0,1304
2	low	0,0927	0,1513	low	0,1432	0,2506
3	middle	0,1879	0,2490	middle	0,2627	0,3501
4	high	0,2550	0,3957	high	0,4601	0,6030
5	very high	0,4329	0,8629	very high	0,8592	0,9571

(a)

Top

Tablet 1			Tablet 2			Tablet 3		
5	1	1	2	2	1	2	5	2
4	3	1	1	2	1	4	4	2
4	2	1	1	1	1	1	4	1

Tablet 4			Tablet 5			Tablet 6		
1	2	2	1	3	5	1	3	2
1	3	3	1	2	1	5	3	1
1	2	1	1	1	1	4	1	1

Tablet 7			Tablet 8			Tablet 9		
4	4	4	3	3	3	3	4	4
4	4	4	3	3	3	4	4	4
4	4	4	3	3	3	4	4	4

Bottom

Tablet 1			Tablet 2			Tablet 3		
2	1	1	1	2	1	2	4	1
3	1	1	2	1	1	3	3	1
2	2	5	1	2	1	1	5	1

Tablet 4			Tablet 5			Tablet 6		
3	2	1	1	3	5	1	1	1
4	2	1	1	1	1	1	1	1
1	3	4	1	1	1	1	1	1

Tablet 7			Tablet 8			Tablet 9		
3	3	3	2	2	2	3	3	3
3	3	3	2	2	2	3	3	4
3	3	3	2	2	2	3	3	3

(b)

Fig. 16.22 ELF magnetic field ranges (cluster classes) obtained by Fuzzy-C-Means algorithm (**a**), and corresponding top and bottom dangerousness maps (**b**)

than the values of the corresponding bottom ranges. In any case, there are no ranges characterized by single values.

The dangerousness maps compared with Fig. 16.4 confirm the clustering results (see Fig. 16.22b). At the top part, we can observe that the highest peaks in the CPU area are classified as very high range (see Tablets 1, 3, 5 and 6). A high emission level is associated with the CPU area and the positions where other inner components are located (see Tablets 1, 3, 6, 7, and 9). The RAM emits a middle range (see Tablet 6). Finally, the less dangerous positions below 0.2 μT emit low and very low ranges. At the bottom part, the highest peaks in the battery and CPU area are classified as very high (see Tablets 1, 3 and 5). Some other positions in the CPU and battery area emit high and middle ranges (see Tablets 3, 4, 7, and 9). All the other components emit a middle-low range (see Tablets 7 and 8).

16.5.2.8 Comparison Among the Methods

From all aforementioned, we can find the most suitable clustering methods for detecting the ELF magnetic field ranges. In particular, the specific problems connected with each clustering result are marked in grey in Figs. 16.16a, 16.17, 16.18, 16.19, 16.20, 16.21 and 16.22a. We observed that K-Means, SOM and agglomerative hierarchical clustering algorithms determine ranges which may be composed of single values. Also, they do not capture or partially capture the difference between top and bottom emission, which is visible from the statistical analysis of the ELF magnetic field values (see Figs. 16.16, 16.18 and 16.21a in grey). Consequently, they are not reliable candidates for finding the ELF magnetic field ranges. On the other hand, the DBSCAN algorithm may detect ranges which are very strict or very large (see Fig. 16.20a in grey), with the consequence that it can overestimate or underestimate the dangerousness of the tablet positions. Also, both DBSCAN and EM algorithms determine ranges which may be overlapped or included one into another (see Figs. 16.19 and 16.20a in grey). The two final candidate algorithms, which are K-Medians and Fuzzy-C-Means, overcome the other candidates in terms of detected ELF magnetic field ranges (see Figs. 16.17 and 16.22a). In fact, they do not find overlapped ranges, or ranges composed of single values. Between the two algorithms, the K-Medians better captures the difference between the top and bottom emission, because all top ranges from low to very high are lower than the corresponding bottom ranges. Accordingly, the K-Medians algorithm is the best candidate to be used for detecting the ELF magnetic field ranges emitted by the tablets. It is also visible from the top and bottom dangerousness maps in Fig. 16.17b, where all high emission areas corresponding to the CPU and battery are correctly classified by the algorithm as dangerous.

In order to confirm the obtained result, a comparison of the dangerousness maps derived by the different algorithms is depicted in Figs. 16.23 and 16.24. At the top part (see Fig. 16.23), we can observe that K-Medians algorithm differentiates from the other algorithms for Tablet 2 at positions T1 and T8, Tablet 4 at position T2, Tablet 5 at position T8, Tablet 6 at position T1, and Tablet 9 at positions T1–T2. In these positions, K-Medians algorithm obtains a higher range than the other algorithms: from low to middle, from very low to low, from middle to high. Observing the ELF magnetic field values and inner components corresponding to these positions in Fig. 16.4, the K-Medians result is fairer than its competitors. At the bottom part (see Fig. 16.24), we can observe that the K-Medians result is similar to the EM result in some positions (see Tablets 4 and 9). The main difference is that K-Medians is able to better-differentiate the dangerous positions from the safe positions (see Tablet 3 at position B4 corresponding to a high peak of 0.31 μT in Fig. 16.4, or Tablet 1 at position B1 of value 0.15 μT which is classified as middle by K-Medians and low by EM).

In order to prove the importance of using clustering, we compare the results obtained by K-Medians algorithm with the results obtained by two typical discretization algorithms: (i) Equal-Width binning, and (ii) Equal-Frequency binning [11] (see Fig. 16.25). The Equal-Width binning splits the range of the values into K intervals of equal width. On the contrary, the Equal-Frequency binning divides the range of the values into K intervals such that each interval contains approximately the same number of values. We can observe that the Equal-Width binning finds one top range which is not connected to any value and one bottom range including only one value

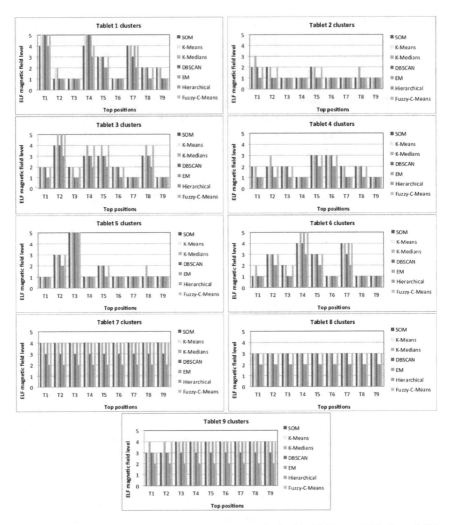

Fig. 16.23 Comparison of the top dangerousness maps obtained by K-Means, K-Medians, SOM, DBSCAN, EM, agglomerative hierarchical clustering and Fuzzy-C-Means for the 9 tablets. For each subfigure corresponding to a tablet, the y axis represents the cluster ID (from 1 to 5), and the x axis represents the top (T1...T9) positions of the tablet

(in grey). In the case of the Equal-Frequency binning, the very high top and bottom range condenses most of the values above the reference limit of 0.2 μT (in grey). Consequently, the different dangerousness levels are flattened. This is mainly because a clustering method is able to follow the natural trend of the data in order to group them into clusters. This is not performed by traditional discretization methods, which may fail in specific contexts like the ELF magnetic field ranges.

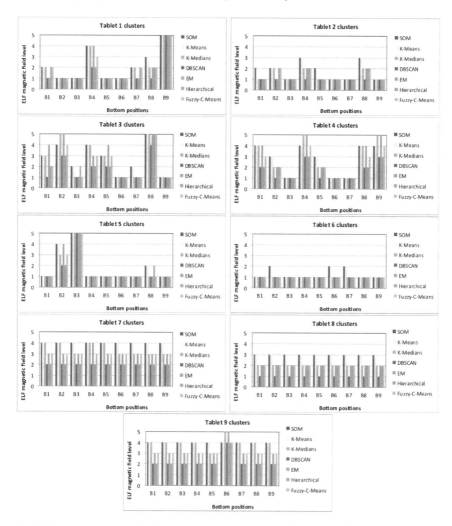

Fig. 16.24 Comparison of the bottom dangerousness maps obtained by K-Means, K-Medians, SOM, DBSCAN, EM, agglomerative hierarchical clustering and Fuzzy-C-Means for the 9 tablets. For each subfigure corresponding to a tablet, the y axis represents the cluster ID (from 1 to 5), and the x axis represents the bottom (B1...B9) positions of the tablet

16.5.3   The foe Results Measurement

The clustering results obtained by K-Medians algorithm are employed for computing the factor of exposure *foe*. In particular, the *foe* results measurement is divided in two parts. In the first part, we separately consider the top and bottom ELF magnetic field exposure. Accordingly, we compute the *foe* factors for the top and bottom emission of the tablets. Finally, in the second part of the experiment, we integrate the top and

Equal-Width binning

	Top ranges	min.	max.	Bin size	Bottom ranges	min.	max.	Bin size
1	very low	0,0100	0,1806	38	very low	0,0141	0,2027	40
2	low	0,1806	0,3512	38	low	0,2027	0,3913	34
3	middle	0,3512	0,5217	4	middle	0,3913	0,5799	3
4	high	0,5217	0,6923	0	high	0,5799	0,7685	1
5	very high	0,6923	0,8629	1	very high	0,7685	0,9571	3

Equal-Frequency binning

	Top ranges	min.	max.	Bin size	Bottom ranges	min.	max.	Bin size
1	very low	0,0100	0,0671	16	very low	0,0141	0,0700	16
2	low	0,0671	0,1300	16	low	0,0700	0,1304	16
3	middle	0,1300	0,2080	17	middle	0,1304	0,2150	17
4	high	0,2080	0,2870	14	high	0,2150	0,3070	14
5	very high	0,2870	0,8629	18	very high	0,3070	0,9571	18

Fig. 16.25 ELF magnetic field ranges (cluster classes) obtained by Equal-Width and Equal-frequency binning methods

Table 16.3 Different values of *foe* factors obtained by measurement in the landscape mode at the top of the tablet

(a)

Tablet	$foe_{top,lw}$	$foe_{top,lb}$	$foe_{top,lt}$
T1	2.2867	2.2867	2.1356
T2	0.8489	0.7733	0.7733
T3	1.1778	1.1778	1.1778
T4	1.6578	1.6200	1.5822
T5	0.8867	0.8111	0.7356
T6	1.3978	1.3978	1.3978
T7	3.2444	3.0933	3.0933
T8	2.0689	1.8422	1.8044
T9	2.4333	2.3200	2.3200

(b)

Landscape			
TOP	$foe_{top,lw}$	$foe_{top,lb}$	$foe_{top,lt}$
T1	middle	middle	middle
T2	very low	very low	very low
T3	low	low	low
T4	low	low	low
T5	very low	very low	very low
T6	low	low	low
T7	high	high	high
T8	middle	low	low
T9	middle	middle	middle

bottom *foe* factors for considering the overall exposure, which is called total *foe*. Next, we denote the 9 tested tablets as T1,...,T9 for making easier the explanation.

From the ELF magnetic field emission at the top of the tablets in the landscape mode, we obtain the *foe* results given in Table 16.3.

The obtained *foe* factor values considering the ELF magnetic field emission in the landscape mode at the top of the tablets are given in Table 16.3a. They are categorized according to the previously established classification. These results are given in Table 16.3b. From the given results, it is worth noting that the tablets T2 and T5 are very safe for a longer use. Nonetheless, the tablets T3, T4, T6 and T8 emit a slightly higher ELF magnetic field than the tablets T2 and T5. Hence, they are also safe for a longer use. In the contrast, the tablets T1, T9 and especially the tablet T7 emit a higher level of ELF magnetic field at the top in the landscape mode. It is worth noting that they should be used with extreme caution over time.

Table 16.4 Different values of *foe* factors obtained by measurement in the portrait mode at the top of the tablet

(a)

Tablet	$foe_{top,pw}$	$foe_{top,pb}$	$foe_{top,pt}$
T1	1.9844	2.0222	1.9844
T2	0.6600	0.6600	0.8489
T3	1.1778	1.1778	1.1400
T4	1.5822	1.5822	1.6200
T5	0.6600	0.6600	0.7733
T6	1.4356	1.3978	1.5111
T7	2.9422	2.9422	3.0933
T8	1.5400	1.5400	1.8800
T9	2.2067	2.2067	2.3200

(b)

Portrait			
TOP	$foe_{top,pw}$	$foe_{top,pb}$	$foe_{top,pt}$
T1	low	middle	low
T2	very low	very low	very low
T3	low	low	low
T4	low	low	low
T5	very low	very low	very low
T6	low	low	low
T7	middle	middle	high
T8	low	low	low
T9	middle	middle	middle

Table 16.4 shows the *foe* factor values considering the ELF magnetic field emission in the portrait mode at the top of the tablets.

Again, the obtained *foe* factor values considering the ELF magnetic field emission in the portrait mode at the top of the tablets (see Table 16.4a) are categorized according to the previously established classification and presented in Table 16.4b. Similarly as in the landscape mode, the tablets T2 and T5 are very safe for a longer use. Furthermore, the tablets T3, T4, T6, T8 and partially T1 emit a slightly higher ELF magnetic field than the tablets T2 and T5. However, they are also safe for a longer use. In the contrast, the tablets T7 and T9 again emit a higher level of ELF magnetic field at the top in the portrait mode. Hence, they should be used with caution over time. If we compare the emission of the ELF magnetic field between landscape and portrait mode at the top of the tablets, it is quite clear that the use of the tablets in the portrait mode is slightly safer for a longer use.

Table 16.5 shows the *foe* factor values considering the ELF magnetic field emission in the landscape mode at the bottom of the tablets. The *foe* factor values considering the ELF magnetic field emission in the landscape mode at the bottom of the tablets (see Table 16.5a) are categorized according to the previously established classification and presented in Table 16.5b. Currently, the tablets T3 and T5 are pretty safe for a longer use. Then, the tablets T1, T2, T6, T8 and T9, although emit a slightly higher ELF magnetic field than the tablets T3 and T5, are also safe for a longer use. In the contrast, the tablets T4 and T7 again emit a higher level of ELF magnetic field at the bottom in the landscape mode. Hence, they should be used with caution over time.

Table 16.6 shows the *foe* factor values considering the ELF magnetic field emission in the portrait mode at the bottom of the tablets.

The *foe* factor values considering the ELF magnetic field emission in the portrait mode at the bottom of the tablets are given in Table 16.6a, while their categorization according to the previously established classification is presented in Table 16.6b. Again, the tablets T2 and T5 emit the lowest level of ELF magnetic field. Hence,

Table 16.5 Different values of *foe* factors obtained by measurement in the landscape mode at the bottom of the tablet

(a)

Tablet	$foe_{bott,lw}$	$foe_{bott,lb}$	$foe_{bott,lt}$
T1	1.8467	1.6200	1.8467
T2	1.1467	1.0333	1.1467
T3	1.0689	0.9933	1.0689
T4	2.2867	2.0600	2.2867
T5	1.0356	0.9600	1.0356
T6	1.2511	1.1756	1.2511
T7	2.7644	2.5378	2.7644
T8	1.4778	1.3267	1.4778
T9	1.6222	1.4711	1.6222

(b)

Landscape			
BOTTOM	$foe_{bottom,lw}$	$foe_{bottom,lb}$	$foe_{bottom,lt}$
T1	low	low	low
T2	low	low	low
T3	low	very low	low
T4	middle	middle	middle
T5	low	very low	low
T6	low	low	low
T7	middle	middle	middle
T8	low	low	low
T9	low	low	low

Table 16.6 Different values of *foe* factors obtained by measurement in the portrait mode at the bottom of the tablet

(a)

Tablet	$foe_{bott,pw}$	$foe_{bott,pb}$	$foe_{bott,pt}$
T1	1.5067	1.5067	1.4444
T2	0.8067	0.8067	0.8333
T3	1.0311	0.9933	0.8333
T4	2.0600	2.0600	1.7778
T5	0.7333	0.7333	0.7222
T6	1.2511	1.1756	1.2222
T7	2.5378	2.5378	2.2222
T8	1.1000	1.1000	1.1667
T9	1.4711	1.4711	1.3333

(b)

Portrait			
BOTTOM	$foe_{bottom,pw}$	$foe_{bottom,pb}$	$foe_{bottom,pt}$
T1	low	low	low
T2	very low	very low	very low
T3	low	very low	very low
T4	middle	middle	low
T5	very low	very low	very low
T6	low	low	low
T7	middle	middle	middle
T8	low	low	low
T9	low	low	low

they are safe for a longer use. Also, the tablet T3 is very safe to be used because it emits a slightly higher ELF magnetic field than the previously mentioned tablets. Furthermore, the tablets T1, T6, T8 and T9 emit a slightly higher ELF magnetic field than the tablets T2 and T5. But, they are also safe for a longer use. In the contrast, the tablets T4 and especially T7 again emit a higher level of ELF magnetic field at the bottom in the portrait mode. Accordingly, they should be used with caution over time. Again, if we compare the emission of the ELF magnetic field between landscape and portrait mode at the bottom of the tablets, we can observe that the use of the tablets in the portrait mode is slightly safer for a longer use.

To show if there exists a difference between the emission of the ELF magnetic field at the top and bottom of the tablets in the same modes (landscape or portrait),

Table 16.7 The results of foe_{ttb} obtained with all the six different ways of working with the tablets

Tablet	Landscape			Portrait		
	$foe_{ttb,lw}$ (%)	$foe_{ttb,lb}$ (%)	$foe_{ttb,lt}$ (%)	$foe_{ttb,pw}$ (%)	$foe_{ttb,pb}$ (%)	$foe_{ttb,pt}$ (%)
T1	**123.83**	**141.15**	**115.64**	**131.71**	**134.22**	**137.38**
T2	74.03	74.84	67.44	81.82	81.82	101.87
T3	**110.19**	**118.57**	**110.19**	**114.22**	**118.57**	**136.80**
T4	72.50	78.64	69.19	76.81	76.81	91.13
T5	85.62	84.49	71.03	90.00	90.00	107.08
T6	**111.72**	**118.90**	**111.72**	**114.74**	**118.90**	**123.64**
T7	**117.36**	**121.89**	**111.90**	**115.94**	**115.94**	**139.20**
T8	**140.00**	**138.86**	**122.11**	**140.00**	**140.00**	**161.14**
T9	**150.00**	**157.70**	**143.01**	**150.00**	**150.00**	**174.00**

we introduce a new measure which represents the ratio of the top and bottom *foe* factors in the same working mode (landscape or portrait). It is computed as:

$$foe_{ttb} = \frac{foe_{top}}{foe_{bottom}}, \qquad (16.34)$$

where foe_{ttb} is determined for all the six different ways of working with the tablets, and foe_{top} is given as: $foe_{top,lw}, foe_{top,lb}, foe_{top,lt}, foe_{top,pw}, foe_{top,pb}, foe_{top,pt}$, and foe_{bottom} is given as: $foe_{bottom,lw}, foe_{bottom,lb}, foe_{bottom,lt}$, $foe_{bottom,pw}, foe_{bottom,pb}$ and $foe_{bottom,pt}$.

Table 16.7 shows the percentage foe_{ttb} which indicates the percentage of ELF magnetic field emission between the top and bottom parts of the tablets in all the six different ways of working with the tablets. Any value higher than 100% indicates that the top emission of the ELF magnetic field in a specific mode of working with the tablets is higher than the ELF magnetic field emission at the bottom of the tablet. To notify it, these values are given as bold in Table 16.7. In the landscape mode, 3 out of 9 tablets (tablets T2, T4 and T5) emit a lower ELF magnetic field at the top than at the bottom of them. On the contrary, the tablets T1, T3, T6, T7, T8 and T9 emit a higher ELF magnetic field at the top than at the bottom, when they are working in the landscape mode. In the portrait mode, the obtained results are almost the same.

At the end, we integrate the way of working with the tablets by combining the positions of hands and fingers at the top and bottom parts of the tablet. In this way, instead of a factor *foe* for the top and bottom part, we define the foe_{total} for all the six ways of working with the tablets. Table 16.8 presents the total *foe* results for the landscape mode. The total *foe* factor values considering the ELF magnetic field emission in the landscape mode of the tablets (see Table 16.8a) are categorized according to the previously established classification and presented in Table 16.8b. The tablets T2 and T5 can be considered as completely safe for a longer use. Then,

Table 16.8 The results of the total *foe* factors obtained for the landscape working mode of the tablet (combination of the top and bottom ELF magnetic field emission)

(a)   (b)

Tablet	$foe_{tot,lw}$	$foe_{tot,lb}$	$foe_{tot,lt}$
T1	2.0667	1.9533	1.9911
T2	0.9978	0.9033	0.9600
T3	1.1233	1.0856	1.1233
T4	1.9722	1.8400	1.9344
T5	0.9611	0.8856	0.8856
T6	1.3244	1.2867	1.3244
T7	3.0044	2.8156	2.9289
T8	1.7733	1.5844	1.6411
T9	2.0278	1.8956	1.9711

Landscape			
FULL	$foe_{total,lw}$	$foe_{total,lb}$	$foe_{total,lt}$
T1	middle	low	low
T2	very low	very low	very low
T3	low	low	low
T4	low	low	low
T5	very low	very low	very low
T6	low	low	low
T7	high	middle	middle
T8	low	low	low
T9	middle	low	low

Table 16.9 The results of the total *foe* factors obtained for the portrait working mode of the tablet (combination of the top and bottom ELF magnetic field emission)

(a)   (b)

Tablet	$foe_{tot,pw}$	$foe_{tot,pb}$	$foe_{tot,pt}$
T1	1.7456	1.7644	1.7144
T2	0.7333	0.7333	0.8411
T3	1.0856	1.0856	0.9867
T4	1.8211	1.8211	1.6989
T5	0.6967	0.6967	0.7478
T6	1.3056	1.2867	1.3667
T7	2.7400	2.7400	2.6578
T8	1.3200	1.3200	1.5233
T9	1.8389	1.8389	1.8267

Portrait			
FULL	$foe_{total,pw}$	$foe_{total,pb}$	$foe_{total,pt}$
T1	low	low	low
T2	very low	very low	very low
T3	low	low	very low
T4	low	low	low
T5	very low	very low	very low
T6	low	low	low
T7	middle	middle	middle
T8	low	low	low
T9	low	low	low

the tablets T3, T4, T6, T8 and partially T1 and T9 are also safe for a longer use. In the contrast, the tablet T7 is dangerous for a longer use and should be used with caution over time.

Table 16.9 presents the total *foe* results for the portrait mode.

The total *foe* factor values considering the ELF magnetic field emission in the portrait mode of the tablets (see Table 16.9a) are categorized according to the previously established classification and presented in Table 16.9b. Again, the tablets T2 and T5 are completely safe for a longer use. Then, the tablets T1, T3, T4, T6, T8 and T9 are also safe for a longer use. Again, the tablet T7 emits the highest ELF magnetic field and it is dangerous for a longer use. Hence, it should be used with caution over time.

If we compare the emission of the ELF magnetic field between landscape and portrait mode using the total *foe* factor, we can observe that the use of the tablets in the portrait mode is slightly safer for a longer use.

From Tables 16.3, 16.4, 16.5, 16.6, 16.7, 16.8 and 16.9, the tablets T2 and T5 can be qualified as the safest ones to be used by the users. Also, it is worth noting that the majority of the tablets is well engineered, which means that they are safe for a longer use, because they emit a level of ELF magnetic field below the safety reference limit. However, the analysis revealed that a user as a buyer should be very cautious during the choose of a specific tablet.

From all above analysis, it is quite clear that choosing a tablet, which is safe for a longer use is not an easy task, because many characteristics of the tablet should be considered. Our analysis showed that many tablets are well engineered for a longer safe use. However, in some circumstances we suggest pausing the work with the tablets after every few hours.

At the end, we recommend a use of the tablets which reduces the exposure to a higher level of ELF magnetic field (safer use of the tablets) as follows: (i) to pause after a longer period of tablet usage, (ii) use a tablet's cover of better quality (materials with lower magnetic field permeability) in order to reduce the close contact of fingers and hands with the tablet, (iii) use of convertible tablets if possible by adding an external keyboard.

16.6 Conclusions

The chapter presented a study for detection of ranges of ELF magnetic field emitted by the tablets, working in normal operating condition. In particular, seven well-known clustering algorithms were experimented: (i) K-Means, (ii) K-Medians, (iii) SOM, (iv) EM, (v) agglomerative hierarchical clustering, (vi) DBSCAN, and (vii) Fuzzy-C-Means. The measurement of the ELF magnetic field was performed for 9 tablets on 9 typical positions at the top and bottom of each tablet. It defined a new measurement geometry for portable devices, i.e. laptops and tablets, overcoming the limitations of the current TCO standard. The measurement process showed that tablet users are exposed in some areas to levels of the ELF magnetic field which can be dangerous for their health. ELF magnetic field values measured at the top and bottom positions were collected into top and bottom datasets. Each of them was processed by the seven clustering algorithms, for finding clusters of values corresponding to ELF magnetic field ranges which represent dangerousness classes for the users.

The obtained ranges were evaluated taking into account the TCO proposed standard safety limit of 0.2 μT. Furthermore, the obtained ranges were compared to find the most reliable algorithm for detecting the top and bottom dangerousness classes. Results demonstrated that K-Medians algorithm is the most suitable for this task. Also, the danger areas of the magnetic field were established in order to warn the tablet users for possible risk assessments. At the end, the results obtained by K-Medians were used for computing the factor of exposure. It quantifies the real level of ELF magnetic field to which the users are exposed in different ways of working with the tablets by considering the top and bottom positions which the user is mostly in contact with.

In the future, we are going to investigate the ranges of ELF magnetic field detected by the clustering algorithms when the tablet operates under the so called "stress condition", i.e. when the tablet is overloaded with heavy programs.

Acknowledgements The authors dedicate this chapter to their colleague and friend Associate Professor Darko Brodić who gave an important contribution to this research topic.

References

1. A. Ahlbom, J. Bridges, R. de Seze et al., Possible effects of electromagnetic fields (EMF) on human health-opinion of the scientific committee on emerging and newly identified health risks (SCENIHR). Toxicology **246**(2–3), 248–50 (2008)
2. C.V. Bellieni, I. Pinto, A. Bogi et al., Exposure to electromagnetic fields from laptop use of laptop computers. Archiv. Environ. Occup. Health **67**(1), 31–36 (2012)
3. J.C. Bezdek, *Pattern Recognition with Fuzzy Objective Function Algorithms* (Plenum Press, New York, 1981)
4. BioInitiative Working Group, Ed. by C. Sage, David O. Carpenter. BioInitiative Report: A Rationale for a Biologically-based Public Exposure Standard for Electromagnetic Radiation (2012). http://www.bioinitiative.org
5. P.S. Bradley, O.L. Mangasarian, W.N. Street, Clustering via Concave Minimization. Adv. Neural Inf. Process. Syst. **9**, 368–374 (1997)
6. D. Brodić, Analysis of the extremely low frequency magnetic field emission from laptop computers. Metrol. Meas. Syst. **23**(1), 143–154 (2016)
7. D. Brodić, A. Amelio, Time evolving clustering of the low-frequency magnetic field radiation emitted from laptop computers. Measurement **99**, 171–184 (2017)
8. D. Brodić, A. Amelio, Detecting of the extremely low frequency magnetic field ranges for laptop in normal operating condition or under stress. Measurement **91**, 318–341 (2016)
9. D. Brodić, A. Amelio, Classification of the extremely low frequency magnetic field radiation measurement from the laptop computers. Measur. Sci. Rev. **15**(4), 202–209 (2015)
10. I. Calvente, M.F. Fernandez, J. Villalba et al., Exposure to electromagnetic fields and its relationship with childhood leukemia: a systematic review. Sci. Total Environ. **408**(16), 2069–3062 (2010)
11. R. Dash, R.L. Paramguru, R. Dash, Comparative analysis of supervised and unsupervised discretization techniques. Int. J. Adv. Sci. Technol. **2**(3), 29–37 (2011)
12. A.P. Dempster, N.M. Laird, D.B. Rubin, Maximum likelihood from incomplete data via the EM algorithm. J. R. Stat. Soc. Ser. B **39**(1), 1–38 (1977)
13. J.C. Dunn, A fuzzy relative of the ISODATA process and its use in detecting compact well-separated clusters. J. Cybern. **3**, 32–57 (1973)
14. M. Ester, H. Kriegel, J. Sander, X. Xu, A density-based algorithm for discovering clusters in large spatial databases with noise, in *Second International Conference on Knowledge Discovery and Data Mining (KDD)* (AAAI Press, 1996), pp. 226–231
15. K.J. Fernie, S.J. Reynolds, The effects of electromagnetic fields from power lines on avian reproductive biology and physiology: a review. J. Toxicol. Environ. Health Part B: Crit. Rev. **8**(2), 127–140 (2005)
16. I.S. Grant, W.R. Phillips, *Electromagnetism*, 2nd edn. (Manchester Physics, Wiley, 2008)
17. A. Gupta, K.G. Mehrotra, C. Mohan, A clustering-based discretization for supervised learning. Stat. Probab. Lett. **80**(9–10), 816–824 (2010)

18. J.G. Gurney, B.A. Mueller, S. Davis et al., Childhood brain tumor occurrence in relation to residential power line configurations, electric heating sources, and electrical appliance use. Am. J. Epidemiol. **143**, 120–128 (1996)
19. Hogg, R., McKean, J., Craig, A., *Introduction to Mathematical Statistics* (Pearson Prentice Hall, Upper Saddle River, NJ, 2005), pp. 359–364
20. IEEE Engineering in Medicine and Biology Society Committee on Man and Radiation (COMAR). http://ewh.ieee.org/soc/embs/comar/
21. T. Kohonen, Self-Organized formation of topologically correct feature maps. Biol. Cybern. **43**(1), 59–69 (1982)
22. T. Koppel, P. Tint, Reducing exposure to extremely low frequency electromagnetic fields from portable computers. Agron. Res. **12**(3), 863–874 (2014)
23. M. Lee, N. Duffield, R.R. Kompella, MAPLE: a scalable architecture for maintaining packet latency measurements, in *ACM conference on Internet measurement conference (IMC), New York, USA* (2012), pp. 101–114
24. Lutron 3D EMF-828. http://www.lutron.com.tw/ugC_ShowroomItem_Detail. asphidKindID=1&hidTypeID=24&hidCatID=&hidShowID=197&hidPrdType=& txtSrhData=
25. MacQueen, J. Some methods for classification and analysis of multivariate observations, in *Fifth Berkeley Symposium on Mathematical Statistics and Probability*, California, USA (1967), pp. 281–297
26. L. Rokach, O. Maimon, Clustering methods, *Data Mining and Knowledge Discovery Handbook* (Springer, US, 2005), pp. 321–352
27. SPECTRAN NF Spectrum Analyzer NF- 5030. http://www.aaronia.com/products/spec-trum-analyzers/NF-5030-EMC-Spectrum-Analyzer/
28. 2012-03-05-TCO Certified Notebooks 4.0, TCO Development, Swedish Confederation of Professional Employees (2012) . http://tcodevelopment.com
29. US Environmental Protection Agency. https://www3.epa.gov/
30. Vannucci, M., Colla, V.: Meaningful discretization of continuous features for association rules mining by means of a SOM, in *European Symposium on Artificial Neural Networks*, Bruges, Belgium (2004), pp. 489–494
31. H. Wang, J. Zhang, One-dimensional k-center on uncertain data. Theor. Comput. Sci. **602**, 114–124 (2015)
32. World Health Organization (WHO), *Extremely Low Frequency Fields* (Environmental Health Criteria, Monograph, 2010), p. 238
33. C. Wu, F. Jeff, On the convergence properties of the EM algorithm. Ann. Stat. **11**(1), 95–103 (1983)
34. Zherebtsov, A.A., Kuperin, Y.A., Application of Self-Organizing Maps for clustering DJIA and NASDAQ100 portfolios (2003). arXiv:cond-mat/0305330

Further Readings

35. J. Han, *Data Mining: Concepts and Techniques* (Morgan Kaufmann Publishers Inc., San Francisco, CA, USA, 2005)
36. C.C. Aggarwa, C.K. Reddy, *Data Clustering: Algorithms and Applications*, 1st edn. (Chapman and Hall/CRC, 2013)
37. B.W. Wilson, R.G. Stevens, L.E. Anderson, *Extremely Low Frequency Electromagnetic Fields: The Question of Cancer*. Battelle Pr (1990)
38. P. Staebler, *Human Exposure to Electromagnetic Fields: From Extremely Low Frequency (ELF) to Radiofrequency* (Wiley, 2017)

Author Biographies

Alessia Amelio received a Ph.D. degree in Computer Science Engineering and Systems from University of Calabria (UNI-CAL), Italy. She is currently a Research Fellow at DIMES, University of Calabria. Dr. Amelio scientific interests fall in the areas of digital image processing, pattern recognition, and artificial intelligence. She is editorial board member of various journals. During her career, she was co-organiser of multiple scientific events, co-authoring more than 100 papers.

Ivo Rumenov Draganov received a Ph.D. degree in Telecommunications from Technical University of Sofia (TU-Sofia), Bulgaria. He is currently an Associate Professor at the Faculty of Telecommunications in TU-Sofia. Dr. Draganov scientific interests fall in the areas of digital image processing, pattern recognition, and neural networks. He is a member of IEEE. During his career, he participated in 4 international and more than 10 national and institutional scientific projects, co-authoring more than 80 papers.

Printed in the United States
by Baker & Taylor Publisher Services